Plants and Environment

A Textbook of Plant Autecology

FRONTISPIECE. *Abies Lasiocarpa* at Upper Timberline in Wyoming. The Lower Branches Form a Dense Layered Mat Which Is Well Protected from Desiccation in Winter by the Snow It Accumulates in Autumn. In a Narrow Zone Immediately Above This Mat All Buds Have Been Sheared Off the Bole by Abrasive Ice Particles Which Are Blown Over the Snow Surface with Great Force. Buds on the Windward Side of the Upper Bole Are Killed Each Year by Ice Blasting or Winter Desiccation. Despite the Outstanding Adversity of This Habitat, the Bole Has Resisted Deformation and Retained a Remarkably Erect Form.

Plants and Environment

A Textbook of Plant Autecology

THIRD EDITION

R. F. Daubenmire
Washington State University

JOHN WILEY & SONS

New York   London   Sydney   Toronto

to Jean

*This book is dedicated
with affection*

Library of Congress Cataloging in Publication Data:
Daubenmire, Rexford F   1909–
 Plants and environment.

 Bibliography: p.
 1. Botany—Ecology. I. Title.

QK901.D3 1974   581.5'22   73-13826
ISBN 0-471-19636-3

Printed in the United States of America

10 9 8 7 6 5 4 3

Preface

Autecology is essentially a consideration of those phases of geology, soils, climatology, zoology, chemistry, and physics which are more or less directly connected with the welfare of living organisms, and a relating of them to the structure, function, and evolution of species. Such an organization of information which cuts across so many discrete fields of science appears to me to have real value in the following ways, and these have been uppermost in my mind during the process of assembling and orienting the material.

First of all, an excursion into the field of autecology of plants may be looked upon as an end in itself, if only for the fact that, like physiography and biogeography, it contributes much to man's interest in his surroundings. The traveler with only a smattering of outdoor sciences appreciates the landscape in a manner utterly incomprehensible to a person not thus trained. Also, most persons have some contact with the living plant even if it is only with a few ornamentals growing about the home, and the satisfaction of understanding something of environmental requirements enriches this contact.

The ecologic approach becomes indispensable wherever plant behavior is studied in relation to the production of cultivated crops or the management of indigenous vegetation, for, unless responses are studied in relation to the natural or cultural environment of the plant, they may be so abnormal as to be of practically no economic value. It is essential that the serious student of the living plant be aware of the numerous and frequently unsuspected influences which are not a part of his study but which nevertheless influence his results. Only a broad grasp of concrete fundamentals can accomplish this.

For those who study plant communities from any angle except that of structure alone, autecology has a fundamental significance, for it provides the basis for the etiology of distribution and of dynamic interrelationships. I feel that the material included in this book is indispensable as a background for anyone who would undertake grasping the complexities involved in the ecology of plant communities. An intelligent approach to that type of investigation demands as a prerequisite a working knowledge of the basic interrelationships between the individual plant and its environment.

Ecology has much to offer in the way of keeping related fields of investigation in contact. The development of all sciences has led to the divergence of innumerable lines of specialization, and, with increasing restriction of the fields, each research worker has become a master of no more than a part of one discipline, yet the fundamental interrelations among the different parts of the universe have in no way lessened. Therefore the maintenance of interconnection among special phases of science has become a problem of increasing importance. Much effort has been wasted by the failure of one scientist to take advantage of information that has become common knowledge to certain others. So often have the benefits of diverse approaches to the same problem been proved that it is now apparent that most of the important problems of any one branch of science are being solved, and will continue to be solved, as a result of the application of other sciences to these problems. Thus the trend in science is toward group research. In research of this type ecologic training has an important role to play in the merging of separate disciplines, portions of which it overlaps. Ecologists themselves are frequently no less specialized than other groups of biologists and are therefore subject to the same criticisms. It should be pointed out that as conceived here, autecology encompasses more than "physiological ecology" by considering morphology ("ecologic anatomy") and genetics ("genecology") along with physiology when striving for an understanding of the ecology of an organism.

A specialist's excursion into ecology may lead to a healthy reexamination of concepts and aims in his own branch of biology, for the most important contribution of autecology as a discipline is to inculcate the holistic outlook. Its most important contribution as a profession lies in planning problems, then synthesizing results after specialists have worked out the details of each phase. Indeed it is only through synthesis that specialized studies acquire their true significance. This role calls for broad training and experience, and an understanding of details without preoccupation with any one of them. A great many problems, and indeed most of those involving the conservation of biologic resources, greatly overreach the limits of any one field of specialization. These opinions concerning the value of formal ecologic study as a borderline science that integrates separate disciplines are reflected in the unusually broad scope of the topics included in this book.

We have recently come to realize that man is doing as much to upset the stability of the earth's surface as did any of the great geologic revolutions of past ages. Changes cannot be prevented, but their course can be altered to minimize their hazard to human survival, and to maintain the most pleasant surrounding possible under the circumstances. Our environmental problems have been largely a consequence of a poor understanding of organism-environment interrelationships.

In keeping with my conception of the synthetic nature of the field, I believe that coursework in ecology should follow at least elementary instruction in plant morphology, taxonomy, physiology, and chemistry. However, in connection with inorganic phases of plant-environment relationships, a compromise with elementary facts is considered desirable in many places, simply because it is impractical to require climatology, pedology, geology, and physics, in addition to botany and chemistry, as prerequisites to a course in plant ecology.

Although the elucidation of basic principles has been the primary objective, special effort has been made to give examples and other consideration to applied ecology, for the subject is indeed the very foundation of all other subsciences dealing with the management of crops and natural biologic resources. Illustrations are taken from a wide area, but at the same time statistics and specific examples have been kept to the minimum consistent with clarity in order to allow the lecturer to develop this phase of presentation in accordance with local conditions which he can demonstrate in the field.

The literature citations are not offered as a complete bibliography of the subject. They are intended to serve three purposes—to give the source of the basic research which supports many of the generalizations in this book, to provide any interested student with initial references to the literature on many of the topics treated, and to indicate places where research techniques are described in more detail than is feasible in a book of this nature. Special attention has been given to recent papers, particularly to those containing summaries of earlier work. Also, in deference to students of intermediate status, the citations are restricted to papers written in the English language.

All special ecologic terms which are believed to be really needed are defined or used in a defining sense when they first appear, so that the proper use of the index makes a special glossary unnecessary.

This third edition has involved more changes than the second. The amount of rewriting considered desirable has varied widely from place to place, though all chapters have been somewhat affected. The bibliography has again been updated and expanded.

I want to thank those correspondents who have suggested specific points for consideration in this revision.

Pullman, Washington   *R. F. Daubenmire*

Contents

Introduction

The Scope of Ecology

The term ecology was proposed in 1885 by the zoologist Reiter. His combination of the Greek roots *oikos* (meaning home) and *logos* (meaning study, or discourse) etymologically implied a study of organisms at home, although he made little further attempt at definition. A year later Haeckel, another zoologist, formulated a simple definition which was in keeping with the original usage and etymology, and which remains today the most widely accepted interpretation of the scope of the word. Haeckel defined ecology as the study of the reciprocal relations between organisms and their environment.

For the sake of convenience ecology is usually subdivided into *animal ecology* and *plant ecology*, although in places these subdivisions merge with each other just as both merge with climatology, geology, and pedology.

Plant ecology in turn may be subdivided into *autecology*, a study of the interrelations between the individual and its environment, and *synecology*, a study of the structure, development, function, and causes of distribution of plant communities. To understand the ecology of a community, the ecologic life histories of at least the most important plants in it must first be understood, so that autecology is of necessity the foundation upon which synecology is built.

Relationship of Plant Ecology to Other Fields of Botany

Pure botany may be divided into four basic fields: morphology, physiology, taxonomy, and genetics. Each of these subsciences may be pursued at length without consideration of the plant's surroundings, but when they take environment into consideration they become part of ecology. Some botanists therefore look upon plant autecology as a viewpoint or aspect of the basic fields of botany, rather than a wholly distinct and circumscribed branch of science. Those who consider it a distinct science are first to admit that its boundaries are not easy to delimit.

In applied fields of botany, such as forest management, range management, plant pathology, and agriculture, ecologic considerations hold a more dominant position. In each of these subsciences a knowledge of the interrelationships between plants and environment provides the fundamental basis for the intelligent management of plant life for the good of mankind, which is their primary aim.

The Soil Factor

Importance of
Soil to Plants

On a dry-weight basis root systems usually comprise no more than one-quarter of a seed plant, but the roots are so finely divided that frequently they occupy a mass of soil greater than the volume of the atmosphere occupied by the shoot. The result is a tremendous amount of surface contact between soil and plant, which is very important in view of the plant's dependence upon the soil for anchorage, water, and nutrients. The soil (*edaphic*) factor deserves much attention on the part of the ecologist largely because, owing to this intimacy of contact, plant and soil are strongly influenced by each other, but also because of the tremendous complexity and dynamic nature of soil. The characteristics of soil are undergoing continual change, the rates of these changes being highly dependent upon a number of environmental factors.

An appreciation of the important part which soil plays in plant growth undoubtedly arose with the very beginning of agriculture. In attempting to control environment for the benefit of his crops, man soon discovered that little can be done to alter climatic factors, but much can be accomplished with the soil. He has learned how to bring about desirable changes in certain of the soil characteristics, and he has developed management practices that compensate for those soil properties less susceptible to alteration.

Soil is even more critical as a factor of plant environment in natural habitats, for the extreme conditions which are encountered there are not tempered by corrective cultivation, irrigation, artificial drainage, fertilizers, etc. In the first textbook of plant ecology, written by Warming in 1895, special emphasis was laid on edaphic factors by classifying plants, wholly aside from their taxonomic relationships,

3

into ecologic groups each of which is characteristic of one type of substratum. Warming designated all plants usually to be found on acid soils as *oxylophytes,* those on saline soils as *halophytes,* on sand as *psammophytes,* on rock surfaces as *lithophytes,* in rock crevices as *chasmophytes,* etc.

Some time ago a comprehensive project designed to show in detail the importance of the soil factor to land plants was undertaken by the British Ecological Society. A series of contiguous plots was established, each composed of a distinct soil type which was transported to the garden for this purpose. Individuals of each species to be studied were planted on each soil type, and the subsequent behaviors of the groups of plants were compared. These experiments (527) have shown that differences in soil may affect plants in the following respects:

Ability of seeds to germinate.
Size and erectness of the plant.
Vigor of the vegetative organs.
Woodiness of the stem.
Depth of the root systems.
Amount of pubescence.
Susceptibility to drouth, frost, and parasites.
Number of flowers per plant.
Dates of appearance of flowers, etc.

By consulting other literature this list could be extended almost indefinitely.

The soil factor is at least as important to aquatic as to land plants. Variations in soil under water strongly influence the distribution of those plants that grow rooted in ponds and lakes, although this factor was overlooked until quite recently (597). Most cryptogams likewise are highly selective as to their substratum. Lichens and mosses often show decided preferences for certain types of rocks or bark, and many fungi are restricted to one or a few types of substrate.

Soil Defined

The word soil has been used above in a very broad but ecologically justifiable sense to include any part of the earth's crust in which plants are anchored: the muddy bottoms of ponds, porous rock surfaces into which cryptogams send their rhizoids, peat, raw gravel deposited by glaciers, etc. More conservatively, soil may be defined as the weathered superficial layer of the earth's crust with which are mingled living organisms and products of their decay. Typically a soil is made up of a *parent material* (the inorganic foundation or mineral framework) into which there has been incorporated an *organic increment* as well as *living organisms,* with the spaces remaining between the solid particles filled with *water* and *gases.* These constituents will be discussed in order.

The Parent Materials

The basic framework of most soils consists of small fragments of mineral matter which have been derived from solid rock by mechanical or chemical types of weathering.

Mechanical disintegration of rock is brought about by temperature fluctuations, by the wedgelike action of freezing water and growing roots and rhizoids, by the abrasive action of particles carried by running water and wind, and by moving ice.

Chemical processes of weathering consist of hydrolysis, oxidation, carbonation, and hydration, by which certain minerals are converted into secondary minerals and solutes, most of which are then moved to deeper levels or are leached away entirely. The role that carbonic-acid secretion of roots plays in carbonation and solution can easily be demonstrated by growing a seedling against a polished marble slab, for the surface becomes etched wherever the rootlets touch it. Chemical weathering is also promoted by certain species of algae, bacteria, and lichens that work their way into self-made cavities in rocks by dissolving away the less-resistant minerals of which the rocks are composed.

Mode of Origin (722)

The nature of a soil is strongly influenced by the history of the fragments from the time they were weathered from solid rock until they became a constituent of that soil. Depending on the mode of origin, most parent materials can be classified as follows:

1. Residual.
2. Transported.
   a. By gravity: colluvial.
   b. By running water: alluvial.
   c. By glaciers: glacial.
   d. By wind: eolian.
   e. By other types.

Residual. Residual parent materials result from the disintegration of rock in place. Because the intensity of weathering is greatest where rock minerals are in direct contact with the atmosphere, the surface layer of a residual parent material exhibits the most complete physical decomposition and chemical alteration. With increasing depth below the surface the mineral particles become larger and less altered chemically, until finally they intergrade with bedrock.

In North America soils developed from residual parent materials are especially predominant in the Appalachians, the Ozarks, and in the southern Great Plains.

Transported. Transported parent materials are composed of, or derived from, mineral particles which have been brought from their place of origin by various agents. Their forms vary widely according to the agents of transport, and, since most of the agents operate intermittently, the parent materials of this class usually occur in layers with abrupt transitions and do not intergrade with the underlying bedrock as is characteristic of residual material.

Colluvial. Colluvial parent materials are those of the transported group in which the particles have been moved by the pull of gravity. They take on a variety of forms.

Fragments from cliffs or steep rocky slopes become dislodged from time to time and may accumulate below as *talus.* This material is characteristically very coarse, consisting mostly of large fragments of rock, and has a rather steep and unstable surface. Small plants becoming established on talus slopes are in danger of being crushed or buried, and large woody plants are often sheared off, pruned, or at least scarred by flying fragments. The percolation of rainwater is so rapid that the moisture quickly penetrates to depths where it is available only to plants with strong taproots, and in dry climates few plants can invade these deposits. In such regions a belt of trees or shrubs with high water requirements is often found around the lower edge of talus deposits where they take advantage of percolation water.

A slow creeping of the surface layers of wet soil or stones down an incline (i.e., *solifluction*) is not uncommon, particularly in high latitudes and altitudes, but it is usually too gradual to attract attention. After such a movement the material is to be classed as colluvial. Some indications that creeping has taken place are: (*a*) scattered bare strips crossing an otherwise well-vegetated slope; (*b*) trees inclined (Fig. 1), or curved at the base (curving is more usually the result of snow action and it is difficult to distinguish between the two kinds); (*c*) saucer-shaped depressions on slopes, which are flanked by ridges along their lower edges; and (*d*) turf forced up in ridges just above or below boulders (690).

Mudflows are riverlike movements of saturated layers of material, which

FIGURE 1. Trees inclined at various angles as the result of soil movement.

are especially favored when a wet surface layer is underlain by frozen subsoil or a smooth rock surface. These flows cover old soil surfaces in their slow movements or when they come to rest. They destroy plant populations by smothering (47) or mechanically injuring roots (Fig. 2).

Landslides are similar in nature and effect to mudflows, except that they take place suddenly, typically involve thick masses of material, and the action results from heavy rain or earth tremors that start material cascading down an over-steepened slope. In mountainous areas there are regions where landslides can be expected at frequent intervals. When landslides are of limited extent (slumps) certain plants may ride the moving mass and continue to grow in the new location even though conditions there are not favorable to the germination and establishment of the species. Slumping is far more common on cultivated slopes than

FIGURE 2. *Pinus ponderosa* reinvading an area on which a mudflow 17 years earlier had covered the original soil and smothered out the forest. Mt. Lassen, California.

where the surface is covered with undisturbed vegetation. However, even deep-rooted trees cannot always prevent gravity-induced movements of large masses of material, and in fact may promote slides by favoring infiltration, adding weight, and swaying in the wind (595).

There is no stratification in a colluvial deposit. Particles of many sizes, including all the different kinds of rock and soil of which the slope or cliff may be composed, are mixed indiscriminately. Because the horizontal movement of colluvial material is so limited, these parent materials are mostly restricted to mountainous regions, but there they often make up a considerable portion of the land surface, particularly where slopes are steep and vegetation sparse.

In cold climates where solifluction is conspicuous, it is the primary force that wears away hills to produce level topography, just as erosion by water (272) performs the same role at lower latitudes.

Alluvial. Materials in this category are deposited by running water in the form of outwash plains, floodplains, river terraces, deltas, and alluvial fans. These deposits have two outstanding characteristics by which they can usually be recognized with ease. First, the individual particles tend to be rounded and smoothed by the action of running water. Second, the strata are usually distinct in that each contains particles of a particular range of size classes, depending upon the speed of the water that deposited the layer (Fig. 3). Coarse materials are laid down by swift water; fine particles are deposited only in relatively calm water.

FIGURE 3. Alluvial deposits overlaid by about 1 meter of loess. The sorting action of running water is responsible for the differing texture of the alluvium. Note the vertical cleavage in the loess. Southeastern Washington.

Alluvium deposited by streams is found bordering most creeks and rivers, especially slow-moving bodies of water such as the Mississippi River. When first formed these deposits are so low that they are frequently flooded at high-water stages, and at such times they receive additional increments of fine sediment, especially if they are well vegetated. At this stage of development they are called *floodplains.* As a stream cuts its channel deeper the floodplain is left above the reach of high water and is then called a *terrace.* Large streams are frequently bordered by series of terraces, the oldest being at considerable heights above the present water level.

Where streams enter a lake or the sea the deposit of silt and clay that settles out in the edge of the still water may build a *delta,* which gradually lengthens until it extends far beyond the original mouth of the stream. The low, marshy delta of the Mississippi River provides an excellent example of this type of alluvial deposit.

Alluvial fans are formed when materials are washed from precipitous mountain slopes and deposited where the velocity of the water is checked abruptly at the edge of the basal plain. They are composed chiefly of stones and gravel along the edge next the mountain, gradually becoming finer in texture with increasing distance from the steep slope. Most of the desert regions west of the Rocky Mountains are characterized by this type of parent material.

Many valleys have nearly level (i.e., aggraded) floors built up of alluvium washed from the adjacent slopes. Many geologists believe that the innumerable beaver dams formerly to be found in most valleys in North America played an important role in the formation of such level valley floors. This view is well supported by the fact that programs of erosion control in wilderness areas frequently include restocking the streams with beaver, which soon build so many series of dams that almost no sediments escape the area.

Glacial.   Glacial deposits (*till*) usually can be recognized by the fact that particles of all sizes from clay to boulders are thoroughly mixed without any kind of sorting (Fig. 4), and that the larger particles are sometimes polished or grooved on one or more flattened faces. Within the layer of till deposited by one glacial advance there is no stratification, but the layers deposited by successive advances of the ice may vary greatly in character. The topography of glacial deposits is also very distinctive. It is usually a flat plain which undulates in such a manner that many depressions lacking surface drainage are formed.

Extensive till plains occur in the colder parts of the northern hemisphere. In North America they cover much of Canada, parts of Alaska, the New England region, and the Mississippi Valley north of the Ohio and Missouri Rivers. Less extensive glacial deposits occur locally in the Cordillera as far south as Colorado and central California. Most of this till was deposited by a series of four continental glaciers which originated at various points in Canada, or in the Cordillera, and carried materials outward in all directions from these centers during the Pleistocene epoch. The deposits tend to be deepest near the periphery of glaciated regions. Near the centers where the ice originated, erosive forces have predominated and the bedrock may be nearly bare.

The physical and chemical nature of a glacial parent material is strongly influenced by the nature of the bedrock in the direction from which

FIGURE 4.   Vertical section through a deposit of glacial till. Particles of all sizes from clay to large rocks are thoroughly mixed. The sharp edges of the rocks differ markedly from the rounded edges of rocks in water-laid deposits.

the ice came. It often happens that two very diverse glacial deposits lie side by side or one above the other, each having been brought by a glacier which approached from a different direction (111).

Unlike other transported materials, plants play no part in the deposition of till.

Eolian.   Wind-transported materials are classed either as dune or loess.

Dunes (Fig. 5) are found in three types of situations. (*a*) They may occur along the shores of seas and lakes as a result of water currents eroding the headlands and depositing the resultant sand particles on the strand in bays, and the wind moving the material back onto the land. Dune areas of this type occur along both the Atlantic and Pacific coasts of North America and around the southern and eastern margins of Lake Michigan. (*b*) Dunes form along river valleys where high waters deposit sand upon floodplains, and the sand is subsequently dried out and moved by wind. Small dune areas with this type of origin are found

FIGURE 5. Unstable body of dune sand.

along the principal streams in arid western North America, along the Pecos, the Arkansas, the Columbia, etc. (*c*) In dry regions the weathering of sandstone and other rocks may produce sand that is subject to blowing because of the sparsity of vegetation. The sandhills of the Great Plains centering about western Nebraska provide an example of dunes having this type of origin.

Although the Sahara Desert is famous for its sand dunes, sand covers only a small part of its area, and sand is not characteristic of deserts in general.

More than any other type of parent material, dune sand tends to be composed of particles of nearly uniform size and chemical composition. The finer particles, silt and clay, have been blown farther away, while the heavier particles, gravel and stones, remain at the sand source, so that the dune sand embraces a rather narrow range of variation in particle size. Usually dune sand is predominantly siliceous, but in tropical regions dunes may be composed entirely of calcareous fragments of coral. In the White Sands National Monument in New Mexico is an area 1300 square kilometers in extent covered with gypsum sand.

Loess is a deposit of particles which have been picked up and transported some distance by wind and therefore is finer textured than dune sand. Such material has usually been derived from rock flour deposited by the melt waters of Pleistocene glaciers then blown onto areas just leeward of the main glacial rivers, as exemplified by the extensive loessal plains of China, by an area overlapping parts of Washington, Oregon, and Idaho, and by strips immediately bordering the lower Platte, Missouri, and Mississippi Rivers. In places these deposits aggregate 70 m in thickness. Sometimes loess is derived in part from volcanic ash.

Wherever it is deposited, loess usually lodges between the shoots of living plants, which in successive generations must develop at successively higher levels as the deposit increases in thickness. Possibly because of the vertical orientation of the root systems which are buried in this manner, loess has a remarkable vertical structure and cleavage (Fig. 3). The horizontal stratification which characterizes most sedimentary deposits is usually lacking here because of the slowness with which the deposits accumulate. Typically loess is a yellowish-buff color. In northern China it provides the coloring material of the Yellow River and the Yellow Sea.

Other types.   Still other types of parent materials may be encountered which do not fit into the above categories. For example, *lacustrine* materials are gray silty sediments that accumulated on the bottoms of lakes. This material is composed in part of dust which was deposited on the lake surface by wind, in part of sediments washed into the lake basin by streams, and in part of shells or excretions of aquatic organisms. Extensive lacustrine materials which accumulated in lakes now extinct occur in the intermountain and Great Lakes regions.

Marine parent materials, such as those along the Atlantic Coastal Plain of eastern North America from New Jersey to Mexico, have been transported in part by ocean currents, in part by innumerable small streams, and in part by wind.

In the vicinity of volcanoes soils may have been derived from various types of ejecta, such as the pumice fields of Oregon.

The classification of parent materials according to their mode of origin is further complicated by the fact that various types of transported materials may come to rest upon one another (Fig. 3) or upon residual materials, or upon a rock surface which later weathers to form an underlying residual material.

Residual soils tend to inherit any mineralogic peculiarities of the parent rock, whereas transported soils tend to be mixtures of particles derived from different kinds of rock, hence are less likely to be nutritionally unbalanced. Also the interpretation of the edaphic factor is more complex in transported than in residual soils, for in the former, roots commonly extend across layers which have different ecologic characteristics.

Textural Classification

The most evident component of soils is the mineral particles that characteristically vary greatly in size. By passing a soil through a series of sieves with different-sized holes, the proportions by weight of the different component size classes of particles, i.e., the *soil texture,* can be determined. Usually the parent material constitutes over 90% of the soil solids; therefore the term texture applies primarily to this constituent.

In 1927 the International Society of Soil Science agreed on the following arbitrary delimitation of particle size classes:

Coarse Gravel
——5.000 mm——
Fine Gravel
——2.000 mm——
Coarse Sand
——0.200 mm——
Fine Sand
——0.020 mm——
Silt
——0.002 mm——
Clay

The U.S. Department of Agriculture uses a system differing from the above in that sands, defined by the size range 0.05 to 2.00 mm, are divided into five instead of two size classes (722).

Analyses of the proportions of different sizes of particles in a soil (79, 407) do not give precise data because the various methods used to disperse the smaller particles strongly affect the results (588). Furthermore, series of five to seven figures representing each soil sample are difficult to interpret. The most successful use of mechanical analyses in plant ecology has been to simply determine the percentage of silt + clay and to use this figure in comparing samples—a method that takes cognizance of the fact that fine and coarse particles have very different properties, as will be pointed out later. An even better means of evaluating textural properties of soils as they are related to plant growth will be discussed in Chapter 2 under the heading "Moisture Equivalent."

Since the proportion of various size classes of particles has an important bearing upon soil properties, and this character varies widely from one soil to another, a system of naming soils which reflects their textural character is highly desirable. On this account the U.S. Department of Agriculture has set up a group of arbitrarily defined textural names (722).

According to this classification loams are any soils in which both fine (silt + clay) and coarse (sand) size classes are well represented. Loams are therefore not necessarily top soils nor dark in color, as they are commonly defined by the layman.

The clay fraction of a soil includes the mineral colloids * with their special and important properties. Since there is a continuous gradation from particles of definitely colloid size to coarse materials wholly lacking colloidal properties, any size limit established for colloids is necessarily arbitrary. However, all the particles within the size range of clay are small enough to possess at least some of the important characteristics of colloids and hence may be spoken of as colloidal.

Whereas larger particles are essentially inert, soil colloids are extremely active physicochemically. The physical basis for this great activity resides in the tremendous aggregate surface area of the plate-shaped crystals of which clay is mainly composed. A gram of such mineral material has approximately 46 square meters of surface!

Mineral colloids also differ from the larger particles in their chemical composition. The larger fragments are for the most part relatively unweathered or inert primary minerals, especially silica, whereas the colloids are mostly secondary minerals, i.e., derived by weathering from minerals very susceptible to weathering. Only these secondary clay minerals admit water into the particles. Water held by primary particles is limited to films over the surface.

The coarser particles are not without some important properties, however. For example, the chemical nature of the inorganic colloids depends upon the chemical nature of the primary rock fragments, including gravel and sand, from which they are derived by weathering.

Importance of Soil Texture to Plants

Relative resistance to root penetration. Soils with high silt and clay content retard root growth, and consequently the extent and degree of branching of the roots are decreased (817). This becomes very important in those regions where relatively long periods elapse between rains. There the survival of seedlings is best where the surface is loose textured, for, unless the root system can penetrate deeply before the surface layer of soil dries out, seedlings that germinate during the moist weather of spring perish during the first dry period of summer (736, 761).

---

* The discussions in this part of Chapter 1 refer primarily to inorganic colloids. Organic colloids and their significance will be treated later.

Infiltration of water. * Rain falling on a coarse-textured soil penetrates almost immediately, so that ordinarily almost none is lost as runoff. In contrast, the rate of infiltration of water into a fine-textured soil is very slow, and, because runoff is greater there, these soils as a group are more susceptible to gully and sheet erosion than are lighter soils. The effectivity of a given amount of rainfall is greatly reduced by runoff, for every drop that runs off an area diminishes the amount of water available to the plants rooted there. The fact that forests extend farthest into arid regions on coarse or stony soils can be explained in part by the rapid infiltration and deeper percolation of rain in such soils, as well as by the deeper root systems of woody plants (593, 859).

Rate of water movement. The rate of movement of water through a soil varies inversely with soil texture; i.e., the finer the state of division, the slower the rate of movement. This is because in a fine-textured soil the tiny interstitial spaces offer considerable resistance to the mass movement of water.

Water moves downward through a sandy soil so rapidly that most of it is soon beyond the reach of shallow-rooted plants. Consequently the most successful perennials on sandy soils (provided the water table is not near the surface) are almost invariably deep-rooted species. On the other hand, where rainfall is low and plants depend upon water rising by capillarity from levels below the reach of their roots the rate of rise may be rapid enough to satisfy plant demands only on sandy soils.

In stratified soils containing one or more layers high in clay content, these layers exert a strong control over the ecologic character of the soil. Such a layer may resist percolation to the extent that during and immediately after rainy seasons the soil above remains saturated (723). The surface of such a body of suspended moisture is called a *perched* or *hanging water table,* and if it is near the soil surface there may be so little aerated soil that only highly specialized plants can grow on the habitat.

---

* Infiltration rates of a series of soils may be compared satisfactorily by using short sections of pipe about 10 cm in diameter and 10 cm long, with one end beveled on the outside to a sharp edge. These are forced into the mineral surface until only about 2 cm are left projecting, with a minimum of lateral movement, then a measured quantity of water (e.g., 100 ml) is poured quickly onto the soil surface encompassed by the projecting rim. The time required for the absorption of this amount of water is determined accurately. Successive aliquots may be added, preferably immediately after the soil surface ceases to glisten with moisture from the preceding aliquot, and results may be expressed as a curve. Obviously a number of replications spaced at random are desirable.

Water-holding capacity.   Water is retained by soils as films which coat the surfaces of the particles, as wedges held in the angles between grains, and as moisture imbibed by the colloids. Surface films tend to have the same thickness over all particles in a soil mass irrespective of their differences in size. In fine-textured soils there are more aggregate surfaces to accommodate films, more angles to hold water, and more colloidal material, and consequently more water can be held per unit volume of soil than in coarser soils. In regions where there are pronounced dry seasons plants are most favorably situated when their roots are in contact with a body of fine-textured soil, because fine soils absorb so much water during each rainy season that the supply is exhausted relatively late in the ensuing dry seasons, as compared to sandy soils. Thus, with the same amount of rainfall, a sandy loam was observed to keep maize from wilting for 20 days, whereas on medium and fine sands maize wilted 12 days after the last rainstorm (732).

The proportion of the rooting horizon that is occupied by gravel, and especially by stones, represents bulk that has almost no water-holding capacity. Stoniness may be evaluated comparatively among different habitats by averaging the depth to which a pointed steel rod, approximately 1 cm in diameter, can be pushed before it lodges against a stone. In Finland it was found that single penetrations deeper than 30 cm should be counted as no more than 30, and the averages so determined for a series of till habitats were directly related to the growth rate of *Pinus sylvestris* (792).

The effect of texture on soil moisture relations is complicated. Although fine-textured soils have a great advantage in holding much water, they (*a*) hold much of it in the upper layers which are highly vulnerable to drying, (*b*) do not admit water readily and so lose more by runoff, (*c*) retard root penetration so that seedlings may not be able to reach deep moisture before the surface dries, and (*d*) tend to be poorly aerated below, thus enforcing shallow rooting which makes the plant susceptible to drouth. Their more constant water supply may favor damping-off fungi, with the result that the seedlings of certain taxa cannot get established on heavy soils (309).

Fertility.   Many of the nutrient ions which plants must extract from soils are held adsorbed by the colloids. This fraction of a soil may be likened to a vast storehouse in which mineral nutrients accumulate and from which they may be drawn as the plant needs them. For this reason it is a general rule that the finer the texture of a soil the greater its fertility (546). Permanent agriculture is possible on sandy soils only with the constant application of considerable amounts of fertilizer, a practice that is the more effective when colloidal matter is likewise incorporated into the soil. When sandy soils are located in regions where irrigation is necessary, care should be taken not to apply an excess of water which would leach away the few colloids and nutrients that such soils contain. In regions of high rainfall sandy soils are usually leached and infertile, resulting in the dwarfing of both roots and shoots of plants with high fertility requirements (330, 343).

Soil structure.   Except where the texture is uniformly coarse, the arrangement of particles in a soil is seldom haphazard. The smaller particles become crowded into the spaces between the larger, and the colloids form coatings over all the larger grains, binding them together into structural units of varying size and firmness, leaving planes of weakness between (722).

If a block of soil in which plants have been growing is excavated with a minimum of deformation, slaked in water for a few hours, and then gently washed onto a sieve which is agitated up and down in water a few times, it will be seen that small, soft lumps of water-soaked soil remain on the screen. * The material that binds the particles together into these lumps, or *water-stable aggregates,* does so with very little force, for they disintegrate under slight pressure from the finger. This loose cementing together of single grains into aggregates is attributable chiefly to colloids (Gr. *kolla* = glue); therefore good aggregation is not possible in coarse, colloid-free soils. To a large extent aggregation is a condition attributable to the activities of soil organisms and roots (12, 848).

The greater the degree of aggregation in a soil the more favorable is the soil for plant growth, for the aggregated condition makes the soil permeable to water, air, and rootlets (315), and at the same time the cemented complexes have high water- and nutrient-holding capacities. Well-aggregated soils are not easily eroded by wind and water, as are soils with single-grain structure. Obviously aggregation becomes more essential with increasing fineness of texture, for many of the desirable characteristics that result from this type of structure are possessed by coarse soils, even though they have single-grain structure. In short, a well-aggregated heavy soil possesses a combination of the favorable features of both coarse- and fine-textured soils.

Aggregates in turn are united into larger structural units. Near the soil surface these tend to be granular or crumblike, below they tend

---

* Determinations of porosity, permeability, or bulk density are other methods of evaluating soil structure (77, 665).

to be blocky, prismatic (Fig. 6), or platy, and are bounded by flat cleavage planes (722).

FIGURE 6. Claypan about 15 cm thick in steppe soil of western Nebraska. Cemented units are in the form of vertical prisms.

In general the morphologic attributes of soils result from different types of colloidal cementation. In certain cases, owing to poor aggregation, colloids are leached to a particular depth by percolating water where they are deposited as a type of claypan. *Because claypans are essentially impervious to water, air, and roots, they often govern to a large extent the types of plants that grow on a habitat, and the growth rate of trees tends to vary directly with the depth of loose soil above any restrictive layer (148) that determines the biologically effective depth of

---

* Brittle hardpans that do not involve high clay content may form as a result of the cementation of particles at a particular level in the soil by calcium carbonate, oxides of iron, aluminum, and silica, or by organic matter. Ecologically they have the same effect as claypans.

soil. In some areas it has proven necessary to blast holes through such layers when it was desired to afforest with deep-rooted trees, but in others forest planting can be used to improve the permeability of a watershed underlain by a clay subsoil (377).

Soil aeration.   In moderately coarse soils, as well as in heavy soils that are well aggregated, there exist large interstitial spaces which facilitate the diffusion of gases. As a result the carbon dioxide produced in a soil by the respiration of soil organisms and roots is able to escape rather easily, and the oxygen used up in this function diffuses into the soil with corresponding ease. In heavy soils, especially those that are poorly aggregated, the deficiency of oxygen and the toxicity of the excess carbon dioxide become limiting factors for many plants. On the other hand, sandy soils in which aeration is exceptionally favorable and temperatures are relatively high, have an inherent disadvantage in that the humus content remains at a permanently low level owing to the rapidity of oxidation.

Quite obviously the influences of fine texture upon aeration capacity are magnified by the high water-retaining property of heavy soils, for the more water a soil holds the less space is available for gases in the interstices. In moist climates masses of poorly aerated clay in the subsoil are often associated with an unhealthy condition of plants. For example, in the Ozarks twenty-year-old apple-tree root systems command 141 cubic meters or more of soil where the subsoil is light, but where it is heavy the space occupied may be less than half as great. Moreover, the trees on the heavy soils have reduced vigor and are especially susceptible to root diseases (743). There are, of course, plants native to such soils that find the conditions very satisfactory.

Soil temperature.   The porosity of coarse-textured soils and of heavy soils that are well aggregated tends to favor a condition of equilibrium between the soil and the atmosphere with regard to temperature. Because of the freer gas exchange and the lower moisture-holding capacity, light soils become warm enough for plant growth to commence distinctly earlier in spring than do heavy soils. Market gardeners generally prefer coarse soils for this reason, because produce which is ready for the market a few days in advance of the main season brings a very desirable margin of profit.

Another aspect of this relationship between temperature and texture is that what has been interpreted as root injury due to low temperature in winter is more pronounced in coarse than in fine-textured soils.

The Organic Increment

The inorganic particles that form the primary framework of soils completely determine the physical and chemical nature of the substratum until humus is added. Although the processes of accumulation and oxidation of humus usually reach an equilibrium before this constituent becomes 10% of the dry weight of the soil, humus is so important that the significance of a parent material as a medium for plant growth is vastly changed by this small organic increment.

Both plants and animals contribute to the organic matter of the soil. Some of the material is derived from dead roots and soil organisms and is therefore well distributed through the soil from the first. On the other hand, much organic matter is deposited upon the soil surface as leaves, twigs, wood, etc., and becomes incorporated into the mineral material only through the activities of organisms. In dense forests as much as several tons of litter may be deposited on an acre of ground annually.

Because roots and soil organisms inhabit chiefly the upper soil layers, and because all materials derived from plant shoots are deposited upon the surface, soils generally show a rapid decrease in organic matter from the surface downward. All loose and largely undecomposed plant debris which has but recently fallen to the ground is called *litter,* or the 01 layer. Beneath this lies partly decomposed debris that is somewhat matted together by fungal hyphae, and this is the *duff* or 02 layer. Under certain types of vegetation both these forms of surface debris are present, but elsewhere the duff, and even the litter, may be essentially lacking, at least at certain seasons.

After the normal biologic processes of decay decompose litter through the above stages, the resultant product becomes incorporated into the mineral soil, imparting a dark color to it. Such finely divided amorphous organic matter as has become mixed with the mineral materials is called *humus,* and the process leading to its formation *humification.* Humus is composed chiefly of those organic residues of the litter that have resisted decay the longest, but it also includes organic wastes synthesized by the soil-inhabiting organisms. Eventually all these compounds are completely converted into carbon dioxide, water, and minerals—a process known as *mineralization.*

The nature of the organic matter on and in a soil is strongly influenced by the chemical composition of the litter from which it is derived. Conifer needles contain acid-forming compounds, resins which are very resistant to decay, and are very low in Ca, Mg, and K. In contrast, the leaves of most deciduous angiosperm trees are nonacid forming and contain little or no resin but considerable quantities of the minerals listed above. *

The rate and degree of incorporation of organic residues into mineral soil vary considerably, depending upon climate and vegetation. In grassland climates litter decays rather quickly and completely so that at any one time there is very little material representing stages of decay intermediate between dead grass leaves and humus. Another peculiar feature here is that the multitude of finely branched, short-lived roots of grasses contribute a great amount of organic matter to the soil and this is mineralized so slowly that throughout most of the depth of root penetration the soil is darkened by humus. In fact, mineralization is so slow that the weight of humus in the soil far exceeds the weight of any one generation of plants, demonstrating that the humus content is the product of many generations of plants.

In forest soils the organic matter is of an entirely different nature and goes through very different stages of decay. Because tree roots are much longer lived than roots of herbs, the amount of organic matter added to the soil by the death of roots is relatively small in proportion to the amount derived from the litter. Another point of contrast is that the litter under forest conditions reaches the humus stage slowly, but after that mineralization is rapid. For these reasons forest soils are low in humus but often have well-developed layers of organic matter intermediate between litter and humus lying upon the surface. The humus content of forest soils is usually less than 100 metric tons per hectare, whereas in grassland soils there may be up to 1200 tons per hectare.

In general, the deeper the mineral soil has been darkened with humus, the more fertile the soil.

Organic-Content Classification

The term *mineral soil* is used for rooting media that are composed primarily of mineral materials. *Peat* and *muck* are largely organic accumulations formed in wet places, the former being so little decomposed as to be still fibrous, and the latter being well decomposed and amorphous (234).

Peat formed on wet ground or under water is named after the type of plant or vegetation from which it was derived, as *moss peat, sedge peat, woody peat,* or *limnic peat,* the last being formed by soft-textured submerged plants. If it is very deficient in plant nutrients it may be called

---

* These generalizations are not without a few conspicuous exceptions. *Thuja* and *Juniperus* are coniferous, but their litter is relatively high in the elements mentioned. *Quercus* is an example of a genus of deciduous angiosperms low in these elements.

bog peat, and if nutrient supplies are moderate *fen peat.* Deposits formed on wet ground, well above the water table, have been classed as *high-moor peat;* those formed near or below the water table and never rising much above it are called *low-moor peat.*

In forests where the soils are low in basic minerals and the litter is unattractive to worms and arthropods, there often accumulates a thick layer of duff, which may contain the bulk of the feeding roots. The ease with which this material dries, and the limited involvement of invertebrates in its decay, cause a very slow type of humification that is characterized by N deficiency (308). Foresters often call these *mor* soils. At other places where soils have higher base content, and the plant litter is attractive to invertebrates, litter is quickly changed to humus with much N becoming available from the excrement of the animals. These are distinguished as *mull* soils. Vascular plants as well as the soil organisms differ strikingly where an alternation of forest types brings areas of mull and mor into juxtaposition. For example, legumes, N-fixing bacteria, and earthworms are common in mull areas but not in mor.

Importance of Organic Matter to Plants

Source of mineral nutrients. All plants take up minerals from the soil and synthesize them into the complex organic compounds of which their tissues are composed (Table 1).

Table 1
Pounds of Minerals Taken Up Annually from One Acre by Different Types of Plant Cover. From Ebermayer's Data as Quoted by Schlich (677).

Plant cover types		Ash	K_2O	CaO	MgO	P_2O_5	SiO_2
Field crops	cereals	208	37	20	11	22	101
	noncereals	240	87	48	18	30	21
Forests	deciduous angiosperm	194	13	89	15	12	56
	evergreen coniferous	104	10	54	8	7	20

When the dead remains of such plants (or the animals which ate them) are returned to the soil, the complex organic compounds are broken down into humus, then mineralized into molecular and ionic forms which become available to future generations of plants. Because these mineral cycles require a long period for completion, a high proportion of the total quantity of biologically essential elements present on a particular habitat is constantly tied up in the existing populations of organisms. Therefore, the ability of the soil to provide adequate nutrients depends upon a return of these elements to the soil not too long after the death of the organisms growing there currently.

During long geologic eras the mineral constituents of living matter pass through an endless series of constantly recurring cycles, but this is not the entire story. Certain elements are continually lost to the atmosphere and to drainage waters, whereas the continued weathering of the parent material, the fixation of N by plants, and the deposition of dust return an approximately equal amount of new materials into the cycle.

Although in general humus is an important source of mineral nutrients and fertility is usually correlated with organic content, peat and muck are notoriously low in P and K despite their high organic contents.

When a virgin soil is first brought under cultivation it is almost invariably fertile, but with repeated cropping the productivity wanes sooner or later, because organic matter is destroyed and the nutrient cycle is interrupted by the removal of minerals from the habitat with each crop. In consequence production decreases steadily for a number of years until a new equilibrium is struck at a relatively low level, depending upon the extent, if any, to which nutrient cycles are augmented by fertilizers. Even where only the leaf litter is annually removed from the forest, as was once the practice in Europe to obtain material for bedding down livestock, the growth of trees is eventually curtailed to a serious extent. Much of this effect, however, can be attributed to the physical rather than to the chemical changes that are brought about (492).

The importance of organic matter as a source of nutrients is strikingly illustrated by the high level of fertility that develops in a new impoundment. The drowned vegetation decomposes in the course of a few years, and during this period the waters support abnormally high populations of algae, invertebrates, and consequently fish (857).

Source of food for most soil organisms. Green plants add more to a soil than they extract from it, for they absorb only small amounts of soluble minerals, and they return these plus even greater quantities of organic materials such as celluloses, lignins, starches, sugars, fats, and proteins. The addition of these high-energy compounds to a soil makes it possible for vast and complex groups of saprophytic soil organisms to develop, and in turn other series of organisms which parasitize these saprophytes develop.

Source of toxins. Agronomists have long noted that the yield of a crop may be strongly influenced by the kind of plant last grown on the field. As an example, crop residues of *Melilotus alba* (sweet clover) yield water-soluble organic compounds which depress germination and growth of *Zea* (maize) (501). Such materials seem to be common, but their ecologic importance tends to be mitigated by the rapidity with which they are inactivated by colloids or decomposed by microbes.

Water-holding capacity. Organic matter is colloidal, and because of this property its water-holding capacity is relatively high. In fact, a given amount of organic matter may hold as much as nine times its own weight of water, a proportion considerably greater than is ever true of clay colloids. Because of this characteristic the addition of organic matter to a sandy soil, either naturally or artificially as manure or compost, offsets the tendency of such soils to be drouthy. However, additions of organic matter to soils under cultivation bring about little or no permanent increase in humus, for cultivation results in rapid oxidation of such residues. The organic content of arable soils cannot be raised artificially except where the content was extremely low before the soils came under management.

Since the organic content is highest near the surface, this part of the soil generally has the highest water-holding capacity. Mismanagement of land in agriculture or grazing, or burning, has the strongest effect upon the surface layers of soil and, by destroying much of the humus in the upper layers, reduces both the water-holding capacity and the nutrient-storage capacity.

Table 2

A Comparison of Certain Properties of Two Forest Soils Developed on the Same Type of Parent Material in New York (126).

Forest type	Horizons[a]	Loss on ignition, %	Exchange capacity, milliequivalents	Exchangeable hydrogen, milliequivalents	Exchangeable bases, milliequivalents	[b]Base saturation, %	pH
Red spruce	O1 + O2	88.6	148.2	128.7	19.5	13.1	3.4
Yellow birch	A2	4.0	8.0	6.4	1.6	20.0	4.6
Balsam fir	B1	20.6	56.0	40.6	15.4	27.5	4.7
	B2	12.0	29.3	21.4	7.9	27.0	4.9
	C1	1.0	2.6	2.0	0.8	23.1	5.0
Sugar maple	O1	79.7	122.5	33.8	88.7	72.4	5.6
Beech	A1	33.3	55.8	29.5	26.3	47.1	5.0
Yellow birch	B1	17.0	30.8	19.7	11.1	36.0	5.1
	B2	12.5	27.2	17.9	9.3	33.8	5.2
	C	2.8	4.7	3.1	1.6	34.0	5.3

[a] The meaning of the symbols which are used to designate different layers of soil will be explained later.
[b] The sum of the adsorbed Ca, Mg, K, and Na.

Soil structure. Humus bears another resemblance to clay in that its presence in a soil helps make aggregate structure possible, since it occurs as a coating over the mineral grains. In fact, it is so much more important than clay in this respect that infiltration capacity commonly bears a direct ratio to humus content.

Humus colloids differ from clay in the manner in which they offset the excessive stickiness of inorganic colloids. When organic matter is mixed into a clay soil, the structure of the soil is vastly improved.

Adsorptive capacity. Organic matter, like clay, can hold by adsorption quantities of nutrients in ionic form, but it differs from clay in that its adsorptive capacity may be a hundredfold. In Table 2 it can be seen that the exchange capacity, a measure of cation-holding capacity which will be explained later, is closely related to the organic content as indicated by loss on ignition.

Mechanical effects of superficial layers of organic debris. Layers of litter and its products which lie on the soil proper are beneficial because they reduce seedling mortality due to frost-heaving, protect the soil surface from compaction by the beating of raindrops, and prevent runoff.

FIGURE 7. Three months after removal of a protective cover of tropical rainforest from this slope, approximately 8 cm of soil has eroded away except where pebbles protected finger-like columns from the beating of raindrops.

Raindrops falling upon bare mineral soil loosen the finer particles and force them between the larger ones, thus making the surface few millimeters relatively impervious. Runoff and erosion (Fig. 7) are also favored by the absence of the damming action of organic debris, which slows down the lateral movement of water over the surface.

Not all the effects of these organic layers are beneficial. They absorb and retain considerable quantities of water, up to an inch of rainfall, which consequently never gets down into the mineral soil where most of the roots are located. The litter may then dry out so quickly that seeds germinating in it desiccate before their radicles can establish contact with more permanent water supplies below. Species such as *Tsuga heterophylla* may germinate and survive chiefly on fallen logs, which retain water much better than do finely divided materials, but many important timber trees require bare mineral soil for seedling establishment. As litter accumulates in grassland fewer species survive, and even these decline in vigor (372, 821).

Soil Organisms

Kinds, Numbers, and Distribution

The principal kinds of living organisms that inhabit the soil can be listed as follows:

Plants	Animals (806)
Bacteria	Protozoa
Streptomyces	Nematodes
Algae (254)	Mites
Fungi (especially molds) (808)	Insects (especially ants and beetles)
Roots, rhizoids, and rhizomes	Earthworms
	Burrowing vertebrates (Moles, gophers, rats, etc.)

The majority of these organisms either feed entirely upon organic residues or are predatory upon organisms that do. This is not true of some of the bacteria, most of the algae, and practically all higher plants. Neither are the majority of burrowing rodents directly dependent upon the soil for their food.

Although most soil organisms are very minute, what they lack in size is compensated for by their numbers. The total animal population of soil and litter in a forest has been estimated as high as 107,000 individuals per square meter. Usually the weight of animals in the soil exceeds the weight of cattle that graze a pasture. Protozoa have been reported as numerous as a million per gram of soil, and the numbers of bacteria and *Streptomyces* may be still higher under favorable circumstances.

In virgin soils microorganisms remain fairly constant in numbers although they exhibit seasonal rhythm in regard to activity and dormancy. In cultivated soils, both cultivation and the periodic addition of organic matter cause great but transitory increases in the saprophyte populations.

Some of the soil organisms remain constantly in rather fixed positions, and others move about actively. The majority of the microscopic forms are held in the colloidal films which coat the larger soil particles. Also a special group of saprophytic organisms remains sedentary in close association with roots, and when the superficial cells of these organs die and are sloughed off, they begin their decay immediately. This zone around a root is referred to as the *rhizosphere,* and studies have shown that its biotic composition varies considerably among various species of plants.

In contrast to the organisms of rather fixed positions described above, there are some that move about continually in search of food, including protozoa, mites, and insects. Also, earthworms migrate with the season, being found at greater depths during dry or cold seasons.

The relative abundance of different classes of soil organisms is considerably influenced by climate and vegetation. For example, fungi are the dominant organisms in the acid soils of forests, whereas bacteria predominate in the neutral soils of grasslands. Species of *Streptomyces* attain their greatest abundance in the saline soils of arid regions.

Chief Roles of Soil Organisms

Decay and the nutrient cycles.   Soil organisms are the chief causes of decay of organic matter whereby elements are returned to the simple forms in which higher plants can use them again. Fungi, arthropods, and other organisms invade fragments of litter sometimes even before they fall to the ground (328). In the subsequent transformations each type of organic substance (carbohydrates, proteins, fats, etc.), as well as each stage in the breakdown of that substance, has its own set of saprophytes which acts only upon that material, so that decay progresses by stages each of which is accomplished by a separate set of organisms and yields a different intermediate substance. For example, proteins are broken down successively into amino acids, ammonium salts, nitrites, and nitrates, each step being the result of the action of a different organism or group of organisms. These cycles are so complicated that the mineralization of a particular annual increment of litter is commonly not completed until after subsequent increments have been added. Thus decay consists of a complex of overlapping cycles.

If no plant material is removed from an area the mineralization process which marks the end of these series of transformations usually releases an abundance of plant nutrients, for nearly all the N and important bases inherent to the area are conserved. However, conditions are not always favorable for decay. The organisms involved in humification and mineralization find optimum conditions only where the soil is warm, moist, well aerated, and not acid. Whenever one or more environmental conditions adversely affect the microorganisms of decay, even if only a single organism necessary in a sequence is affected, nutrient cycles are completed only at a very slow rate and the soil becomes depleted of essential elements, and the complex organic compounds in which they are locked accumulate.

Accumulations of organic matter such as peat are evidence that some factor has interfered with complete decay, the quantity of undecayed materials being inversely proportional to the rate of decay. Bogs and fens are usually cooler than the 22°C level below which bacterial activity is sharply reduced. On upland habitats also, temperature is a factor of widespread significance in humification, for, other factors remaining equal, the colder the climate the higher the humus content of soils.

Aside from their role in liberating nutrients from humus, soil organisms also have a considerable influence on the rate of solution of K, P, and Ca from soil minerals. This is largely a result of the dissolving action of carbon dioxide which the living cells secrete.

Production of toxins.   In poorly aerated soils the processes of decay of organic debris are entirely different from those in aerated soils. Reduction rather than oxidation is the predominating chemical process in the former, and certain of the resultant products (aldehydes, organic acids, vanillin, etc.) are strongly toxic to many species of higher plants (681). In aerated soils also, the products of tissue decay may be toxic to certain living plants, depending on the kind of microbes and the nature of their substrate (289).

Production of growth-stimulating substances (400).   A wide variety of heterotrophic soil organisms, including both bacteria and higher fungi, are known to produce 3-indol acetic acid and other growth-promoting substances. Although little is known of the actual significance of this factor, it is quite possible that these excretions, if produced in the soil at the root surfaces, constitute an important factor in the subterranean environment of higher plants.

Nitrogen fixation (733). Not only are elements already in the organic cycle perpetuated as such by microorganic activity, but one element, N, is continually being brought into the cycle by fixation from the rather inert gaseous form in which it exists in the atmosphere. Several groups of bacteria (*Azotobacter* in aerated soils of extratropical areas, *Beijerinckia* in the tropics, *Clostridium* in unaerated soils, *Rhizobium* in legume roots, *Streptomyces* in non-leguminous roots) and a number of blue-green algae are able to use N_2 and thus bring it into the organic cycle. The number of N-fixing * organisms in soil generally bears a direct correlation with fertility. Algal crusts that occur on the soils of arid regions appear to be quite important in fixing N (692), and in *Oryza sativa* (rice) paddies, where blue-green algae play a crucial role in fertility, these plants can fix as much as 78 kg N/ha/6 weeks (755).

Competition with higher plants for nutrients. Freshly fallen litter usually has 20–70 times as much C as N, but, since sugar and starches are oxidized, whereas N is merely transformed into microbial protoplasm, this ratio is narrowed during humification. The bacteria that decompose cellulose require N to synthesize new protoplasm following each division of a cell, and, being more efficient than roots in absorbing it, they tend to use up the N released in protein decomposition about as fast as it becomes available. As a result of the competition between roots and microbes, absolutely virgin soils of high fertility frequently give a negative test for nitrates. It is only after the C/N ratio has been reduced to about 20:1 that the microbial demand for N declines so that a surplus becomes available for vascular plants (572), and when the ratio drops to about 12:1 it tends to remain constant.

This inherent tendency toward N deficiency in soils can easily be aggravated by the addition of any organic matter high in cellulose but very low in N. Under such conditions certain of the bacteria involved in protein decay tend to shift their activity to carbohydrate decomposition, and this broadens the C/N ratio still further by reducing the rate of liberation of bound N. As a result the amount of available N in a soil heavily charged with carbohydrate may drop so low that it becomes a limiting factor in humus decomposition, and the higher plants growing on such a soil suffer from N deficiency. In agricultural practice an unfavorable C/N ratio is prevented by incorporating nitrogenous material (such as green manure) into the soil whenever straw or other carbohydrate-rich material is plowed under. In the forest an abundance of N-impoverished litter and duff sometimes accumulates to the extent that

---

* Nitrogen fixation, the utilization of N_2, is not to be confused with *nitrification,* which designates the oxidation of ammonium salts to nitrites, then nitrates.

tree growth is very slow and species that cannot use N in the form of ammonium salts are eliminated. Such a condition is relieved only when the organic accumulations are destroyed by fire or when logging alters the microclimate and allows more rapid oxidation.

Through competition, soil microbes may also critically reduce the supplies of other nutrients, especially Mn, to the point of creating a deficiency for higher plants.

Soil mixing. The larger soil organisms are responsible for considerable mechanical mixing and weathering of soil. The wedgelike action of roots and other underground plant organs widens the fissures in rocks and compacted soil so that they are more susceptible to other forms of weathering. Soil grains are rotated, and may be temporarily compressed between thickening roots so that infiltration is reduced (38). Trees that are blown down from time to time tend to produce pit-and-mound microrelief in a forest. This involves much mixing of mineral and organic materials (731), and on slopes it may be the principal agent of erosion. Rodents, insects, and worms turn over great amounts of soil in the aggregate, with a result that this material is repeatedly exposed to the physical and chemical agents of weathering. Worms are especially important in this respect. As much as 25 tons of material may be drawn through their alimentary canals and deposited upon the surface of a hectare of soil in a year, and at this rate in 65 years their casts should form a layer 2 dm deep on the old soil surface. Wherever these organisms abound * the physical character of the soil is strongly influenced by them (672).

Burrowing rodents also bring great quantities of subsoil to the surface, as is especially shown by the maze of "rodent eskers" which appear when winter snows melt to expose the materials that the animals have excavated and distributed along their intraniveal runways during the winter (Fig. 8). On the west slope of the Sierras of California it has been estimated that gophers bring approximately 25 kg of subsoil to the surface per hectare per year (293). In addition these animals cut and bury much vegetation, of which some is used for food and some for nesting materials. The latter, together with excreta buried below the surface, greatly increases the organic content and hence improves the soil.

Improvement of soil aeration. Those macroscopic organisms that in-

* Earthworms demand for their food litter that contains much mineral matter. Conifer needles and oak leaves are so low in Ca and Mg that earthworms are few or lacking in areas dominated by these trees, and among other types of litter strong preferences are exhibited.

habit the soil greatly improve its structure by facilitating aeration. Root decay leaves the soil riddled with channels, and the burrowing of worms and other animals creates innumerable passageways by means of which oxygen can enter and carbon dioxide leave a soil readily. Root tips elongating through soil readily follow any such channels they encounter. Water penetration is also promoted by the system of interlacing channels, sometimes to the extent that an undesirable amount of loss by percolation takes place. Field tests have shown that infiltration and subsequent percolation are much more rapid under grasses than under tap-rooted herbs, presumably because of the greater abundance of roots in the grasses (599).

FIGURE 8. Working beneath winter snow cover, small rodents bring much mineral material to the surface. Although this activity thins the plant cover, it also keeps the soil porous and transfers organic matter below the surface.

Improvement of aggregate structure. Bacteria and blue-green algae, both of which are abundant in the soil, are especially characterized by a mucilaginous excretion in which each cell is usually enveloped. This mucilage plus other organic excretions of the cells is very effective in cementing soil grains into aggregates. A similar function is performed, although in an entirely different manner, by the hyphae of soil fungi which bind soil particles together.

Injury of higher plants by soil organisms. Many soil-inhabiting organisms live, or can live, by eating or otherwise parasitizing the subterranean organs of plants. The larvae of June beetles frequently cause considerable damage to lawns, and similar injury by insects must be common in nature where, for the most part, it passes unnoticed. Wireworms are serious pests to crops and bring about no little damage to tree seedlings. Damping-off fungi likewise kill many seedlings, both in natural habitats and in propagating beds. Soil-borne diseases of crop plants, such as root rot, crown gall, potato scab, and nematodes, cost millions of dollars annually.

Removal of water and solutes by roots. Roots have several special features which facilitate their role as absorbing organs. Other conditions being equal, the ability of a root system to absorb is directly proportional to the amount of surface exposed, and this is very great because of the abundant rootlets and still more numerous root hairs. The roots of a 4-month-old rye plant were found to have a combined length of 240 km and a total surface of 639 m^2 (200). Roots of *Prosopis* have been found 53.3 m below the soil surface (619); the roots of *Pinus palustris* extend 26.6 m from the trunk. Generally the roots of grasses do not extend as deeply or as widely as those of trees, but in the tropics they may penetrate to 5.5 m (361).

Roots tend to dry the soil, at least during the growing season, and thus have a pronounced influence upon weathering, leaching, mineralization of humus, and other processes. By absorbing minerals that are later cast on the soil surface as litter, roots play a crucial role in perpetuating the vertical cycling of nutrients.

Soil Moisture and Air

Soil particles, together with the bodies of soil organisms, may lie as close together as their shapes permit, but there always remain between them many small angular cavities connected irregularly by still smaller openings. All these interstices, collectively referred to as *pore space,* comprise a fairly constant volume (usually 40–60% of the total soil volume) which is filled with water and gases in varying proportions. In a "dry" soil water occupies a very small percentage of this space; in a "wet" soil the converse is true. The relative amounts of water and air contained in the pore space are very important and will be discussed in detail later.

The specific gravity of mineral soil particles varies but slightly from the mean value of 2.65. However, the dry weight of a given volume of undisturbed soil (its *bulk density* or *volume weight*) is usually but 1.4 to 1.8 times as great as the weight of an equal volume of water. The difference between real and apparent specific gravities is due chiefly to the amount of pore space (and somewhat to humus content) and therefore constitutes a measure of this important characteristic. Coarse-textured soils generally are more compact and less well aggregated than fine soils and in consequence tend to have higher bulk densities and less pore space. Except in sands, low bulk densities in a given type of soil are most favorable for plant growth.

Within a particular soil type, bulk densities vary, depending upon factors which alter the degree of aggregation and compactness. The trampling of cattle or people, and the passage of machinery, are effective agents of soil compaction, and comparisons of bulk densities between virgin and disturbed areas nearby serve as a criterion of the extent to which soil structure has deteriorated (354, 529). When the bulk density reaches a critically high value, roots cannot elongate to tap further supplies of water and nutrients, and the drainage of any excess water is greatly retarded (39). However, it will be pointed out later that variations in the relative proportions of large and small pores are generally greater and of more ecologic importance than variations in the total pore space.

Soil Solutes

All the water contained in a soil, together with its dissolved solids, liquids, and gases, is termed the *soil solution*. Of the 15 or more elements essential for plant nutrition all but C, H, and O are derived exclusively from the soil, and it is believed that most substances taken up by the roots must be contained in the soil solution.

Plants cannot live normally when any of these nutrient elements are present in amounts too small, or in combinations too complex for absorption. Neither can they live normally if the soil solution contains a harmful excess of one or more solutes such as acids, bases, salts, organic toxins, or even fertilizers. It is only when the soil solution has neither too little nor too much of any solute that the soil can be considered fertile. Fertility, however, is always a relative matter, for a soil fertile to one kind of plant may be infertile to another with higher or different nutrient requirements, and the same soil is less capable of satisfying nutrient requirements under conditions of excessive or deficient moisture.

Adsorption and Exchange of Ions

Except for N, the nutrients required by the average plant have their ultimate origins in parent materials, from which they are liberated as the parent materials weather to clays. The chemistry of rocks that form parent materials is therefore important. Among the igneous rocks is an acidic series (granite, granodiorite, rhyolite, gneiss and schist) which generally produce infertile soils, and a basic series (basalt, andesite, gabbro and diabase) which are much better sources of plant nutrients. Sedimentary rocks include limestone which usually gives rise to highly fertile soils, and sandstone that produce soils the fertility of which varies directly with the clay content of the rock.

As humus develops in a soil it becomes an additional secondary source of supply, since the essential nutrients used by one generation of plants are passed on to the next by decay of organic remains. Rain, especially after having come in contact with foliage, is still another source of nutrients (383), and sometimes carbohydrates as well. Plants protected from rain may contain as much as 25% more ash than plants not so protected (540), and moss growth beneath tree canopies may be markedly increased by solutes leached from the foliage above (745).

Of the total quantity of plant nutrients which a chemical analysis shows to be present in a soil, the vast bulk is usually combined in mineral or organic compounds too complex to be immediately useful to the plant. As the weathering of minerals and the mineralization of humus slowly render these nutrients soluble, they may be (a) washed out of the soil by leaching, * (b) absorbed at once by soil organisms and roots, or (c) if ionized they may be adsorbed and temporarily held out of solution on the surfaces of the colloidal micelles.

The micelles of colloids are negatively charged and therefore attract swarms of cations which become loosely attached to their surfaces. Each micelle has its charges balanced by holding sufficient numbers of hydrated cations of Ca, Mg, K, NH_4, Na, Fe, H, etc. This electrical union of cations with a micelle takes the cations out of solution so that the micelle with its swarm of cations is analogous to a molecule of an insoluble salt. However, the adsorbed and the free ions are in equilibrium,

---

* This phenomenon of leaching and percolation is best studied by growing plants in large containers arranged so that the percolate can be collected for analysis. Such an apparatus is called a *lysimeter* (417).

so that as free ones are used up or lost others are released. Therefore the colloids with their adsorbed ions constitute a vast storehouse of available plant nutrients.

The composition of these cation swarms exerts a powerful influence upon both the chemical and physical attributes of a soil. In temperate and cold climates with moderate or high rainfall, H ions are formed in abundance in the soil. They are of no direct use to plants, and when they dominate the colloidal complex there is nothing to prevent the loss by leaching of the important basic ions, especially Ca, Mg, * and K, and fertility is consequently low. This condition may arise as the result of differences in the abilities of ions to displace each other, the order being H, Ca, Mg, K = NH_4, and Na. The degree of saturation with basic ions can always be reduced by cation exchange wherever there are enough H ions to bring about this displacement. Once displaced the bases are carried away in drainage waters. Colloids are said to be unsaturated when the adsorbed bases have thus been reduced to a very low level. In Table 2 it will be noted that

Exchange Capacity = Exchangeable H + Exchangeable Bases

and these figures provide the basis for calculating percentage base saturation. In acid mineral soils (pH 5.0 and lower) Al plays a role similar to that of H in moderately acid soils as described above.

Because chemical analyses of soil do not show the relative proportions of available and unavailable forms of most critical elements, they are of limited value. A much more significant measure of soil fertility, as far as basic nutrients are concerned, is a determination of the quantity of exchangeable bases held by the colloids. † Fertility can also be deter-

---

* In most good agricultural soils Ca plus Mg comprise 90–95% of the adsorbed bases.

† Metallic cations which are instantaneously released from colloidal adsorption on treatment with certain salt solutions are called *exchangeable bases*. Soil samples are commonly treated with a neutral salt such as ammonium acetate, which is then filtered off (277). Due to mass action all the adsorbed basic ions are replaced by the more strongly adsorbed ammonium ions and, along with cations in solution, appear in the filtrate as acetates. The acetates can be analyzed and the results expressed as milliequivalents (m.e.) per 100 g of dry soil. Of the three most important clay minerals, montmorillonite has a cation capacity of 80–150 m.e., illite 15–40 m.e., and kaolinite 3–15 m.e. per 100 g. The cation capacity of humus may exceed 300 m.e. per 100 g.

If, after leaching as above, the amount of adsorbed ammonium is determined by analysis of the residue, the total *cation exchange capacity* of the soil is known, and by using this figure in connection with the first obtained above it is possible to calculate the *percent base saturation* (Table 2).

For studies in which the relative amounts of different basic cations are not important, total exchangeable bases can be estimated by a rapid technique involving the use of a pH meter and suitable reagents (104).

mined by controlled experiments in which the response of plants to various fertilizer applications is observed (340, 409), by determining the specific electrical conductance of the soil solution (27), and by chemical analyses of plant tissues to determine the amount of nutrients which have been taken up from different soils (541, 773).

The discussion thus far has taken only the cations into consideration, but some important elements exist in the form of anions, for example, N, P, and S. These occur in the soil solution along with small amounts of cations, and it appears that P and S can also be adsorbed and stored on the colloids owing to their amphoteric nature. In comparison with cations, anions are held less firmly by the colloids and therefore are more subject to loss by leaching.

Plants absorb some solutes as undissociated molecules, some as free ions; and some ions may be exchanged directly between plant and soil micelle without going into free solution. In any event, because of the energy relations involved, an ion cannot be absorbed unless the protoplast simultaneously releases another ion of like charge. This phenomenon of constant ion exchange between plant and soil may in part account for the universal secretion of carbonic acid by roots, for this acid is always abundant in nature and readily yields ions of both charges.

Nutrient Deficiency

In cultivated soils, ions of N, P, and K are the ones most commonly present in suboptimal amounts; hence they are the chief constituents of most commercial fertilizers. In humid regions the loss of Ca by percolation makes this a fourth ion of great importance in maintaining fertility, but this problem is rather easily overcome by the periodic addition of crushed limestone (chalk) to the soil. S and Mg are also frequently limiting, and many cases have been reported where the minor elements, those such as Cu, B, Zn, Mo, and Mn, which are needed only in minute traces, exist in concentration suboptimal for plant growth or in unavailable combinations.

Although in nature soil fertility is not continually reduced through the annual removal of material from the area, as is true of cropped lands, nutrient deficiency is definitely a factor of considerable importance under certain conditions. When a raw parent material first becomes available for plant colonization, it contains practically no N, P, or S, all of which ordinarily become available to plants through the mineralization of humus. Yet plants colonize such parent materials, and before long these nutrients appear and accumulate. Undoubtedly the pioneer plants on such habitats have low requirements for these elements but absorb and accumulate the minute amounts which become available to them through impurities in rainwater (383, 399) or from the weathering of minerals. For example, the liverwort which colonizes freshly deposited volcanic ash in Alaska can thrive on media as low in N as it is possible to synthesize (290). Also there is a good possibility that certain of these pioneer species may be able to play the role of an invader of relatively infertile substrata chiefly because their roots, by means of vigorous CO_2 excretion, have superior abilities to extract ions from difficultly soluble minerals. It is quite likely that the orderly process of development of vegetation on certain raw parent materials is governed more by the rate of increase in fertility than by any other single factor (10).

Nutrient deficiency may arise through the action of natural vegetation upon a soil originally more fertile. Plants such as coniferous trees may synthesize so much acid-forming material, which is deposited as litter, that soil colloids eventually become saturated with H, and nutrient bases are lost by leaching. As this change is brought about, the vegetation, especially the subordinate undergrowth, changes from a rich type which is characteristic of fertile soil to an impoverished one of species that endure low fertility.

Waters vary considerably in their nutrient content, and, since this factor governs the quantity of algae and other water plants that they can support, fertility consequently determines the kinds and abundance of fish and other animals that depend upon the plants for food. Fertilized and well-managed ponds have been made to yield over 720 kg of fish per hectare per year! However, this response commonly is of short duration unless fertilization is regularly repeated, for the soluble nutrients soon are converted into insoluble compounds (515, 857).

Water derived from melting snow may be nearly free of solutes, commonly containing only a few parts per million. As a stream flows from its source it accumulates more and more solutes and loses water continually by evaporation. This process gradually increases the solute content, which commonly reaches 800 ppm, and is referred to as the *aging of water*. Obviously the type of rock and parent material with which waters come into contact govern the kinds and amounts of solutes they will contain (842).

Where water moves horizontally below the ground surface, yet within the rooting zone, fresh supplies of nutrients constantly pass through. Vegetation tends to be taxonomically richer with the plants growing more vigorously on such sites, which are common on lower slopes (280).

Waters or soils with suboptimal concentrations of nutrient solutes are called *oligotrophic* (Gr. *oligos* = little, *trophikos* = nourishing), and those with near-optimal solutes are *eutrophic* (Gr. *eu* = well). Fens are eutrophic; bogs are oligotrophic.

Plant reactions to nutrient deficiency are quite varied. By means of water cultures, physiologists have long been able to demonstrate the symptoms exhibited by plants suffering from complete or nearly complete lack of each essential element. These symptoms consist of combinations of morphological and pigmentation characters usually so distinctive that a trained observer can tell at a glance the nature of the deficiency. Also the infertility of a soil is reflected in the size and health of animals feeding upon plants that grow on it.

Roots remain sparingly branched where the nutrient levels are low but branch profusely when they encounter fertile areas of soil (817). Nutrient deficiency generally decreases the amount of shoot even more than that of the root (46).

Many plants fruit prolifically only when the nutrient level is relatively low. In certain species of *Arisaema,* staminate plants are associated with infertile and carpellate plants with fertile soil, and the change from one sexual state to the other can be brought about experimentally by increasing or decreasing fertility (675). Such plants are to be considered potentially monoecious, with soil fertility controlling sex expression.

The inherently slow growth of many plants may well be an important evolutionary adaptation to low fertility, since their slow rates of nutrient uptake are in harmony with the limited rates at which the soil can supply them (348).

Acids and Bases in Soils: Soil Reaction (714)

Solutions are acid wherever they contain more H than OH ions; they are basic where the reverse is true. If both types of ions are present in equal concentrations the *reaction* is said to be neutral.

Reaction is usually expressed by a pH scale, which is the negative log of the H ion concentration. From neutrality, written as pH 7.0 (an abbreviation of $10^{-7.0}$), numbers represent increasing alkalinity up to pH 14, or increasing acidity down to pH 1. The concept of pH originated in connection with simple ionic solutions. Later, as it was applied to soils there arose a strong conviction that the same kind of determinations made on soil suspensions have a different significance and should not be treated as logarithmic. The contention is supported by several phenomena. (a) The pH of a freshly stirred suspension changes as the solids settle out, suggesting that the electrode is not only affected by ions free in the water but by those adsorbed (565). (b) Random measurements of environment in a small area exhibit a normal frequency distribution, but do so with regard to pH only when the values are used arithmetically. (c) When pH is plotted against related soil properties which appear to be arithmetic (e.g., percentage of adsorbed Na), a straight line fits the data when ordinary graph paper is used. Therefore, pH values of soils are now usually treated as an arithmetic series in computations.

Most types of soil and organic matter contain certain types of colloids or solutes that have the property of keeping the reaction of the soil solution fairly constant. These substances are called *buffers*. When bases are formed in or are added to a soil the buffers liberate H ions in sufficient quantities to keep the reaction constant, whereas OH ions are freed if acids are added. From the standpoint of plant nutrition a highly buffered soil is most satisfactory for it prevents violent fluctuations in the nutritive balance of the soil, which would frequently tax the ability of the plant to make adjustment.

To a certain extent the plants themselves behave as buffers. Experiments have shown that through exchanges of ions between soil and root the pH is always adjusted in the direction of the optimum range of the species. This may be accomplished directly by an unequal absorption of acidic and basic components of the soil solution or by secreting carbonic acid from the roots, and indirectly by producing litter which releases either predominantly basic or predominantly acid residues.

Measurement of pH (646). Titration is of no value in connection with pH, for acids that ionize very little give high normality upon titration, even though they free relatively few of the biologically important H ions. Furthermore, adsorbed H ions would be neutralized by titration, and these too are inactive. Thus measurements of total acidity are far less significant than measurements of the relative proportions of H and OH ions.

The concentration of free H ions may be measured by the use of colorimetric testing sets which are relatively inexpensive and at the same time accurate to about 0.2 pH. In this method various indicator dyes are added to water extracts of the soil and the color the extract assumes is compared to a standard color chart from which pH is read directly (136).

Electrometric apparatus, especially the glass electrode, is even more convenient and more accurate, although greater accuracy is of questionable value in field work. The method employing the glass electrode is the most highly recommended. Although electrical apparatus is more elaborate and more expensive than the colorimetric outfits, convenient instruments that are relatively simple to operate in the field are available, and readings are reliable to 0.1 pH.

The pH of soils may be tested immediately upon collecting the samples, or the samples may be air-dried and stored in a chemical-free room. Air-dry soils may be screened, but should never be triturated in a mortar when they are to be used for pH determinations. Soils need to be moistened to make electrometric determinations, and if they have been dried the moistened soil should be allowed to equilibrate for 24 hours. Although acidic mineral soils may be mixed with as much as 2 parts of water, and peat with 7 parts of water, without affecting their pH, certain basic soils undergo hydrolysis with so much dilution and give considerably higher values than obtained under field conditions. Consequently these should be moistened with just enough water to give them the consistency of thick cream. Soil suspensions should be stirred before the electrode is inserted, and then the electrode should be agitated briefly in the suspension.

Field soils contain much CO_2 which is dissipated on sampling, and this gives an unnaturally high pH test. To counter this, many workers use 0.01 molar $CaCl_2$, rather than distilled water, to moisten soils for testing.

Sources of acids and bases.   Acids can be formed in soils in the following ways. (*a*) Parent materials derived from light-colored igneous rocks that have much silica and oxides of Na and K tend to be acid owing to the chemical character of the products of rock decay. (*b*) Acids are the products of metabolism of soil organisms. Depending upon the types of organisms and prevailing soil conditions, oxalic, lactic, acetic, nitric, sulfuric, and other acids may be formed in addition to the ubiquitous carbonic acid. (*c*) Certain types of plant litter yield considerable quantities of acidic materials when they decompose. In this way most species of Pinaceae, Ericaceae, *Sphagnum, Polytrichum,* etc., strongly decrease the pH of the soil upon which they grow, unless it is already strongly acid. (*d*) The metabolism of aquatic plants has a marked rhythmic effect upon the reaction of the water. The pH is lowered at night as the result of the liberation of CO_2, in respiration, but during the day green aquatics absorb this gas in photosynthesis and the trend in pH is reversed. (*e*) Smoke and fumes from industrial plants, especially those containing sulfur dioxide, frequently lower the pH of soils in the vicinity.

Basic soil reaction is the result of an abundance of salts of types that hydrolyze to yield strong bases. In humid regions the salt involved is almost always calcium carbonate; in arid regions both calcium and sodium carbonates are common. These basic salts may have either of two origins. (*a*) Limestone decomposes to liberate great quantities of calcium carbonate. Also, there are dark-colored igneous rocks, low in silica but high in oxides of Ca, Mg, and Fe, which yield basic compounds as they decay. (*b*) Drainage waters may carry basic salts formed where rocks are weathering into poorly drained depressions from which the water evaporates, leaving the carbonates concentrated.

Spatial and temporal variations in soil pH. The sea is a medium of constant and uniform pH of about 8.1, but on land pH varies widely from one habitat to another, and even within the same habitat it differs at different levels in the soil. Surface soils are nearly always more acid than subsoils because of the greater abundance of acid-forming organic matter and the stronger leaching action in the upper levels. The bases replaced by H in the upper soil levels may accumulate at lower levels or in depressions in the topography, so that in hilly regions the hilltops commonly are more acid than the base-enriched valleys. On mountain slopes soils generally increase in acidity upward because of the vertical increase in organic content and in leaching intensity.

In warm, dry climates soils are usually circumneutral to strongly basic because there is insufficient rainfall to leach away the bases as soon as they are released in weathering, and few acidic materials are produced there by natural processes of decay. The soils of cool, moist climates characteristically vary from slightly to strongly acid.

Also, pH is known to exhibit a rhythmic seasonal change in the same soil, but owing to the abundance of buffers this cyclic fluctuation is slight and therefore usually unimportant.

Relation of pH to other soil characteristics. The calcium status of a normally drained soil is fairly well correlated with its pH. Above about 8.3 soil usually contains free * calcium. The range 8.3 to 6.0 indicates saturation with bases, among which calcium is usually the most abundant. Soils with about half their cation capacities saturated with bases and half with H tend to have a pH of about 5.0. At 3.8 and lower the colloids are practically saturated with Al and H, and only those plants can grow that can endure extreme calcium deficiency.

Both H and Ca not only are antithetic in regard to soil chemistry but they are also biologically very active and have very different effects upon plants. Many species of moist regions grow only on basic soils containing free $CaCO_3$ or on circumneutral soils from which the Ca has not yet been lost from the colloids. These are called *calciphytes*. They disappear from an area as the soils become acid and Ca is lost, or they become confined to springs or stream margins where the soil solution may remain circuneutral. *Oxylophytes* are plants associated with acid soils. They may possibly be excluded from alkaline soils owing to the insoluble state of Fe or P, and at the same time require little Ca. *Leguminosae* are

---

* Soils with approximately 1% or more of free calcium carbonate will effervesce when treated (after wetting) with a few drops of 20% HCl. In agricultural soils this much lime is detrimental to certain crops.

mainly confined to neutral or alkaline soils because their root nodule bacteria require such environment.

All plants that have been studied have been found to need at least small amounts of Ca for various metabolic roles, but there is every gradation of tolerance and/or requirement for this ion among species. In the colder part of the temperate zone, where soils developed from noncalcareous parent materials soon become strongly acid, the floras of these strongly acid and of limestone soils are often strikingly distinct (238).

An apparently paradoxical situation is frequently encountered in which plants characteristic of acid soils are found growing on shallow residual soils derived from limestone, where their roots are mainly confined to a thin surface layer which is decidedly acid. Presumably the seeds of such oxylophytes have germinated on decayed logs or other bits of nonbasic debris, and then the plants were able to enlarge the initial mass of Ca-free soil through their own activity. There are many instances where small initial differences between two habitats favor different plant populations, and these in turn exert such divergent influences upon the habitats as to render them markedly different.

The pH of a soil indicates the availability of other nutrients as well as of Ca. At approximately 6.5 all minerals are sufficiently soluble to satisfy plant needs, but with increasing deviation in either direction certain nutrients become less soluble (Fig. 9). Also nitrification is impaired below 6.0 and above 7.7. pH is not a quantitative indicator of fertility, since clay of low pH may still supply basic ions in abundance owing to its high cation exchange capacity.

Still another aspect of soil chemistry, not indicated in Fig. 9, is that at extremes of pH the balance among nutrients becomes unfavorable, and certain materials become so excessively soluble as to be toxic. At low pH, for example, Al, Fe, Mn, Zn, and Cu may be toxic. The H and OH ions themselves are directly injurious below pH 4 or above 9.

An indirect influence of pH on higher plants is felt through its effect upon soil-borne diseases. Although the principle must also operate in nature (187), studies prompted by economic considerations provide practically all our sources of evidence. These researches have shown that some important crop diseases can be controlled where the pH tolerances of host and parasite differ. For example, low pH can be employed to control the organism causing root rot of cotton (*Phymatotrichium omnivorum*), potato scab (*Streptomyces scabies*), and root rot of tobacco (*Thielavia basicola*). On the other hand, high pH levels are effective in controlling the fungi that cause damping-off, the rhizoctonia disease of potatoes (*Rhizoctonia solani*), and club root of cabbage (*Plasmodiophora brassicae*).

If the electric charges of the micelles of soil colloids are not satisfied with certain types of cations, the residual negative charge makes them repel each other when Brownian movement causes them to collide. But when the micelles become dominated by adsorbed divalent cations, especially Ca or Mg, the negative charges are neutralized to the extent that the micelles come together to form small units called *floccules*. Flocculation (electrical attraction) is a less permanent condition than aggregation (cementation) but seems to be a prerequisite for it, and hence for a porous condition in fine-textured soils. Under high acidity the amounts of Ca and Mg are inadequate to cause flocculation, and H, though positively charged, seems unable to perform the same function if organic colloids are abundant. On the other hand where soils are strongly basic owing to excesses of adsorbed Na or K, they likewise are deflocculated. Thus, under either high or low extremes of pH a soil has all the undesirable features associated with single-grain structure, and unless flocculated, colloidal micelles are readily washed from the surface to deeper levels.

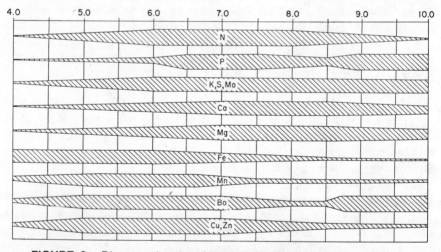

FIGURE 9. Diagram showing the relation of soil pH to the relative availability of nutrient elements required by plants. It should be noted that owing to other environmental variables, a favorable pH range cannot be construed as certain indication of adequate supplies of a given element, and vice versa. Furthermore, the significance of extreme conditions of pH differs for crop plants and for other plants native to such soils. [Modified from Truog (765).]

In acid forest soils the principal microorganisms of decay are acid-tolerant fungi. Bacteria and earthworms are sensitive to acidity and are abundant only in circumneutral soils. Low pH levels consequently interfere with nitrification and with the functions of some important classes of decay organisms, thus allowing residues to accumulate. These are chiefly intermediate products of incomplete decomposition and are normally nontoxic, but when they accumulate in great abundance they become decidedly toxic.

Since soil reaction is so closely correlated with so many other concomitant variables, it is a valuable indicator of these correlated conditions, and many plants have restricted ranges of pH under which they will grow (727). However, it should be quite clear that to assess individually the relative importance of each of the chemical, physical, and biological properties of the soil which vary concomitantly with pH is a difficult if not impossible task, and pH is probably not as directly related to many biologic phenomena as its close correlation may lead one to assume. Still, no other single value for a soil tells so much about its ecologic character.

Control of pH. Where agriculture is practiced on acid soils, a common method of offsetting most of the unfavorable conditions associated with low pH is to make applications of crushed limestone from time to time. This replenishes the soil with Ca, neutralizes the acids, and thus retards loss by exchange and leaching of other basic nutrients. In east central North America approximately 4 metric tons of limestone per hectare are required every 15 years to balance the loss of Ca by leaching (775). Crushed dolomite is another highly valuable amendment where available, for it supplies the important base Mg in addition to supplying Ca and counteracting acidity.

Changes in the reverse direction, from a basic to a neutral or acid condition, can be accomplished through the use of acidifying amendments such as sulfur, sulfates, *Sphagnum* peat, litter from certain trees (e.g., *Quercus, Pinus, Picea, Abies*), or bark residues from tanneries. Under certain conditions, to be discussed later, artificial drainage can be used to reduce the basicity of sodium soils. None of these methods is cheap enough to apply on an extensive scale, so that ordinarily the utilization of basic soils is economically possible only by choosing crops that are at least base-tolerant.

Whenever commercial fertilizers need to be applied, soil reaction can at the same time be altered by choosing particular combinations of chemicals which will yield the necessary nutrient ions and also react in the soil to produce the desired change in pH.

Residual Soils with Unbalanced Mineral Constituents

Residual soils developing from special kinds of rocks often support a thin or discontinuous plant cover composed of relatively few taxa, many of which are peculiar to these soils. Such areas are frequently referred to as "barrens." Either certain of the physiologically essential elements are present in critically low concentrations, or certain elements are unusually abundant and so soluble as to be toxic. The plants tolerant of such conditions are dwarfed and widely spaced, in comparison with contiguous soils with better nutritional balance.

Some of the rocks that characterize barrens in different parts of the world are limestone (calcium carbonate), dolomite (calcium carbonate containing appreciable magnesium carbonate as well), magnesite (magnesium carbonate), serpentine (498, 628) (hydrated magnesium silicate, usually deficient in Ca if not in Mo, N, K, or P also, but with excessive Ni, Co, and Cr), gypsum (hydrated calcium sulphate), and calamite (hydrated zinc silicate). The influence of these minerals on plant distribution or chemical composition has often proven useful in locating new sources of ores (120).

Occasionally deep soil mantles are simply deficient in one or more nutrients so that the kinds of plant or their rates of growth are strikingly affected. Thus in Nevada an islandlike stand of *Pinus ponderosa* surrounded by steppe has been explained on the basis of P and N being deficient for the steppe vegetation (668). In Australia contiguous areas of scrub, low forest, and tall forest appear to owe their distinctiveness to very low, low, and adequate supplies of P (46).

Mine Wastes as Parent Materials

Ecologically similar to the above are problems presented by mine wastes that accumulate where ores are processed or where strip-mining is practiced (680). These materials are highly unfavorable for plant life owing to nutrient imbalance, or more seriously, to highly toxic amounts of heavy metals (Pb, Zn, Hg, Cu, etc.) often combined with low pH, sometimes as low as 2.5.

The steady increase in area of these deposits, and in the biologic devastation of streams flowing from them, has stimulated much research into methods of their revegetation. However, their reclamation has barely started when the difficult process of developing a new plant cover has been achieved. Thus these habitats are products of short-lived industries that produce long-lived derelict land. Environmentally conscious communities variously require mining companies to level the surface, conserve topsoil and redistribute it over the waste, encorporate garbage into the surface, etc.

Salinity

European ecology of the nineteenth century was notable for its controversies as to whether the physical or the chemical properties of soil are the more important. In 1910 Gola, by his "osmotic theory of edaphism," settled this question by demonstrating that either property may be controlling, depending upon its relative intensity. His studies led him to conclude that, when the soil contains 0.5% or more by weight of soluble salts, chemical attributes are the more important in determining its ecologic character; otherwise, physical differences (especially colloid content) are of more importance. This theory proved to be an important landmark in the development of our understanding of the relationships between plants and soils.

Halophytes and their distribution.   A number of economic plants, such as *Beta vulgaris* (sugar beet), *Prunus communis* (almond), and *Medicago*

FIGURE 10.   Zonation of plants around a saline basin, Grand Coulee, Washington. The wettest and saltiest soil in the center of the basin is devoid of macroscopic vegetation. Surrounding this is a broad belt of salt grass *(Distichlis stricta)*, then a narrow belt of greasewood *(Sarcobatus vermiculatus)*. In the distance is nonhalophytic vegetation dominated by sagebrush *(Artemisia tridentata)*.

sativa (alfalfa), can be made to grow on saline soils and therefore may be classed as facultative halophytes, but true halophytes are those plants that occur naturally only on soils or in water too salty for the average plant. Many experiments have shown that halophytes do not require salinity; for unknown reasons they can complete their life cycles only in saline environments.

Relatively few species, especially of mosses and ferns, have been able to adjust their physiology to the conditions prevailing in saline habitats. Among those that are so adapted the ranges of salinity tolerated varies somewhat with species, so that different species of halophytes indicate by their presence differing degrees of salinity. Thus the vegetation of the strand as well as that around the periphery of inland salt basins (Fig. 10) often exhibits a remarkably definite sequence of zones. It was once thought that these zones of vegetation reflect principally the degree of salinity, but research has shown that the poor aeration which accompanies increasing moisture and salinity is also an important factor (230), so that the correlation between plant distribution and salt content is far from perfect. In general, the greater the salt tolerance of a species, the wider the range of salinity of the soils on which it grows, i.e., the degree of maximal salt tolerance is more definite than the minimal (248, 611).

The osmotic potential. In thermodynamic terminology pure water at atmospheric pressure has zero free energy. As solutes are added, its declining free energy is expressed as negative bars (1 bar = 0.98692 atmosphere) of *osmotic potential*. Water tends to move in a direction of decreasing potential. Therefore as the concentration of salts increases about a root, it can take up water only if it can keep the concentration of solutes in its cells greater than that of the soil solution, i.e., maintain a more negative osmotic potential. Once within the plant, tension caused by transpiration then draws the water upward.

In most well-drained soils the water potential lies between −2 and −30 bars. Halophytes commonly must develop potentials to about −50 bars, with an extreme record of −202 bars for *Atriplex confertifolia* growing in very salty soil. In such plants the osmotic potential in the roots is kept low by freely taking up salts without risk of ill effects, but when *glykophytes* (i.e., plants normally found only in nonsaline habitats) adapt themselves to somewhat saline conditions, osmotic potential is lowered by converting starch to sugar.

Determination of salinity and osmotic potential. Because the salts in sea water consist of about 78% NaCl and 11% $MgCl_2$, variations in its salinity can be approximated by titrating for chloride ions (193), or by interpreting hydrometer readings in terms of pure NaCl solutions.

For soils the most satisfactory method of evaluating salinity is based on the fact that the greater the concentration of electrolytes in the soil the greater its electrical conductivity at a particular moisture content (651). To standardize the dilution factor, samples are moistened uniformly to "saturation"—a paste (twice field capacity) that will barely settle to a level surface if jarred; then the water is extracted by vacuum filtration and its conductivity measured as mmho/cm with a Wheatstone bridge. Salinity is not very detrimental to any crops or native plants when the conductance is less than 4 mmho/cm, but relatively few crops can be grown between 4 and 15, and at higher values only halophytes will grow. Usually no attempt is made to convert conductivity values into other units, but salt content can also be determined as parts per million, as milliequivalents per liter, or as bars of osmotic potential. Mmho $\times$ 0.36 is an approximation of osmotic potential when given a negative sign.

The water potential of either soil or plant tissue can be determined with a thermocouple psychrometer, which tests the relative humidity of the air about a sample of soil or tissue enclosed in a small chamber, after equilibrium has been established (640, 841).

Alternatively the osmotic potential of plant tissue can be determined by calculations based on the freezing point of expressed sap (478), or of living tissue (789).

Much of the earlier work on sap concentration was done by the plasmolytic method, in which the investigator, by trial and error, finds the highest concentration of some aqueous solute, usually sucrose, which will not cause a plant cell to plasmolyse (227).

Salinity in arid regions.   Almost all rocks liberate considerable quantities of soluble salts as they weather. In regions of high rainfall these salts are leached into streams and carried away from the area, but where evaporation exceeds precipitation, there is so little movement of water through the soil that salts formed by rock weathering remain in rather high concentration. In years of above-average precipitation excess salts tend to wash from high parts of the topography toward depressions, and if the depressions are undrained they gradually accumulate solutes. The closer the water table is to the surface of the soil, the stronger the concentration by evaporation, and the higher the salt content (248). Depending on the mineral composition of the rock at the source of the salt, the chief ions concentrated in this manner are Ca, Mg, Na, and SO_4, with K, HCO_3, CO_3, and NO_3 less commonly dominant.

During dry weather the salts are brought upward as water evaporates from the surface layer of soil; in rainy weather they are temporarily leached downward again. Therefore the vertical variations in degree of salinity fluctuate widely during the year. In addition there are wide variations in salinity within short horizontal distances even during the same season. It is important to keep in mind that conditions in the least saline part of a root system are most important to a plant.

In dry regions saline conditions are often brought about through ill-advised irrigation practices. If an area originally not saline is irrigated without allowing drainage water to escape, salts accumulate even though the water used is fairly pure, for evaporation removes water from the land but none of its solutes. Also, salts may originally have attained significant concentrations only at some distance below the soil surface, but with an abundance of water they are dissolved and drawn up to the surface where they accumulate as the water evaporates. Permanent agriculture on saline irrigated land is possible only if sufficient water is applied to maintain a downward and outward movement to such an extent that the quantity of salt contained in the irrigation water is at least equaled by the amount removed by drainage on a yearly basis.

FIGURE 11.   Salt flats around the margin of Great Salt Lake, Utah. *Allenrolfea occidentalis* in the distance; in the foreground *Distichlis stricta* reproducing by rhizomes.

It is widely believed that the decline of agriculture in ancient Mesopotamia was a consequence of salinization resulting from not understanding this principle.

Two aspects of solute excess must be distinguished. Saline soils are those that have accumulated sufficient salts, mainly chlorides and/or sulphates, to interfere significantly with the growth of glykophytes, but are not basic in reaction. Technically such soils have been defined as having a conductance of the saturation extract exceeding 4 mmho/cm, but with Na comprising less than 15% of the adsorbed cations. The pH is usually below 8.5 and the colloids are so well flocculated that the soils are porous and fluffy. Owing to the white crust of salt that is drawn to the surface during dry weather (Fig. 11), these soils in the western United States have long been referred to as "white alkali." *Solonchak* is a more widely used term.

Alkali soils are those with enough Na (or K) to be injurious to glykophytes (i.e., 15% or more of the adsorbed cations), but with negligible free salts as indicated by conductivities less than 4 mmho/cm. As Na comes to dominate the colloids, Ca and Mg become suboptimal. The colloids are deflocculated so that drainage and aeration are poor, and clay washes down to accumulate as a pan below. High Na content also raises the pH above 8.5, and its hydrolysis yields NaOH which is highly corrosive to humus and to living tissue, especially at the base of the roots. During the dry season dissolved humus is drawn to the surface and appears as a blackish to reddish crust. In the western United States these soils are often called "black alkali," but *solonetz* is a more internationally used term.

Poorly drained soils may give basic reactions as a result of either $CaCO_3$ or Na_2CO_3, and both salts will effervesce with HCl. A pH above 8.4 almost certainly indicates high Na_2CO_3 concentration. Also, since Na salts hydrolyze to form strong bases, the pH of a paste and of a 1:5 dilution differ appreciably if the colloids have a high Na:Ca ratio. If dilution increases the pH as much as 0.4, Na is probably present in harmful amounts.

Saline-alkali soils combine detrimental (for glykophytes) solute concentration with a great deal of Na, as defined above. They are well flocculated, and usually have a pH above 8.5. Owing to their dark colors they are usually classified under "black alkali" soils.

Salinity of waters and coastal regions.   The salinity of the ocean (dissolved solids about 3.5%, osmotic potential about −24 bars) is not high enough to be considered a retarding influence upon life, as is witnessed by the tremendous variety within its biota. Most biologists believe that life arose in the ocean, but in all probability the waters were essentially fresh when primitive forms were evolving, and they have become salty so gradually that adaptation to salinity has been able to take a leisurely course. At present, however, the salt content of sea water is definitely injurious to land plants other than halophytes.

Inland salt lakes develop a much higher degree of salinity. Great Salt Lake, Utah, one of the saltiest lakes in the world, contains as much as 28% salt in some bays. Nevertheless fourteen species of organisms, exclusive of bacteria, have been identified in this lake.

Salt basins are occasionally formed when an arm of the sea becomes separated from the remainder and the salts, which are about 80% NaCl, are concentrated by evaporation. The Salton Sea basin in southern California was formed in this manner.

Much more abundant are salt marshes on the lowlands bordering the seas (611, 630). They may be rather uniformly salty where subject to regular tidal action, provided that there is no seepage from the land to keep the salts from sea water washed out, or they may be *brackish* if the salt water is diluted with fresh water. Sometimes habitats are alternately fresh and brackish with the ebb and flow of the tide, and the plants growing there must have highly specialized physiology to be able to withstand such sudden and extreme changes in the water potential of their water supply.

Physiologic effects of salinity (522).   The action of dissolved salts on plants is partly physical or osmotic, and partly chemical or specific, depending on the kinds of ions.

In order to maintain turgor, the concentration of osmotically active substances of any protoplast must exceed that of its water supply. Most glykophytes begin to exhibit reduced growth when the water potential of the soil solution drops below about −2 bars, and even salt-tolerant crops fail completely by the time the water potential of the soil reaches about −47 bars, but halophytes can continue to lower their water potential to compensate for salinities far beyond this. Their tissue fluids may be so charged with salt that they have a salty taste. Some of them (*Distichlis spicata, Spartina* spp., *Glaux maritima,* etc.) excrete superfluous salt by means of special glands, and salt incrustations may form on the foliage of those species that guttate freely (119). Secretion directly through the cuticle has been reported for species of *Tamarix* and *Armeria*.

When crop plants are grown in saline soil their root systems are dwarfed, absorption and transpiration are reduced, and they use less water in accumulating a given amount of carbohydrate. Because salts so evidently interfere with the absorption of water by glykophytes, saline soils have long been considered "physiologically dry," though physically wet, for these plants. Several decades ago the concept of physiologic drouth likewise embraced halophytes, but research has shown that it does not apply to them. Halophytes experience no difficulty in absorbing water from highly concentrated solutions, as is demonstrated by their high transpiration rates and by guttation, which may be observed in many species. They are an ecologic group of plants characterized by an ability not only to adapt themselves easily to high concentrations of certain ions in their water supply, but also to absorb water with ease under these conditions. In fact, this is the only manner of defining halophytes, for a number of them can be grown easily under nonsaline conditions (774). The importance of this osmotic aspect of their natural environment and their close adjustment to it is indicated by the fact that the osmotic potential of halophytes varies directly with the salinity of their water supply over a wide range of concentrations.

Plants on saline soils germinate and grow chiefly during the rainy season when the soil solution has been diluted and the salts moved down below the root zone. The characteristic shallow rooting of many halophytes is a decided advantage in this connection, and it may be an equally important factor from the standpoint of aeration, for saline soils are frequently waterlogged for at least a part of the year. When saline soils are used for agriculture, the best irrigation system is one that keeps the soil at the maximum moisture content possible without bringing about a waterlogged condition. When saline soil is favorably moist, the water potential is only about half as low as when plants growing on such soil wilt for lack of water, at which time the potential may drop as low as −200 bars (515).

Although its significance is not clear, a large percentage of halophytes exhibit succulence * (Fig. 12). Dissolved salts in the soil solution, especially chlorides, stimulate succulence in many nonhalophytes and govern the degree of succulence in a number of halophytes that have been studied. The salt-tolerant *Asparagus officinalis* is much more succulent when grown on saline than on nonsaline soils—a fact of some economic importance.

In the majority of both halophytes and glykophytes germination is greatly retarded and seedling survival is difficult under saline conditions. Certain halophytes, e.g., *Salicornia,* have more than overcome this handicap by physiologic adaptation, for the presence of a little salt is much more favorable to their germination and survival than ordinary nonsaline conditions. Nevertheless, the degree of salinity is an important limiting factor for halophytes (130). Their successful germination is usually re-

---

* Succulence refers to the possession of soft, thick organs which contain much water owing to an enlargement of one or more of the parenchymatous tissues, coupled with a reduction of the sizes of intercellular spaces.

FIGURE 12. Succulent halophytes along the Pacific coast of Baja California. *Mesembryanthemum aequilaterale* above and right; *Abronia maritima* below and left.

stricted to those seasons when the salts are most dilute, and this is consequently the season when salt analyses provide the most critical information. When the seeds of *Medicago sativa* (alfalfa) or other salt-tolerant crops are planted in saline soil, heavy applications of irrigation water are made so as to dilute the soil solution as much as possible during the germination period. Later, after the seedlings become established, the plants are less sensitive to salt. A number of halophytes of seacoasts have developed the habit of *vivipary,* which is the germination of seeds before the fruits have abscised (Fig. 13). It has been hypothecated that this character may be an adaptation by which salt interference with germination is avoided, but in one species of *Rhizophora,* at least, this is not true (519).

That the effects of salinity are largely osmotic is shown by the fact that equal growth reductions can be obtained with equal concentrations of different salts, and by the fact many gardeners will attest to, than an over-application of mineral fertilizers will kill glykophytes. However, each of the salts responsible for salinity tends to have certain specific effects upon plants, with Cl and B being more detrimental than others (56). In general plants can tolerate concentrations of different salts in direct proportion to their abundance in nature (90). For example, ordinary plants will tolerate saturated solutions of the common $CaSO_4$, but even a dilute solution of CuSO$_4$ is toxic. Because of this correlation between tolerance and the abundance of ions in nature, and because differences in toxicity are not related to physical or chemical laws, tolerance appears to be a matter of physiologic adaptation acquired through time, and plants have had more opportunity to acquire a degree of tolerance for the more common salts.

Another specific character of the ions is their effects upon the toxicity

FIGURE 13. *Rhizophora mangle,* a mangrove shrub growing in brackish water at the edge of a lagoon in southern Florida. Note the elongated hypocotyl projecting from the fruit, which will shortly pull the remainder of the embryo away from the fruit. The stiltlike prop roots are also conspicuous.

of one another. For example, in the presence of a little $CaSO_4$ alfalfa will tolerate a 300 times greater concentration of $MgSO_4$ than it will in the absence of Ca. In general, Ca seems to mitigate the toxicity of other ions of saline habitats, except for those which dominate alkali soils.

Assessing the importance of alkalinity in the field is complicated by the fact that high pH is associated with poor aeration from deflocculation, and also with unfavorable states of several essential nutrients.

Soil Development

The Dynamic Concept of Soil

The weathering of a parent material under the influence of any particular combination of climate, vegetation, and relief progresses along a definite series of stages which may be looked upon as passing from youth to maturity. When, after a lapse of many years, maturity has been attained, the soil is in a state of dynamic equilibrium with those factors that have determined its course of development. Credit for outlining the principles of this dynamic aspect of soils is due the Russian soil scientists, and this fact explains the frequency of pedologic terms derived from their language.

The Soil Profile

Any weathered parent material, even though it may not have had time to attain the state of equilibrium which characterizes maturity, is made up of a series of superimposed layers or *horizons,* each of which has been affected differently by the weathering processes, and which collectively are referred to as the *soil profile.* The various horizons in a profile may differ in color, structure, consistency, thickness, reaction, and chemical composition. Usually, though not always, the horizons can be distinguished with the unaided eye.

The development of horizons is brought about largely through the action of rainwater, which leaches materials from a surface layer and deposits most or all of them at a slightly greater depth. This process consequently results in three major horizons (Table 3). The weathered *A* and *B* horizons constitute the true soil or *solum* and are usually between ½ and 2½ meters in thickness. The material below the *B* horizon constitutes the *subsoil* in the true sense, although this last term is often used indiscriminately for any depth below the very surface of the ground. Both *A* and *B* horizons, and sometimes the *C* as well, are subdivided where it is possible to do so (Table 2). In general, the more nearly a soil has approached maturity, the more distinctive its horizons.

Table 3

The Soil Profile. Horizons, Indicated by Letters and Numbers, May Be Lacking Individually or in Groups. [After Smith and Moodie (719).]

Plant remains lying on the mineral soil.	*O1* Recently fallen leaves, twigs, and other litter, so little decomposed that origin is apparent. *O2* Duff that has been derived from litter; particles have lost their identities and are commonly matted together.

A horizon: Mineral material that contains most of the humus in the profile, and has lost clay, or Fe, Al, and organic colloids by leaching.	*A1* Thick and very dark with humus; with well-developed crumb or granular [a] structure.	*or* Thin to absent.
	A2 Not quite so dark, and with little crumb or granular structure.	*or* Light-colored and structureless owing to loss of clay, Fe, Al, and organic colloids.
	A3 Transitional, but more like *A* than *B*.	

B horizon: Mineral material that has become cemented with Al, Fe, and organic colloids, or has developed blocky to prismatic structure as a result of enrichment with clay from above.	*B1* Transitional, but more like *B* than *A*. *B2* Maximal clay, Fe, Al, or organic colloids, hence maximal development of blocky to prismatic structure or cementation. *B3* Transitional, but more like *B* than *C*.

Subsoil, designated according to its character as follows:
C Material similar to that from which at least the *B* horizon was derived, but lacking blocky or prismatic structure; may contain deposit of $CaCO_3$.
D Material dissimilar to that from which the B layer was derived.
G Glei, gray structureless material with compounds in reduced condition owing to long periods of waterlogging.

[a] For technical definitions see *Soil Survey Manual* (722).

Under a particular type of climatic influence, all parent materials tend toward the same general type of mature or climax profile. However, the details of the mature profile are determined by the kind of vegetation which grew on the soil, the physiographic processes of erosion or deposition, drainage, and the nature of the original parent materials.

Effects of Climate on Soil Development

The major bioclimatic provinces of the earth's surface tend to be arranged in a zonal sequence which is governed primarily by the pole-equator temperature gradient. Therefore each takes the form of a circumpolar belt occupying a particular range of latitude. Mountainous regions interrupt this normal pole-equator sequence of zones by producing aridity, and in the region of their influence there occur secondary systems of zones which are governed more by aridity than by temperature and are consequently centered about the mountains. These zones have practically no relation to the latitudinal series of zones.

The climate of the eastern half of North America, which is essentially free from the dominating influence of any major mountain system, has abundant rainfall from Greenland to the West Indies. Because precipitation exceeds evaporation and transpiration, the excess rainwater sinks down through the soil and escapes from the area as streams. Here, and wherever else the net movement of soil water is downward, certain soluble nutrients are continually lost and frequently the development of the soil toward maturity is accompanied by a reduction in fertility with respect to at least certain of the essential nutrients.

In this part of the continent the most important geographic variable in climate is temperature. Under the extremely cold climate of areas bordering the Arctic Ocean, soils undergo a type of development referred to as *gleization.* Southward from this boreal region to approximately the southern limit of the United States the soil-forming processes are of a type called *podzolization,* and in the warm regions still farther south *laterization* is the characteristic weathering process. Both the podzolic and the lateritic regions of North America can be subdivided into minor provinces which likewise form temperature belts.

The interception of rainfall by the Cordilleras of western North America results in arid or semiarid climates on the basal plains to the lee of most of the mountain ranges. Within the sphere of influence of these mountains, aridity is of more importance than temperature. Consequently the soil and vegetational provinces here are arranged in concentric zones about the mountains, and these zones correspond more closely with effective rainfall than with temperature. Because little or no water escapes being drawn back to the surface and evaporated or transpired, most of the soluble nutrients freed by mineral weathering cannot escape, and their accumulation results in high fertility. The characteristic process of soil evolution which predominates in this arid province, whether warm or cold, is *calcification.* The relationships among climate, weathering processes, soil groups, and vegetation are summarized in Table 4.

Table 4
Some Relationships of Climate to Vegetation, Weathering Processes, and Great Soil Groups.

Climate		Weathering Process	Characteristic Vegetation	Great Soil Group
Moist	Cold	Gleization	Tundra	Tundra
	Cool	Podzolization	Taiga	Podzolic Brown podzolic Gray wooded
			Temperate forest	Gray-brown podzolic Prairie podzolic Red-yellow podzolic
			Chaparral	Noncalcic brown
	Warm	Laterization	Rainforest	Lateritic
Arid (cool to warm)		Calcification	Steppe	Chernozem Chestnut Brown Sierozem Gray desert
			Desert	Red desert

Gleization. In arctic regions, and wherever else soils are cool and continually wet but not saline, there develop sticky, compact, and structureless *B* and *C* horizons. These *glei* horizons are gray, bluish or olive in color because poor aeration brings about a reduced condition of iron compounds; at the same time it usually favors surface accumulations of peaty material. Where water tables are not continuously high, the glei layer is commonly mottled with rust-colored areas. The particular series of physical, chemical, and biological changes which produce such a profile, wherever it may occur, are embraced in the term gleization.

In arctic and subarctic regions where the subsoil is continually frozen, a circumpolar belt of soils with a glei horizon is found. These are called *tundra soils* after the nature of the vegetation which occurs in these regions. The glei layer is commonly streaked with iron and organic matter, and is subject to mixing by irregular pressures set up when the soil freezes each winter.

The *alpine meadow soils* found under corresponding vegetation on the summits of high mountains at low latitudes are high in organic matter but do not have a glei horizon and therefore do not belong in this soil group.

Podzolization. Where the dominant plants of wet climates have low nutrient requirements and grow on very infertile soils, the litter cast upon the soil undergoes a slow type of decomposition in which organic acids are produced in abundance. As a consequence the rainwater that percolates through the mineral soil is acid and thus dissolves out the free carbonates and then the adsorbed basic ions, both of which are carried away in streams. This process, podzolization, culminates in a distinctly acid condition due to the accumulation of H and A1 ions upon the micellar surfaces. Acidity in turn brings about the solution of sesquioxides of iron and aluminum (Fe_2O_3 and $A1_2O_3$) together with a deflocculation of the colloids. Both the oxides and colloids are carried downward only a short distance where they are precipitated as the *B* horizon. Silica, being practically insoluble in acid solutions, is left behind and eventually may comprise as much as 80% of the mineral material of the *A* horizon.

In those coniferous forests characterized by the genera *Picea, Abies,* and *Tsuga,* which extend from the arctic timberline down to approximately the latitude of the Great Lakes, and even farther south at high altitudes along the Cordillera and the Appalachians, true *podzols* predominate. These are highly acid soils of low fertility in which the dark *A1* horizon, usually very thin, if distinguishable, is underlain by an ashy-gray (podzol is the Russian word for this color) *A2* horizon of siliceous sand 1–45 cm thick (Fig. 14). This *A2* horizon is more acid than either the *A1* above or the *B* below. The *B* horizon is dark brown * owing to its accumulation of iron oxide. Where the parent material is sandy, the minerals of the *B* horizon may be cemented together into a hard *ortstein* layer, but where the parent material is finer-textured, blocky cemented units usually form. The very acid litter and duff may attain a thickness

---

* Munsell color charts in book form are widely accepted standards for describing soil colors (607).

of 45 cm and are usually thick enough to interfere seriously with the germination and establishment of seedlings. Where the *A2* is thin or discontinuous, and the *B* weakly developed and conspicuously brown to yellowish-brown, the profiles are segregated as *brown podzolic.* If the litter and duff are thin, the *A1* clearly defined, the *A2* exceptionally thick but not so pale, the *B2* less intensely colored, and the profile neutral to slightly acid, the soils are classed as *gray wooded.*

Even in virgin condition these soils are relatively infertile, and their crop-production capacity drops rapidly unless Ca, P, and other fertilizers are used regularly. Ca is limiting because it is lost by leaching, and

FIGURE 14. Podzol profile in a stand of *Pinus sylvestris* with an understory of *Vaccinium,* near Saranac Lake, New York. The A horizon, from the base of the grass plant at the right to the bottom of the white layer, is 7 cm thick.

P is a problem because it becomes cemented in insoluble form in the *B* horizon, often attaining this condition so quickly after fertilizer is applied that the plants cannot benefit from it.

In climates slightly warmer than those with the three great soil groups described above, are the *gray-brown podzolic* soils. This group occurs also in the mountains at elevations below the podzols, especially along the Pacific coast. Some characteristic trees of the gray-brown podzolic soils are *Acer saccharum, Fagus, Pinus ponderosa, Pseudotsuga,* and *Sequoia.* This vegetation returns more bases to the soil than does that of the podzol regions, and the warmer temperatures favor decomposition processes which do not liberate acids so abundantly. Accordingly the soil remains moderately fertile, and the sesquioxides are not so completely translocated.

In the winter-deciduous forest the leaf litter is often high in base content. Earthworms and other agents of decay bring about its decomposition so rapidly that it often disappears by late summer, and the resultant humus is deeply incorporated into the soil. In these soils the *A1* is weakly acid, and the *A2* light colored but not ashy-gray and only moderately thick. They support an exceptionally wide range of crops because of a combination of favorable temperature, abundant moisture, and good fertility.

Along the western edge of the winter-deciduous forest of eastern North America, chiefly occupying a triangular area from northern Missouri to central Minnesota to western Indiana, the soils have long developed under grassland vegetation, but in the last few millennia the climate has become sufficiently moist that the area is now capable of supporting forest. Long domination by grassland vegetation produced the deep dark humus layer characteristic of grassland profiles, but the rainfall has been high enough to favor mild podzolization as indicated by slight acidity, a light-colored *A* horizon, the loss of carbonates, and a slight accumulation of clay in the *B* horizon. Before cultivation the vegetation of this area was essentially like that of the chernozem region along its western border, but the soils are classed as *brunizem* or *prairie podzolic* since carbonates have been leached away.

These are probably the most fertile soils in the world. Though used chiefly for growing maize, they will grow a wide variety of crops. Relatively small areas of true grassland soils in the western states have also been classified in this group although the potentialities of the soil in regard to native vegetation and cultivated crops differ greatly from the midwestern area.

In chaparral areas of the southwestern United States occur *noncalcic brown* soils, characterized by brownish surface horizons overlying reddish material of heavier texture, which becomes leached of lime, yet remains neutral.

Laterization.   In warm regions with high rainfall parent materials undergo a distinctive type of weathering called laterization. Here iron and aluminum oxides do not leach downward, whereas silica is converted into silicic acid which is lost by leaching along with bases. Red clays, mainly kaolinite of low nutrient capacity, are residual in this process, and they are generally too infertile to sustain agriculture without the continual application of fertilizer. These clays are peculiar for their high aggregation and consequent remarkable permeability. Since temperature and moisture favor decomposition all year the humus content is extremely low. Thus most of the nutrient supply is tied up in the living organisms.

Most true *laterites* occur in the tropics. Here both the *A* and *B* horizons are reddish (*later* is the Latin word for brick) and barely distinguishable. They harden irreversibly following the removal of their forest cover which then allows alternate drying and wetting, or even without disturbance they may harden as nodules or vescicular masses. Clearing such land essentially ruins it in terms of human needs, resulting in a permanent replacement of forest by savanna.

From southern Missouri to Tennessee to Virginia and southward in the United States are soils intermediate between lateritic and podzolic soils, and usually classed as *red-yellow podzolic*. The *A1* horizon is poorly developed, the *A2* is yellowish, and the *B* is either red or yellow and heavy textured. These are soils of a forest region where *Quercus* is the characteristic forest dominant. On vast acreages cropping has become no longer profitable, owing to long cultivation and sheet erosion.

Calcification.   In the development of forest soils (podzolization and laterization) the earliest phases of weathering are characterized by the loss of Ca and Mg by leaching, but in grassland and desert regions precipitation is insufficient to leach away these ions. Furthermore, in grasslands the percolation waters are in intimate contact with humus which decreases gradually to the depth of the grass root systems, and there is ample opportunity for the adsorption of cations as they are released. Those ions of Ca and Mg which escape adsorption are leached only a short distance before they are precipitated, depending upon the average depth of rainfall penetration. The resultant accumulation of carbonates (Fig. 15) is one of the two most outstanding characteristics of the calcification process. The other is the immobility of colloids, which, being Ca-saturated and therefore highly aggregated, cannot be moved downward as in podzolization, with the result that clay content is relatively uniform throughout the profile, except in areas where it forms in place in the *B* horizon. The pH varies but little from neutral.

In North America the largest area of calcified soils developed under grassland climate and vegetation lies east of the Rockies from Canada to Texas. Along the eastern edge of this region where prairie meets forest, rainfall is highest, gradually decreasing westward to the moun-

FIGURE 15. Soil profile under steppe in Montana. The whitish lime layer starts 5–6.5 dm below the surface. The depth of the humus-rich *A* and *B* horizons coincides well with the depth of root penetration. Each mark on the rod is 10 cm long.

tains. The vegetation and soils vary accordingly from one edge of the region to the other. *Andropogon scoparius, Stipa spartea,* and *Sorghastrum nutans* are dominant along the eastern edge, while *Agropyron smithii, Stipa comata, Bouteloua gracilis,* and *Buchloe dactyloides* are the most important species along the western edge.

Pedologists recognize three major soil zones across this region. The *chernozem* soils occupy the easternmost zone where precipitation is adequate for agriculture, and the blackish *A* horizon is 0.6–1.5 meters deep, with a whitish accumulation of lime at the bottom of the *B* horizon. These are highly fertile, and although they are ideally suited to only a few crops such as wheat and maize, they are agriculturally very important.

Just west of the chernozem belt is the *chestnut* soil region in which the dark-brown *A* horizon is only 0.4–0.9 meter thick and where agriculture frequently suffers heavily during dry phases of climatic cycles. Finally, along the western margin of this great grassland belt lie the *brown* soils with a brown *A* horizon no more than 0.3–0.6 meter deep. In this third zone rainfall is so low that for the lack of irrigation facilities agriculture is profitable only during the wet phases of climatic cycles, and the best permanent land use is cattle grazing.

In the southerly portions of the prairie, chestnut and brown soils tend to have a slightly reddish tinge owing to weak laterization, and are called *reddish-prairie, reddish-chestnut,* and *reddish-brown* soils.

Vegetation in which *Artemisia tridentata* is conspicuous, from southern Nevada to central Wyoming to southern British Columbia, is associated with brown, *sierozem* and *gray desert* soils. Except for the brittle hardpan of $CaCO_3$, locally called *caliche,* the horizons are extremely faint, and Sierozems are gray even at the surface. Gray desert soils are distinguished by being calcareous in the upper profile, whereas sierozems are not. Agriculture is extremely limited without irrigation.

Red desert soils occur in the hot desert country dominated by shrubs and cacti which occurs in southern Arizona and adjacent parts of California and Nevada, in western Texas, and in adjacent Mexico. These soils have extremely low humus content. The annual production of litter is scanty; roots are widely spaced through the soil; and temperatures favorable for decomposition prevail over much of the year. The profile typically has a pinkish cast, and, like gray desert and sierozem soils, has a caliche (703) layer. Agriculture here is not attempted without irrigation.

Effects of Vegetation on Soil Development

To attempt to draw a clear distinction between the effects of climate and vegetation upon the nature of the soil profile is very difficult, for microclimate, * vegetation, and soil form an interrelated dynamic complex. When one member of this complex is altered the others likewise change and a new equilibrium is established. The broader aspects of climatic influence upon the development of soil profiles have just been discussed briefly, and some examples will now be cited which emphasize the influences of vegetation.

Grassland-type profiles tend to develop in forest climates wherever the forest is kept cleared away so that herbaceous vegetation remains in possession of the soil. Thus old lawns in podzol climates exhibit as many characteristics of chernozems as of podzolic soils. Also, when a forest is cleared and replaced with grain fields, the organic matter may increase as the result of crop residues, despite the usual tendency for cultivation to decrease humus.

On the other hand it is possible for grassland soils to develop some of the characteristics of a forest soil profile. This phenomenon can be observed where a windbreak or "timber claim" has been established in an area which was originally grassland.

The effects of different forest types upon the soil may be so distinct that the replacement of one forest type by another produces marked changes in the profile within a few decades. In the New England region many abandoned farms have grown up into white pine forests, and these in turn are replaced by other forests dominated by deciduous dicots: beech, sugar maple, yellow birch, ash, etc. This transformation is accompanied by a striking evolution of the soil profile, the podzol formed during the pine occupancy changing to the gray-brown podzolic type in a few decades (Fig. 16). A parallel phenomenon has been described in the southeast (147).

In all instances mentioned above the changes from one vegetation type to another bring about changes in the microclimate and in the chemical nature of the organic increments which the plants add to the soil. Under grass cover soils are always warmer than under trees, and with deciduous trees rainfall- and heat-interception are not so uniform throughout the year as with evergreens. The important differences in the organic residues of grasses, conifers, and deciduous dicot trees were emphasized earlier. Obviously it is nearly impossible to evaluate the relative importance of these two kinds of vegetation influences upon

---

* *Microclimate* refers to strictly local combinations of atmospheric factors which, owing to uneven topography, plant cover, etc., differ from the *macroclimate* as measured in locations where these modifying factors have negligible influence. Within each area embraced by one macroclimate there exists an intricate matrix of microclimates, at least some of which differ sufficiently to be of ecologic importance.

soil. The important conclusion to be drawn is that vegetation exerts a strong influence upon soil development because no two vegetation types produce exactly the same microclimate at the level of the soil, nor do they yield the same kind of organic matter (585).

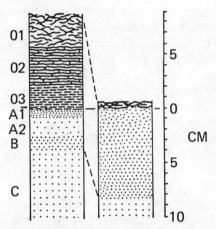

FIGURE 16. Soil profiles showing changes from a podzol (left) to a brown podzolic (right), which took place within 18 years after the dominant *Pinus strobus* was cut from a mixed forest so that a pure stand of deciduous angiosperm trees remained. [After Fisher (245).]

The fact that different kinds of plants have different effects upon soil properties is of great practical importance in forestry. Silviculturists have found that the infertility of podzol soils, and the resultant slow growth of the conifers growing upon them, can be alleviated by thinning the stands (altering the microclimate, and green-manuring from cut roots) and by encouraging admixtures of deciduous dicots (improving the quality of the organic matter). Sufficient information is available in temperate latitudes that the principal species of forest trees can be rated as to their relative efficiency in accumulating basic nutrients in their litter.

The same principle finds less application in agriculture solely for the reason that there soil fertility is more easily maintained by the addition of fertilizers. However, a major aim of crop rotation is the enrichment of the soil by one or more of the crops, commonly legumes, which are included in the sequence. When crop residues are returned to the soil definite and specific changes are effected.

In arid regions certain shrubs selectively absorb Na, K, or Mg salts until their foliage is highly charged with them and the area affected by their litter becomes highly basic (243). In wet climates single trees of highly influential taxa are sometimes the cause of markedly stronger islands of podzolization (69).

Effects of Erosion and Deposition on Soil Development

It will be necessary at this point to differentiate between two types of erosion. *Normal erosion* is the term generally applied to the gradual wearing away of the land surface under natural conditions. *Accelerated erosion* is a much more rapid removal of surface materials which may result from man's alterations of vegetative or soil conditions; it is additive to normal erosion.

Normal erosion tends to produce an undulating land surface with a system of stream courses connecting the depressions so that the entire surface is well drained. This is accomplished chiefly by means of a slow migration of particles downslope as they are moved by successive rains. In arid regions wind erosion during the long dry seasons is also an important factor.

This phenomenon is normally so slow that the surface minerals are not removed until after they are weathered to such an extent that they no longer have much potentiality as a source of nutrients. So much time is involved that the profile maintains its characteristic thickness by a gradual extension of weathering into the subsoil, which keeps pace with the surface degradation.

On flat land, however, there is little or no removal of surface materials, and the exhausted minerals accumulate as an infertile layer as a result of what may be called suboptimal erosion.

On hilltops and steep slopes the upper part of the *A* horizon is removed so rapidly even in the absence of accelerated erosion that the solum cannot attain the depth characteristic of gentle slopes. The effect of this above-average rate of erosion is heightened by the fact that increased runoff is accompanied by decreased percolation, drier soil, a thinner plant cover, less leaching, and less vegetational influence. Under accelerated erosion in hilly agricultural land the *B* horizon or even the subsoil may soon appear at the surface of hilltops.

Deposition is the universal complement of erosion. In contrast to the effects of erosion, deposition causes soil profiles to be deeper than average because surface increments of new materials are constantly added, and upon weathering these become continuous with older layers of similar origin beneath. Floodplains, for example, receive an additional increment of alluvium during each high-water stage. As a result of this progressive burial of older layers, floodplain profiles are deep and cannot begin to approach maturity until the combined effects of the building up of the surface plus the downcutting of the stream convert the flood-plain into a terrace which is above high-water level.

Effect of Lack of Drainage on Soil Development

Poorly drained depressions are common features of the earth's surface, especially in regions that are physiographically young. In such habitats leached and enriched horizons cannot form in the usual manner, owing to the lack of requisite downward movement of water, and the development of soil and the vegetation are more strongly determined by poor drainage than by climate. However, those that develop in arid and semiarid regions are separated as *halomorphic* (saline) from those of moist climates which are called *hydromorphic* (nonsaline).

Halomorphic soils. From a developmental standpoint the kinds of soils discussed under the topic of salinity are believed to be related. Salt accumulation begins as aridity develops and water flows toward basins from which it evaporates. This first stage results in saline (*solonchak*) soils. The colloids resist Na which remains mostly in the soil solution, so that Ca and Mg maintain flocculation, prevent downward migration of clay, and thus prevent profile differentation.

If the proportion of Na in the soil solution continues to increase, it eventually begins to increase also as an adsorbed ion, thus leading to the saline-alkali condition.

Saline-alkali soils persist until drainage begins to develop, then since the soluble salts leach away before the adsorbed Na is lost, they are converted into alkali (*solonetz*) soils. Deflocculation now allows clay to migrate downward to produce vertically oriented, round-topped columns 10–15 cm in height, which comprise the *B1* horizon. Sparse vegetation and the dry, windy climate often conspire to remove the leached *A* horizon, with the result that "slick spots," shallow pits floored with blackish clay, appear. Although the vegetation is still limited to a few species having specialized physiology, it is not as distinctly halophytic as that of solonchak soils.

Further leaching, in the presence of a normal abundance of Ca and Mg, then takes away the excessive amounts of adsorbed Na and a normal zonal profile develops.

Halomorphic soils (as well as hydromorphic soils, rendzinas, etc.) owe so much of their character to influences unrelated to climate that they appear in two or more climatic zones, as defined on the basis of vegetation and average soil conditions. These form the category of *intrazonal* soils.

Hydromorphic soils. Hydromorphic soils are found in poorly drained depressions in moist climatic regions. Here drainage, though feeble, is sufficient to keep the soil solution from being dominated by inorganic salts, as is true of corresponding habitats of arid regions.

Frequently hydromorphic soils become covered with a layer of peat so deep that plant roots seldom reach the mineral surface, and such profiles are referred to as *bog soils.* In the plant associations characteristic of these soils sedges, sphagnums, and such woody plants as willows, alders, ericads, and conifers dominate.

In many poorly drained habitats of moist regions peat does not accumulate on the surface, but instead the mineral framework of the soil becomes impregnated with an abundance of organic matter, rendering the upper part of the profile black. These are the *humic glei soils,* and their humus content is largely attributable to the abundance of short-lived roots of grasses and sedges which characterize them. On account of their high humus content and unleached character they become excellent agricultural soils, although the glei layer and a high water table usually make artificial drainage necessary.

The term *planosol* is often used for any soil in temperate climates with internal drainage impeded by a thick, compact or cemented *B* horizon that lies below an *A* horizon which is leached but still fairly high in organic matter. Flat, sandy land in the subtropical southeastern United States is often distinguished as *ground-water podzol.* Here negligible surface layers of organic matter are underlain by a thick ashy-gray *A2,* and this in turn by a dark well-cemented *B2.* The native vegetation is dominated by *Pinus, Quercus, Nyssa,* and *Taxodium.* In the tropics under savanna vegetation occurs *ground-water laterite.* This is much like ground-water podzol, but with perhaps wider annual fluctuation in the position of the water table.

Time and Soil Development

Most of the soil-forming processes tend to operate very slowly. The constant physiographic remodeling of the earth's surface proceeds rapidly enough by comparison so that the profile developing in a particular body of parent material frequently does not attain a state of equilibrium before the surface is again altered by erosion or deposition. In consequence, examples of truly mature solums are not always easy to find, whereas unaltered parent materials and immature profiles abound in most regions. In the terminology of dynamic pedology, such parent materials as have not yet had time to develop weathered horizons are referred to as *azonal.* The *A* horizon, darkened with humus, lies directly on the *C* horizon.

The term *lithosol* is applied to azonal soils in which the parent materials are very shallow and stony, and lack much horizon differentiation either as a result of recency of origin or excessive erosion in relation to weathering. It is also used for excessively stony soils that have well-differentiated horizons. *Regosol* is used for deep, nonstony deposits of loose material with faint profile development.

Effects of Parent Materials on Soil Development

In very cold or very dry climates weathering is extremely slow, so the character of the soils there is rather permanently determined by the nature of the parent material.

In less extreme climates soils on different parent materials show the influences of the same soil-forming process, unless they are subject to excessive erosion or deposition. This, however, does not mean that the mature profiles on all parent materials will be identical. For example, in dune sands of podzolic regions distinct layers of iron-cemented ortstein form at depths considerably greater than the less compact *B* horizon of loamy soils. Furthermore, in these sands there can never be much downward movement and concentration of inorganic colloids, as in finer-textured parent materials. Thus, profiles developed in dune sand remain distinct from those developed on other types of parent materials even though all undergo podzolization.

In the northeastern United States there are considerable areas of *brown forest* soils, in which the sola are so thin that the tree roots are in intimate contact with limestone, with the result that bases remain relatively abundant in the profile and earthworms are especially active. Here a thick, brown *A1* lies directly on the calcareous parent material.

A similar type of soil is the *rendzina,* which is characteristic of islandlike areas of grassland interspersed over forests of the southeastern quarter of the United States. These differ from brown forest soils especially in the shallowness and blackness of the *A1* horizon, possession of an *A2* but lack of a *B* horizon. Soil nutrition is strongly unbalanced by the predominance of Ca throughout the profile.

The interpretation of soil profiles is frequently complicated in that stratification of the parent materials may be confused with, or at least modify, genetically developed horizons.

As interest in ecology developed during the past century, strong emphasis came to be placed on the importance of parent material. In Europe soils derived from sandstone and granite rocks usually have levels of base saturation far below those of other soils derived from calcareous rock, and especially northward, plant distribution is closely correlated with the availability of Ca. Early plant ecologists saw the close correlation between plant distribution and parent rock, and therefore looked upon the latter as the most fundamental basis for soil classification. At the end of the century, however, Russian pedologists demonstrated that even on diverse parent materials, the processes of soil formation are the same under uniform climate. Soon climatic influence became the predominant theme in soil classification. In the past half century, however, it has been realized that although there is a tendency toward convergence between soils on contiguous parent materials, the latter, if highly contrasted at first, seem never to lose their individuality. Hence today there is a better balance in thinking on the relative importance of climate and of parent materials, with other soil-forming factors, namely vegetation, topography and time, also coming in for a good share of attention.

Effect of Man on Soils

Previous mention has been made of irrigation practices that have led to severe salinization, to accelerated erosion, to the reduction of humus in consequence of cultivation, and to the disastrous consequences of cultivating lateritic soils. Elsewhere the *A* horizon has been modified so that an *A1p* layer, in which plowing has mixed the natural horizons, must be recognized in profile description.

Chemical alterations have been brought about by adding pesticide residues. Although arsenical sprays are no longer used, some soils contain this poisonous element in high concentrations in consequence of its repeated use in the past. More recently a wide variety of organic herbicides, insecticides, fungicides and fumigants have come into use, and some of these persist in the soil for a long time, especially where humus is abundant. Although there is no question but that crop production has been greatly enhanced by these chemicals, we are largely ignorant of the cost resulting from alterations of the soil biota. Nitrifying organisms suffer especially, and the germination and growth of certain vascular plants are adversely affected (216, 858).

The use of $CaCl_2$ are NaCl on highways to minimize ice hazard to traffic salinizes broad strips on either side, so that in these areas salt-tolerance is a factor that must be considered in planning the landscaping of highways (24, 807).

Trails in parks become eroded and the soil so compacted that adjacent vegetation deteriorates. The soils of campgrounds and picnic areas, usually sited among trees, become so compacted that these areas must be given several years respite from use at intervals of a decade or so. The amount of use tolerated varies considerably with the type of soil and species of trees, so these factors need consideration even as the locations of the sites are determined.

The U. S. Department of Agriculture
System of Classification

For use in its soil surveys, the Department of Agriculture has devised a binomial system of soil nomenclature which reflects both genetic and textural characteristics (722, 775). A *soil series,* the first part of the binomial, embraces all soils which are alike in every respect except texture, i.e., alike in origin of parent material, climate, and horizon sequence. A *textural class* name is appended to the series name to indicate the texture of the upper part of the solum. Some examples of binomials of recognized soil types which are in use are: *Fargo clay, Miami silt loam,* and *Norfolk sand.*

2

The Water Factor

Importance of
Water to Plants

In the physiology of plants water is of paramount importance in many ways. As the closest approximation of a universal solvent, it dissolves all minerals contained in the soil. It is the medium by which solutes enter the plant and move about through the tissues. By permitting solution and ionization within the plant it greatly enhances the chemical reactivity of both simple and elaborated compounds. It is a raw material in photosynthesis. It is essential for the maintenance of turgidity without which cells cannot function actively, and indeed it is necessary for the mere passive existence of protoplasm, for very few tissues survive if their water content is reduced as low as 10%. The fact that water can to a remarkable degree absorb much heat from warm surroundings with relatively little change in temperature tends to slow the rate of temperature changes in protoplasm and thus make uniform the temperature conditions affecting the rate of biochemical reactions.

The water in the soil is continuous with that in the plant, and the entire system is in constant upward movement since the shoot loses water to the atmosphere at almost all times. Nearly all this water moving upward in the plant is lost in transpiration, only about 0.1–0.3% of it being tied up in chemical compounds. From the ecologic standpoint the course of water upward through the plant is of relatively minor importance, but the intake and loss of water are of great concern, for these processes are strongly conditioned by environment.

Atmospheric Moisture

Invisible Vapor: Humidity

The invisible water-vapor content of the air is usually expressed as *relative humidity,* which is not a statement of the actual moisture content of a volume of air, but a statement of this amount as a percentage of the maximum quantity that the air can hold at the prevailing temperature. Warm air can hold more water vapor than cold air. In fact the capacity of air for holding invisible water vapor doubles with each increase of 11°C in temperature. It follows that, when a body of moist, warm air is cooled, the relative humidity approaches 100% (even though the actual water-vapor content of the air remains constant), and if further cooling takes place the *dewpoint* is reached and the excess vapor is condensed into droplets of liquid. Thus, a cubic meter of saturated air (R.H. 100%) at 20°C, when cooled down to 9°, will lose by consensation half its water content as visible droplets. At the changed temperature, with only half its former *absolute humidity,* the relative humidity is still 100%.

Relative humidity normally undergoes a daily rhythm, changing from low during the day to high at night when the air cools. In rainforest habitats the diurnal "low" may remain above 80%, whereas in deserts it may drop below 10%.

During the day relative humidity usually decreases with distance above the ground, this gradient being especially pronounced where there is a dense plant cover. The high humidities that prevail within dense vegetation are proportional to the lowered temperature, showing that temperature is a more important governing factor here than the giving off of water vapor in transpiration.

Measurement of Relative Humidity (544)

The most common instrument used to measure relative humidity, ordinarily designated as a *psychrometer,* consists of a pair of thermometers, the bulb of one being covered with a thin muslin wick which is wetted just before the instrument is to be used. The pair of thermometers (*sling psychrometer*) is then whirled in the air, or air is forced past them by mechanical means (e.g., *hand-aspirated psychrometer*), and owing to evaporation of water from its wick the *wet bulb* will assume a temperature below that of the *dry bulb* which indicates true air temperature. The lower the relative humidity, the more readily water will evaporate from the wick and the greater will be the *wet-bulb depression,* i.e., the difference in temperature between the two thermometers. When relative humidity is 100% the two thermometers give identical readings. At other times, by means of the two temperatures obtained, the relative humidity can be read directly from tables prepared especially for the purpose (533). In freezing weather the ice coating on the wet bulb should be kept very thin.

A second type of instrument is based upon the ability of human hair to absorb moisture and shrink, or dry out and stretch, and in this way respond quickly to even small changes in relative humidity. Several strands of hair are harnessed so as to actuate a lever, at the tip of which is a pen which makes a continuous record of changes on a revolving drum. Such instruments are calibrated with the wet- and dry-bulb apparatus and need to be visited but once a week. They are called *hygrographs,* and the records obtained from them *hygrograms.*

Hygrographs, as well as temperature-recording apparatus to be described later, must be housed in properly constructed shelters to obtain reliable data. The shelter should be louvered on four sides, have air vents in the bottom, have a double roof with a free air space between the layers, and be painted white.

Useful adaptations of the two basic types of instruments described above have been devised for working under special conditions (117, 747). Also, special methods and apparatus for controlling humidity for experimental purposes have been devised (851).

Visible Vapor: Cloud and Fog

Cloud and fog consist of water droplets, or sometimes tiny ice crystals, which result from the cooling of air to a temperature below its dewpoint. They differ only in location. Clouds usually form when the cooling is due to a movement of air upward from the land surface into colder levels of atmosphere; hence a cloud is nearly always separated from the earth, except when it moves horizontally and comes in contact with a mountain top. On the other hand, fog is caused by the cooling of air at or near the land surface and is therefore ordinarily continuous from the surface upward, at least while forming.

Fog can be formed by (a) warm air passing over cold water currents in the sea, (b) by warm air rising up a sloping land surface to high elevations, and (c) by the nocturnal radiation of heat, resulting in a rapid cooling of the soil, at times when there is no appreciable air movement. In the last instance the cause is similar to the first, where air comes into contact with cold water. Examples of these three types of fog, respectively, are the "fog belt" along the Pacific coast from central California northward, the fogs that form on intermediate mountain slopes, and the morning fogs of valleys which are quickly dissipated when the sun rises.

When further cooled, both cloud and fog condense into particles or droplets large enough to precipitate out of the atmosphere. On the other

hand, should clouds be forced downward to warmer levels of the atmosphere, or foggy air be warmed up, the visible vapor evaporates into
the invisible form, after which the relative humidity begins to drop.

Importance of Atmospheric Vapors to Plants (404)

Effect on the intensity of solar radiation.   Radiation is the propagation
of energy through space, and the term *solar radiation* is used for the
heat, light, and other rays that are transmitted from the sun to the earth.
Water vapor in the atmosphere intercepts much of this energy before
it reaches the earth's surface, with visible vapor reducing photosynthesis
and transpiration far below normal rates. The slow-growing dwarfed trees
of fog-shrouded tropical mountains may owe their stature largely to the
limited supply of solar energy available to them (366).

Effect on evaporation and transpiration.   With all other factors remaining
constant, an increase in relative humidity reduces the rates of evaporation and transpiration because the vapor-pressure gradient between the
atmosphere and moist surfaces is lowered. In one set of experiments,
transpiration was found to increase about sixfold as relative humidity
dropped from 95% to 5%.

Certain species of plants are very sensitive to dry air and consequently
are found only in habitats where the humidity is always high. They are
called *hygrophytes.* The filmy ferns (*Hymenophyllaceae*) of the shady
interiors of tropical rainforests provide good examples of vascular plants
belonging to this ecologic category. Among cryptogamic hygrophytes,
damping-off fungi are especially important because of the high seedling
mortality they cause. Glasshouses are ordinarily watered in the morning,
then well ventilated during the day to keep the humidity down, and in
this way damping-off is fairly well controlled. Moist air favors many other
fungi which likewise become serious pests on higher plants only under
high humidities. Protracted cloudy weather or a series of showers within
a short period commonly cause rusts and other parasites to spread
rapidly through field crops. Beneath a cover of snow the needles of
conifers are commonly attacked by hygrophytic fungi (319, 418) such
as *Phacidium infestans,* which causes snow blight, and *Neopeckia coul-
teri* or *Herpotrichia nigra* which cause brown felt blight. Organic debris
lying on the ground is vigorously attacked by fungi while a snow cover
keeps the humidity high.

Source of soil moisture.   When fog moves horizontally, or a cloud comes
in contact with the earth's surface, minute water droplets are deposited
as they pass through finely divided materials such as foliage. Such
precipitated moisture may be absorbed directly, and on the rainless coast

of Peru it serves as the sole source of water for plants. In less arid regions the water is deposited so copiously that drops form on foliage and fall to the ground, materially augmenting the supply of soil moisture. This water, *fog drip,* gives character to the vegetation of intermediate mountain slopes in the American tropics (the "cloud forests"), and is important in determining plant distribution along the west central coast of North America (Fig. 17) (832). The amount of water falling under a tree canopy may be considerably greater than that falling on open ground, especially when measured under isolated trees or on the wind-ward edge of a forest. When it is very pronounced the phenomenon has a marked effect upon the distribution of low plants in relation to the trees (575). So objectionable is fog in human communities along the southeast coast of Hokkaido that special strips of forest are main-tained along the seashore to comb a large part of the water out of fogs moving landward (359). Belts of *Araucaria* trees were planted along the crests of ridges on the island of Lanai when it was discovered that there trees can comb an extra 75 cm of rain out of the Trade Winds (223).

Instruments for comparing potential fog drip at different places can

FIGURE 17. Looking westward from the dry climatic region of Mount Diablo, California, toward the region along the Pacific coast which has much more favorable moisture conditions owing to the fog belt that is visible in the distance.

be made by suspending screenwire cylinders above rain gages with a conical roof above to keep out rain.

Direct use by plants. Desiccated mosses and lichens absorb moisture from a humid atmosphere without preliminary condensation. In general the abundance of these plants, especially species that grow on rock or bark surfaces, is in direct proportion to the humidity of the climate. Certain tropical orchids and bromeliads growing on tree branches, and some desert plants, can also take up water directly from the air when the relative humidity rises above 85% (610, 832).

Whenever loss of heat by radiation cools a surface below the dewpoint, water vapor from the air will condense on it as a film of dew, even if relative humidity at a height of 1.2 m is no more than 60%. Turbulent air prevents the required temperature gradient, but gentle air currents thicken dew films by bringing fresh supplies of air into contact with the surfaces (189). Such dew forming on leaves may be absorbed through the cuticle of normal epidermal cells, or through specialized cells. Lysimeters have demonstrated annual dew increments on soil equal to 23 cm of rain, but in any one night dewfall probably never amounts to as much as 1 mm in depth (2). Tiny shallow-rooted desert annuals may depend more upon dew than upon rain. For deep-rooted plants dew appears to do no more than shorten the diurnal period of transpiration, and thus conserve soil water.

Alpine snowfields have been shown to gain an average of about 5000 liters of water a day during a summer month, owing to active condensation at night followed by feeble evaporation in daytime (23). This is a matter of considerable importance where snow melt in the mountains is the major source of irrigation water on the adjacent basal plain.

The quantity of dew condensing is best measured with a lysimeter. The duration of dew films can be recorded with clock-driven instruments (746, 749).

Evaporative Power of the Air

The atmosphere is nearly always dry enough to allow evaporation from the surfaces of plants, from the soil, and from bodies of water. So great is its capacity for holding moisture that approximately three-fourths of the precipitation falling on the land surface is drawn back into the air before it can run off the continents back into the oceans. In the United States this amounts to 10–150 cm of water annually (301).

The evaporative power of the air is increased by high temperature and wind as well as by low relative humidity. If any two of these factors remain constant, the evaporative power of the air varies directly with the third. Normally it is influenced by all three simultaneously (Fig. 18).

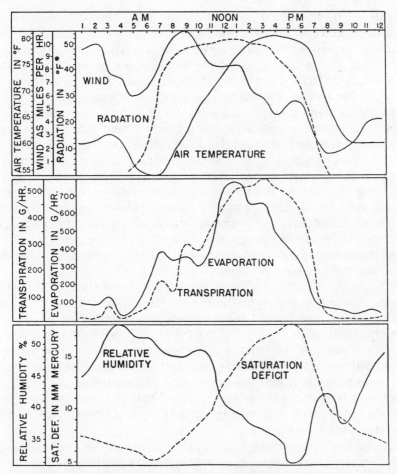

FIGURE 18.   Diurnal march of the principal atmospheric factors and of the transpiration of alfalfa *(Medicago sativa)* at Akron, Colorado, August 11, 1914. Asterisk indicates the difference between a black-bulb thermometer exposed to the sun and an unmodified instrument in a louvered box. [After Briggs and Shantz (94).]

Measurements of the evaporative power of the air are very important because they integrate the influences of all the strictly atmospheric factors that promote evaporation from the soil, and most of the external factors that affect transpiration. In fact, the expression "evaporative power of the air" is often used interchangeably with "transpiration stress." However, transpiration is subject to the effects of structural and functional features peculiar to the plant and is strongly influenced by light, so that evaporation rates cannot be assumed to indicate transpiration rates (Fig. 18). While the stomata are wide open the transpiration curve tends to follow the evaporation curve until the mesophyll develops so strong a water deficit that its ability to deliver water to the atmosphere diminishes, and at that time the stomata usually begin to close. With the stomata nearly closed there is no apparent relation between the rates of evaporation and transpiration, transpiration being controlled chiefly by the dimensions of the stomatal openings. Transpiration continues through the cuticle after the stomata close, but the rate under this condition is very low in relation to the evaporative power of the air. The effect of stomatal closure on transpiration is clearly shown by the sharp reduction in transpiration when a leaf wilts or when nightfall closes the stomata. In the first instance stomatal closure may not be accompanied by any change in atmospheric factors, and in the second the nocturnal reduction in evaporative power is by no means proportionate to the degree of transpiration reduction (Fig. 18). Light increases transpiration more than evaporation, whereas wind increases evaporation more than it increases transpiration.

Many studies have shown that the evaporative power of the air increases rapidly with elevation above the ground (Table 5), so that plants growing side by side may exist under widely different evaporative conditions when they differ in height of their foliage. This vertical stratification is largely a result of vegetation influence. As a barren area becomes clothed with vegetation the shade, reduced air temperature, interference with wind movement, and the water vapor given off by the leaves all conspire to reduce the evaporation rate (Table 6). The rate just above the tips of the leaves is several times greater than that just below them in certain plant communities.

Because the diurnal period of lowest relative humidity coincides with the period of highest temperature and brightest light, transpiration stress is greatest when these factors attain their greatest combined intensities at midday or in early afternoon (Fig. 18). However, the fact that riverbank vegetation and well-irrigated crops are not much affected by atmospheric drouth, of whatever intensity or duration, indicates that many if not most vascular plants can tolerate relatively high transpiration stress as long as their roots remain abundantly supplied with water. There is also good experimental evidence that atmospheric drouth is of much less importance than soil drouth, except insofar as atmospheric drouth contributes to soil drouth by promoting water loss from plant and soil (114, 176, 557).

Indirectly the evaporative power of the air is important to plants through its influence upon the effectivity of precipitation in replenishing soil moisture. When the sun shines on a wet soil it loses water by evaporation more rapidly than a free water surface would, because of the large amount of total evaporative surface presented by the minute irregularities, and because the heat input is concentrated in the very surface. However, as the surface layer dries out the rate of water loss diminishes rapidly even though the subsoil remains moist. Surface evaporation can desiccate a normal soil to a depth of 2–3 dm; any quantity of precipitation which is inadequate to penetrate deeper than this can be drawn back into the atmosphere directly and is therefore available to only very shallow-rooted plants, and to these for only a short period. The drying of soil layers deeper than about 3 dm (except sand or soils that crack deeply) is due almost entirely to absorption by roots.

Table 5
Vertical Gradients of Evaporative Power of Air and Controlling Factors. Data Obtained in Prairie Vegetation in Iowa Between 7:30 A.M. and 5:30 P.M. in Early September (266).

Height above ground, cm	Relative evaporation [a]	Mean wind velocity, mph	Mean relative humidity, %	Mean temperature, °C
237	170.7	11.01	37	31.5
114	130.3	7.68	40	31.4
10	100.0	2.19	46	31.1

[a]Piche evaporimeter.

Table 6
Comparison of Certain Habitat Factors in a Virgin Pine Forest and in an Adjacent Clearing in Northern Idaho (444). Data for the Month of August.

Factor		Forest	Clearing
Air temperature, °C	Maximum	25.9	30.0
	Minimum	7.4	3.9
	Range	18.5	26.1
Mean relative humidity at 5 P.M., %		38.8	35.2
Mean daily evaporation,[a] ml		14.1	36.1
Mean soil temperature at 15 cm, °C		12.8	17.0
Mean soil moisture at 15 cm, %		32.0	43.2

[a]Livingston atmometer mounted 15 cm above the ground.

Evaporation from an open water surface can be closely approximated by calculations based on (a) duration of sunlight, (b) air temperature, (c) relative humidity, and (d) wind speed (612). Since relative humidity and wind data are so few, another method of approximation has been worked out which uses only the first two of the above elements, although accuracy here is reduced (756). From such calculations one can in turn approximate the transpiration from short green vegetation, providing it covers the ground completely and has optimal soil moisture. Such results are useful in controlling irrigation as well as in watershed management (612).

Measurement of Evaporative Power of the Air (544)

The evaporative power of the atmosphere is usually measured by one of three types of evaporimeters, an open *pan,* which is the standard instrument of the U.S. Weather Bureau, the Livingston *atmometer* (468), which has been widely used by ecologists in North America, or the *Piche evaporimeter,* that is commonly used in other parts of the world.

A pan exposing a free water surface has the very desirable feature of giving values for evaporation in the same units as precipitation, i.e., in millimeters or inches of depth. However, the volume of water in the pan, the height of the rim projecting above the water surface, the radius of the pan, its color, and other factors have great influence upon the results obtained. To be of real value all such instruments should be constructed, installed, and operated in identical manner, but unfortunately this has not always been the case. Additional disadvantages accrue from wind splashing out water and carrying debris into it, birds drinking from the pan, and algae fouling the water. However, it is possible to construct simple evaporimeters that solve all these problems (544).

Livingston's atmometer consists of a thin sphere of unglazed white pottery filled with distilled water which soaks through the walls readily and evaporates into the air (Fig. 19). By means of glass tubing and rubber stoppers this sphere is connected with a reservoir bottle, and the amount of water lost by evaporation is ascertained from time to time by measuring the amount necessary to refill the bottle up to a scratch mark on the neck, or by weighing the apparatus. Evaporation easily pulls the water up from the reservoir against gravity, but when the outside of the sphere is wetted with rain, gravity would pull water into the system and the previous evaporation record would be lost if it were not for a 5–8 mm layer of mercury installed in the delivery tube and held in place by two plugs of loosely packed lamb's wool (Fig. 19). Upward pull draws water around the mercury, but downward pressure flattens it and stops flow; thus the mercury layer acts as a rain valve.

For determining evaporation rates during short intervals an atmometer may be connected to a burette which permits readings to 0.01 ml. An appropriate length of glass tubing and rubber connections must be used to keep the porous sphere at a height somewhat above the water level in the burette; otherwise gravity will force water into the sphere and greatly increase the rate of water loss.

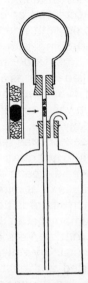

FIGURE 19.   Diagram of an atmometer mounted on a reservoir bottle and equipped with a mercury-drop rain valve.

Since no two instruments can be expected to lose water at the same rate, each must be calibrated. * Research workers can purchase new calibrated instruments with which they can calibrate others which they use. The calibration of a group of instruments simultaneously is accomplished by distributing them, along with a calibrated one, around the periphery of a round table fixed on a bicycle wheel mounted in a horizontal position which is slowly turned by means of a breeze from an electric fan (566). Thus, all are exposed to the same evaporation stress, and by comparing their water loss over a period of time a coefficient can be calculated for each, which, when applied to the data, will adjust them to the universal standard and thus render them comparable to similar data obtained by other investigators. If a new calibrated instrument with a coefficient of 0.70 loses 150 ml during a test, and a used instrument loses 140 ml during the same period, the coefficient of the latter is $(150 \times 0.70)/140 = 0.75$.

* The arbitrary standard is kept by the Livingston Atmometer Co., Box 305, Pauma Valley, Calif. 92061, U.S.A. which is the sole manufacturer and distributor of spherical pottery atmometers.

Since coefficients tend to change slowly while instruments are in use, they must be determined both before and after use, and the two determinations taken as bases for interpolating other coefficients to be applied in connection with each record obtained during the period of study. When atmometers become discolored or covered with algae they may be renovated by drying, rubbing with fine sandpaper until the surface is once more white, then scrubbing with a brush and distilled water to remove particles from the pores. Under no circumstances should any chemical be added to the distilled water as a deterrent to algae. Insertion of a short section of air-filled rubber tubing in the sphere, and use of a slightly more complicated rain valve, are desirable as insurance against damage where light frost is likely (469, 849).

Atmometer data are expressed only as milliliters loss per unit of time, and they cannot be converted into any other quantitative value. Therefore the atmometer is valuable only as a means of comparing the evaporative power of the air at different places or at different times. Unlike any other evaporimeter, this spherical instrument is not biased in favor of zenithal radiation. Another unique advantage lies in the universal standardization, so that data obtained by different workers are comparable.

The Piche evaporimeter consists of a disc of blotting paper clamped across the mouth of an inverted and calibrated glass tube filled with distilled water. The size, color and thickness of the blotting papers are not standardized, and the blotters need frequent renewal (556).

Like relative humidity, the evaporative power of the air varies at different elevations above the ground (Table 5). Therefore the position of the instrument must be carefully chosen with respect to the objectives of the problem at hand.

Evaporation data are valuable in many studies, but in their interpretation it must be kept in mind that the daily march of evaporation differs considerably with different types of instruments. In fact, different types of evaporimeters exhibit greater differences in water loss than different types of plants! The term "evaporative power of the air" therefore has no quantitative meaning except when used comparatively in connection with a particular type of evaporimeter.

Precipitation

Causes of Precipitation

There is always sufficient water vapor in the air for precipitation, but the proper meteorologic conditions to cause precipitation are frequently absent. Depending upon the causal meteorological phenomena involved, precipitation has three major types of origin.

Cyclonic (or *frontal*) *precipitation* is caused by eddies of warm air several hundred miles in diameter rising in vertical spirals about centers of low atmospheric pressure. These storms move easterly along the edge of cold polar air masses that cap both the northern and southern hemispheres. As the ascending mass of air in a cyclone rises it expands, mingles with cold air, and thereby is greatly cooled. Precipitation easily results from this phenomenon. Storms of this type are long-lived but move almost continuously following definite storm tracks across a continent (Fig. 20).

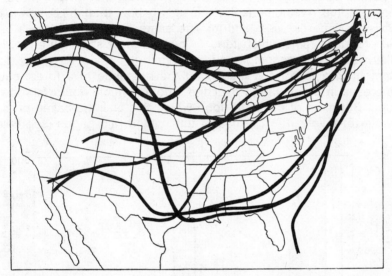

FIGURE 20.   Principal paths followed by cyclonic storms moving across the United States, based on a study of 1160 storms. [After Van Cleef (779).]

Orographic precipitation is caused by currents of air rising over an elevation, even if less than 200 meters high. Expansion and cooling is the basic cause of condensation here as with other types of precipitation. In general, rainfall of this type increases up mountain slopes only to an altitude of about 3000 meters at most, and beyond this there is a progressive decrease toward the summit (Fig. 21). This phenomenon can be explained in part as an approach toward exhaustion of the moisture content of the air, or possibly by the fact that the higher the altitude the more chance the air masses have of crossing the mountains through low divides without passing up over the highest points. Because of the strong control which even minor topographic irregularities exert over precipitation, attempts to estimate this statistic from rainfall maps or from measurements made at stations a few miles distant in mountainous country may be subject to considerable error (317).

Convective precipitation occurs in summer when the land surface becomes very hot under strong insolation, and the lower strata of the atmosphere are in turn heated considerably. As the air next to the ground becomes heated its density decreases below that of the upper atmosphere so that an unstable condition is brought about. Sudden violent overturns result, and these often project warm air vertically to altitudes where it is chilled to the condensation point. Cumulus clouds are the upper ends of columns of overturned air formed in this way, the upward movement having resulted in conditions that caused no more than the condensation of invisible vapor to visible form. The same phenomenon on a larger scale produces rains which tend to be heavy, of short duration, local in distribution, and accompanied by lightning.

Many claims have been made that the presence of a forest causes an increase in precipitation over the amount which would fall on bare ground. But most of the data upon which this claim is based probably reflect error involved in the use of unshielded rain gages (see below),

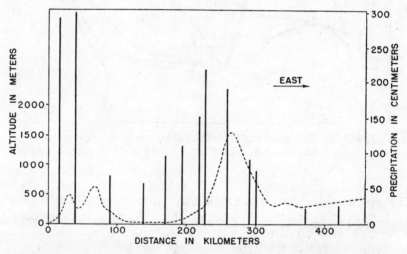

FIGURE 21. Topographic profile (broken line) and mean annual precipitation (heavy vertical lines) at a series of stations centered about 47° 50′ N latitude, extending from the Pacific Ocean across the Olympic and Cascade Mountains in Washington. The diagram shows (*a*) the approach effect west of the Cascades, (*b*) the existence of a zone of maximum precipitation on the middle western slope of the Cascades, (*c*) rainshadows on the landward sides of both mountain ranges, and (*d*) the lack of a very close relationship between altitude and precipitation.

with the trees serving somewhat as a shield and increasing catch rather than amount of rainfall. Precipitation forest may actually decrease convective precipitation slightly owing to its influence in preventing the land surface from becoming excessively hot.

Measurement of Precipitation (544)

Rainfall is measured with instruments called *rain gages,* which consist essentially of a funnel with a vertical collar at least 5 cm high which delivers water into a collecting reservoir. A calibrated stick can be inserted in the reservoir to measure the water collected, or the volume can be measured by pouring it out into a graduated cylinder and dividing the quantity by the area of the funnel.

Perfectly satisfactory rain gages can be constructed very cheaply by soldering a 5 cm vertical collar around the rim of a metal funnel, then soldering the spout of the funnel into a hole stabbed in the top of a metal kerosene can. With all types of rain gages the rim of the collar should be kept perfectly round.

Cylindrical precipitation gages, especially with orifices of large diameter, tend to deflect wind upward as it strikes the instrument, thus also deflecting snowflakes or drops that would otherwise continue their gradual descent and fall into the gage. Unless special precautions are taken the error is always important, and may result in catching less than half the precipitation that falls (335). Funnels of small diameter (down to 3 cm) (268) with narrow stems are not subject to significant turbulence error.

A large funnel-shaped shield installed about the cylindrical type gage and raised but slightly above the orifice allows the gradual descent of particles to continue without interruption so that the gage catch is similar to that of the ground (Fig. 22) (811). On sloping ground the instrument should be tilted so that the orifice and its shield conform to the slope of the soil surface. The catch must then be divided by the secant of the angle of slope to make the results comparable with those of a level orifice on level topography (306).

Tall objects near a gage intercept precipitation during a storm, therefore the gage should not be placed within a distance of 3 times the height of the tallest nearby object. Standard weather stations rarely use shielded gages and the instruments are often placed too near tall objects, so that their data must be used with great caution (544).

Where it is not necessary or feasible to read instruments after each precipitation, light mineral oil may be poured into the reservoir to form a film at least 4 mm thick over all water subsequently collected so that none evaporates. An instrument so set up to accumulate precipitation

FIGURE 22. A precipitation gage equipped with an Alter-type shield, and well elevated above the ground to avoid coverage by deep snow. The slats are hinged to a large hoop and drawn in slightly with a chain at the bottom. Their agitation by the wind prevents snow accumulation with subsequent capping over.

over a period of time is called a *storage gage*. The instrument can be weighed at intervals, or both oil and water can be poured out into a graduated cylinder, and, after the upper end of the water column is read, the oil alone is decanted back into the reservoir to function in the same manner during the subsequent period. In freezing weather a 34% solution of $CaCl_2$, 1.5 times the quantity of the expected catch, is placed in the reservoir with the oil. Metal surfaces should be protected against the salt solution with a coating of asphaltum.

Forms of Precipitation and Their Importance to Plants

Although rain is of tremendous importance to plants as a source of soil moisture, in general it is of little direct significance. An exception is the cracking of thin-skinned fruits such as cherries, plums, and tomatoes if wetted when ripe and full of sugars.

Snow may be injurious or beneficial to plants in a number of ways. Its weight may break off tree branches, especially when a heavy snowfall is followed by wind, or when a cold rain falls on vegetation already burdened with a great weight of snow. Saplings, especially slender-stemmed individuals growing under crowded conditions and those with asymmetrical canopies, may be bent from their normal vertical positions by the weight of snow held in their branches. Once bent over, the crowns are in a position to intercept still more snow so that the tree is bent prostrate or is broken off. The loss of one tree in this manner may create a break in the canopy which causes other individuals to grow asymmetrically and thus be predisposed to bending so that eventually quite a number of individuals are snow-thrown (168).

At high latitudes and altitudes snowdrifts persist exceptionally late on certain habitats where topography favors accumulation, but exposed south-facing slopes are characterized by little accumulation and early melting. Between these two extremes there is a wide difference in the length of the season when the ground is free of snow, and many plants show strong preferences for particular degrees of snow cover. The sedges and other plants which are confined to *snow patch* habitats prosper under the short growing season and cold, wet soil condition, and if necessary they can occasionally endure a summer when there is no snow-free period because of exceptional depth of accumulation the preceding winter (Fig. 23). Other species appear to demand more heat and a longer growing season, and are therefore confined to habitats where the snow cover melts early. Near alpine or arctic timberline, trees are commonly confined to habitats of moderate snow accumulation. On the one hand, inadequate snow cover exposes their seedlings to injury by wind, and on the other, excessive accumulation results in growing seasons too short for seedling establishment. Evergreen foliage often shows the protection afforded by moderate snow cover (Frontispiece), but deep snows often stunt growth, owing to their persistence into the growing season, and cause compressed, crooked stems (Fig. 24). The vertical distribution of thallophytes and bryophytes on tree trunks may be closely related to average snow depth (Fig. 25).

FIGURE 23. An area on the lee slope of a high ridge where forest is inhibited by a deep bank of snow that accumulates annually and persists until early summer. Various species of *Carex* dominate the snowpatch vegetation. Northern Idaho.

FIGURE 24.   A tree of *Abies lasiocarpa* exhibiting nearly 100 growth rings that was cut from an area where the forest is dwarfed by deep annual accumulations of snow. In addition to retarding growth, the snow has created an environment unfavorable for the retention of the lower branches, and its weight has distorted the main axis. Northern Idaho.

Deep snow may press seedlings down to the ground so as to favor their parasitism by fungi which are active at the ground surface (418). The lower branches of older evergreen conifers, if covered with snow for long periods, are especially subject to attack by the brown felt fungi mentioned earlier.

Where snow accumulates on steep slopes topographic conditions may favor sudden devastating avalanches which recur with such frequency as to keep the avalanche tracks nearly denuded of brittle-stemmed plants (256). Forest planting may be used for control, especially through the establishment of trees on windward slopes to prevent transfer, with subsequent build-up on the lee slopes. Elsewhere a slight creep of snow may deform trees while in the seedling stages, before

FIGURE 25. *Parmelia olivacea,* a lichen confined to portions of *Betula* trunks which are not covered by snow in winter. (*Left*) Tree in forest interior. Abisko, Sweden. (*Right*) Tree on a bluff where wind prevents snow accumulation.

they attain sufficient mechanical strength to resist the lateral pressure (Fig. 26).

Snow that melts at times when the soil is not frozen adds to the supply of soil water. The gradual melting of mountain snow feeds streams descending to the lowlands during the summer, so that in arid regions the snow on adjacent mountains constitutes the chief reservoir upon which irrigation practice depends. Winter surveys of the depth and water content of snow in such regions make possible a good forecast of the supply of irrigation water which will be available in the following season.

Hail and sleet are too infrequent to be significant as sources of soil water, but where they occur they may cause considerable temporary damage to the aerial parts of plants, especially in the case of tender crops.

In east central North America glaze storms, in which rain freezes as it strikes objects, build up thick coatings of ice on tree boughs to the extent that considerable breakage may result from the great weight of ice (455).

FIGURE 26. *Populus tremuloides* showing the effects of creeping snow in deforming the stems while they were still too young and weak to resist its movement.

Some Geographic Aspects of Precipitation

Most of the precipitation that falls on land results from the condensation of vapor derived from the surfaces of oceans. Consequently the amount of precipitation that falls on a particular area depends somewhat upon its proximity to an ocean and a landward movement of air masses, the latter tending to become progressively depleted of their moisture as they advance inland. Landward breezes forced abruptly to high altitudes drop an average of 11.45 m of precipitation a year at a station on the island of Kauai, Hawaii. But on the desert at Iquique, Chile, descending air that is warmed yields an average of only 1.3 mm per year, and up to 13 consecutive years may pass without measurable precipitation.

When the direction of these prevailing winds differs with the seasons, they exert an especially important control over the precipitation regimen.

In crossing well-elevated land masses warm layers of air are forced to rise so that orographic precipitation occurs. Mountain slopes usually have a relatively wet side facing the ocean from which the moisture is derived, and a relatively dry slope to the leeward (Fig. 21). Vegetation differs on seaward and landward slopes in direct ratio to this degree of difference in rainfall. Obviously mountain ranges with their longitudinal axes at right angles to the direction of the prevailing winds cause much more precipitation than those oriented parallel to the wind.

The effect of mountains in increasing precipitation extends some distance in a windward direction, rainfall usually increasing before altitude begins to increase noticeably. This has been called the *approach effect.* On the lee slope, and extending outward for a distance over the adjacent basal plain, there is usually a region of very low rainfall referred to as a *rain shadow* (Fig. 21). This phenomenon results from the descent and warming of air masses that have had most of their moisture extracted on the windward slope, and it is most pronounced where winds are from a constant direction. The rain-shadow effect may begin at the crest of the mountain or well down the lee side, and it is often sufficiently pronounced to obscure the normal altitudinal gradient of precipitation.

At stations in deep valleys or canyons precipitation may far exceed the amount characteristic of locations at similar altitudes on the adjacent basal plains because rainfall is stimulated by the surrounding heights. This is called the *canyon effect.*

The presence of cold or warm ocean currents near the margins of continents affects the potential precipitation on the adjacent land. Warm currents give up much moisture to the air and also raise the temperature of the air until it has a great capacity for water vapor. The high precipitation along the west central coast of North America is largely attributable to the warm Japan current which approaches the continent from the southwest at this point.

Latitude also has a considerable influence on precipitation. Rainfall is potentially highest in equatorial regions and lowest in polar, for in polar regions the air is too cold to hold much moisture. Furthermore, cold air tends to settle rather than to rise and expand.

On approaching dry regions, the relative variability of annual precipitation increases rapidly as the total drops below about 45 cm per year (154). Precipitation of excessive magnitude in any one month or year commonly elevates the mean, giving a highly erroneous concept of the amount of water usually available for plants. For this reason, *median* monthly or annual figures are more indicative of "normal" conditions than are means.

Effectivity of Precipitation (189)

Although all soil moisture is derived from precipitation, not all precipitation is equally effective in increasing soil moisture (189). In the first place, the slower and more gentle the showers the higher the percentage of water that soaks into the soil in relation to that lost as runoff. Second, the greater the quantity of water falling during any one rainy period, the more of it sinks below the reach of direct surface desiccation. Thus, in dry climates a number of showers totaling up to several inches of rain in summer may have no effect in raising the soil moisture content because the individual rains are too light and too widely separated for successive increments to have cumulative effect. The longer and more severe the drouth the greater the quantity of rainfall that will be required subsequently to break it. Finally, the severity of evaporative conditions after a rainfall has a very important bearing upon the duration of favorably moist conditions. The more cool and humid the climate, the more effective a given amount of rainfall.

The effectivity of precipitation as a source of soil moisture for plants is best measured by direct studies of the degree of penetration and duration of moisture in the soil. However, this method is laborious, and detailed studies are appearing but slowly. In the meantime efforts have been made to find indirect means of evaluating this factor of plant environment. Some of the methods of expressing the climatic effectivity of precipitation are as follows.

A simple, yet reasonably adequate method is to discount precipitation (P) in proportion to the warmth of the climate (T):

$$\text{Effective annual precipitation} = P \text{ in mm} / (T \,^\circ C + 10)$$

The 10 is added only with a view to avoiding negative values and provide an index of climate even when the mean temperature is as low as $0\,^\circ C$. To compare monthly values, which are far more useful ecologically, the appropriate modification is:

$$\text{Effective monthly precipitation} = (P \times 12) / (T \,^\circ C + 10)$$

A similar but slightly more complicated expression is the Precipitation/Saturation deficit Ratio. Precipitation is again expressed in millimeters. The saturation deficit is obtained by subtracting the actual vapor pressure from the maximum possible vapor pressure at the prevailing temperature, the result being expressed in millimeters of mercury. * The special merit of this method lies in the fact that evaporation rates parallel

---

* Determine the saturation deficit as follows. Look up the maximum water-vapor pressure for the prevailing air temperature in dewpoint tables (533). Convert inches to millimeters by multiplying by 25.4. Multiply this pressure by relative humidity expressed as a decimal. Subtract the actual vapor pressure thus obtained from the maximum vapor pressure to get the saturation deficit, i.e., $SD = VP - (VP \times RH)$.

The hygrothermograph, an instrument which makes a continuous record of air temperature and relative humidity on a paper chart, is uniquely suited to studies where saturation deficit data are needed.

variations in saturation deficit more closely than relative humidity or temperature, because saturation deficit expresses the absolute capacity of the air for additional moisture, whereas relative humidity indicates no more than the degree of saturation. For example, the evaporative power of the air at 60% R.H. is not the same at different temperature levels, as indicated by the following comparisons:

R.H., %	T, °C	SD, mm
60	10	12.73
60	12	7.02
60	30	3.68

These data bring out the fact mentioned previously that at 20°C the capacity of the air for water vapor is twice that prevailing at 9°, and relative-humidity data obscure this important difference. Since the relative wetness of air and its drying power do not necessarily vary together, some biologic phenomena may be correlated more closely with one than with the other.

There is much to recommend the P/SD ratio as a measure of effectivity of precipitation, but the paucity of relative-humidity records is a major obstacle.

A still more complicated method, devised by C. W. Thornthwaite (756), is based on these premises: (a) temperature is closely related to the supply of solar energy available for evapotranspiration,* (b) a given monthly temperature mean is associated with higher evapotranspiration when days are longer, (c) wind and humidity are of minor importance and may safely be ignored in estimating evapotranspiration, and (d) the soil has a capacity to store approximately 10 cm of rainfall, so that biologically significant drouth does not begin at once with the onset of hot, rainless weather. Calculations are minimized through the use of tables and nomograms (589), so that it is not difficult to evaluate monthly precipitation as excessive or deficient in terms of what is needed to maintain favorable levels of soil moisture. H. L. Penman's method includes wind, relative humidity and hours of sunshine in addition to air temperature, but is only slightly more accurate despite the increased complexity.

The last two methods above not only provide month-by-month evaluations of climatic moisture in terms of potential evapotranspiration (i.e., water that would be lost from a dense and active plant cover well supplied with soil moisture), but also of actual evapotranspiration, which drops

---

* *Evapotranspiration* is the total loss of water vapor to the atmosphere by vegetation and soil, on a unit area basis.

considerably below the potential rate as soil moisture reserves are depleted during dry weather. By a continual bookkeeping technique monthly deficits and surpluses can be balanced and the status of currently available soil moisture can be predicted rather accurately (533). Thus the appropriate time to irrigate crops can be determined as it approaches without actually sampling the soil. However, irrigation need can be anticipated much more simply by following the course of evaporation from a blackened atmometer, which is more sensitive to solar radiation than the usual white instrument (304).

Importance of Seasonal Distribution of Precipitation

The statistical methods for evaluating precipitation effectivity which have been outlined above ordinarily yield different values for each month of the year. Some investigators, desiring to obtain single-figure criteria of precipitation effectivity, have chosen to average the twelve monthly figures for each station. This is definitely an oversimplification and is not to be recommended for any climatic data, because the same averages can be obtained by different combinations of monthly values, and the differences are ecologically very important.

The time of year when moisture is deficient is always a critical factor (406). For example, the steppe region centering about eastern Washington and the steppe of the central North American plains are semiarid by any climatic criterion, the annual averages of climatic statistics showing no definite and consistent differences between them. However, in the former the precipitation is confined chiefly to winter, whereas in the latter the winters are dry, most of the rain falling in spring and early summer. The floras of these two steppe provinces have very few species in common. In the one, only those plants are successful which can complete their annual growth and set seed in spring before the winter accumulation of moisture is exhausted. In the other, the inception of growth and the season of flowering come distinctly later in the season, coinciding with the period when soil moisture is most favorable.

Nearly every difference in rainfall, either in the annual total or in the season distribution, is accompanied by a difference in the natural plant populations. For a plant to be successful in a region, the sequence of phases in its life cycle, with their varying demands, must match the climate to the extent of allowing good vegetative growth as well as adequate reproduction. For this reason the introduction of economic plants into climatically unsuited regions would result in almost complete failure but for irrigation, cultivation, and other means of artificial control of the environment which compensate for many climatic adversities.

Furthermore, man's efforts to extend the ranges of economic plants are aided by the fact that it is not always necessary that the life cycle be completed in order to get a crop.

Aside from its influence in determining whether a particular plant can live in a particular region, the time of year when moisture is suboptimal has an important bearing upon the quality of growth. Thus, the amount of rainfall during and immediately preceding the growing season has an important influence on the amount of wood laid down by trees in dry climates, and the time of application of irrigation water can increase or decrease the yield of crops (406). A dry summer may be beneficial in ripening winter wheat and at the same time greatly reduce the yield of spring wheat which is not quite so far along in its cycle of development (412). In addition, the chemical composition of plants is decidedly affected by rainfall and other weather variables.

Drouth

When a dry season is a regular and characteristic feature of a climate, no matter how severe it may be, native plants are all so thoroughly adapted that they ordinarily show no ill effects from it, but sometimes rainfall drops far enough below the normal amount to affect the plants adversely, and such an event is referred to as a drouth. The effects vary from slight reductions in size, vigor, and yield, to outright killing of the plants (535). Drouth is common in all climates, but in moist climates a relatively short rainless period will bring about as pronounced an effect as a much longer period in a semiarid climate. Low rainfall is not the only factor involved in drouth, for climatic dryness is invariably associated with higher temperatures which cause plants to use up soil moisture more rapidly. Therefore the statistical methods described for evaluating the precipitation effectivity of climates are useful as a basis for evaluating the dryness of weather for any one month in relation to its norm.

Although the theory of climatic cycles has been the subject of much debate, the facts are indisputable that precipitation tends to be abnormally high for a number of years, then to vary in the opposite direction rather consistently for a time. Our knowledge of this phenomenon rests partly upon direct observation and careful records, but this source has been supplemented considerably by an indirect method based on tree growth. In some but by no means all temperate and tropical areas precipitation is correlated with the thickness of annual increments of xylem formed by exogenous woody plants as long as the individuals grow where the chief source of soil moisture is precipitation falling on the immediate surroundings. Therefore, it has been possible, by measuring and comparing xylem layers in old trees, to learn something of cycles of drouth for many centuries past, although there are many pitfalls in interpreting this information accurately (4, 255, 740).

The information gained by the study of alternating periods of drouth and heavy rainfall has proved to be important to mankind in many ways. Climatic fluctuations have produced alternating periods of profitable agriculture, and crop failure accompanied by land abandonment, on the dry plains of central North America, and this fact is fundamental in any phase of land-use planning (143). Natural plant populations on these plains are quite responsive to these fluctuations, drouth-tolerant and drouth-intolerant species benefiting alternately (818). Shelterbelts established in steppe regions during periods of above-average rainfall suffer heavy mortality during drouths, and studies of the relative abilities of different species to endure these drouths have provided valuable information to guide further shelterbelt planting (195). Because the survival of tree seedlings is so much better during the wetter phases, it has been proposed that manpower be devoted chiefly to reforestation during these periods, then shifted to fire protection during drouth periods (528). The pattern and intensity of alternating drouth and flood periods are also of great interest in hydrology, * where engineering must be at once adequate to conserve subnormal runoff and to control severe floods. Man's ability to cope with ecologic problems arising from climatic cycles is definitely limited by the fact that they are not rhythmic, hence he cannot predict their timing. However, the maximum ranges of intensities can be estimated with fair accuracy.

Soil Moisture

Precipitation Losses to the Atmosphere

Earlier the fact was mentioned that the evaporative power of the air exerts a strong influence upon the effectivity of precipitation in keeping the soil moist. An analysis of the mechanics of this is now in order.

Some of the water precipitated out of the atmosphere is evaporated before it reaches the ground. If the period of precipitation is long enough, sufficient water is evaporated to bring the relative humidity of the atmosphere up to 100%, which is the condition usually to be observed immediately after a rainfall.

Anyone who has taken refuge under a tree during a shower knows that the crowns of trees intercept and hold considerable amounts of

---

* Hydrology is the science that deals with the interrelationships among precipitation, runoff, evaporation, ground-water supplies, etc. (537).

precipitation before allowing much to penetrate through. In addition, shrubs, herbs, surface mats of cryptogams, and layers of litter and duff take a toll of moisture before it reaches the mineral soil. Snow as well as rain is intercepted. The total quantity thus held varies with the density of the vegetation, whether it is in leafy or leafless condition, the form of the precipitation, and the amount and rate of precipitation. As much as 100% of light showers may be intercepted, but with a heavy rain the shoot surfaces accumulate water to their capacity and then let the remainder drip off or run down the stem so that the percentage of interception drops very low. Some of the water that does not drain off under the influence of gravity can be shaken off by wind, but a certain amount can be removed only through evaporation. Of the total precipitation falling during a season usually no more than a third is prevented from reaching the ground before it evaporates back into the air. There is surprisingly little difference among tree, shrub, and grass vegetation with respect to the amount of water they intercept. Although a small amount of intercepted precipitation is absorbed, and the remainder greatly reduces transpiration until it evaporates, the bulk represents a loss to the potential water supply of the soil.

The windward slope of a ridge may get little moisture as a result of wind transfer to the lee side, where it results in a limited area of relatively lush vegetation. Fences and windbreaks create snowdrifts that result in inequalities of moisture distribution that differ in pattern from those favored by natural topographic irregularities.

The contrasting influences of windbreaks in the northern and southern parts of the central steppe of North America illustrate this point to good advantage. Northward, where much of the precipitation falls as snow, windbreaks accumulate snow and build up moisture at the expense of adjacent fields to the extent that the growth of the trees remains vigorous during all but severe drouths. Southward there is little snow, and the trees intercept only rain, decreasing the amount of moisture the soil beneath the trees would otherwise receive. In this region windbreaks suffer frequently from drouth. Crops on the lee side of a windbreak may clearly show the effects of a miniature rain shadow.

The morphologic features of the individual plant, as these divert water toward or away from the axis, strongly influence the microdistribution of precipitation. Thus plants like *Tsuga canadensis* (795) may create a relatively dry spot centered on its stem, whereas grasses tend to funnel water inward so that stemflow wets the soil considerably deeper under the axis than elsewhere (274). In arid regions especially this has a bearing on the plant's own survival, and on seedling establishment. In wet climates it can mitigate or aggravate leaching. Furthermore, the chemical composition of foliage drip, however it is distributed, is strongly altered while the drops are in contact with leaves or stems. Small amounts of minerals contained in the drops initially may be taken in through the epidermis, whereas larger amounts of nutrients (especially K), toxins (e.g., phenols, terpenes), etc. are usually dissolved from the epidermis or periderm.

Much of the water that enters the soil is evaporated quickly again. However, contrary to popular opinion, surface evaporation ordinarily does not reduce the soil moisture content below 2–3 dm whether the soil is cultivated or not (781). Upon this fact hinges the practice of cropping and fallowing on alternate years in semiarid regions. All the moisture stored by the soil below the upper 2–3 dm is conserved, provided the growth of weeds is prevented, so that each crop has nearly all the benefits of two seasons' precipitation. However, since the upper 3 dm of a dry loam may retain over 5 cm of rainfall, only continuously wet periods furnishing moisture in excess of this amount can result in the storage of water below the reach of surface evaporation. Thus, it is clear that, wherever rains are widely spaced and not much water falls each time, a high percentage of the annual precipitation is lost by surface evaporation. Under average conditions in Saskatchewan, a one-day rain amounting to 9 mm in May or June is returned to the air by evaporation within 10 days. In July and August a 12 mm rain would be vaporized in the same time (353). All the above facts can be utilized to good advantage in irrigation practice. If the soil moisture content is allowed to approach exhaustion each time before water is applied, there will be a minimum number of irrigations and less loss by evaporation from the surface 2–3 dm each season.

The influence of vegetation in shading the soil, reducing wind velocity, and giving off vapor by transpiration all tend to reduce losses of moisture from soil or snow cover by direct evaporation. However, the vigorous absorption of soil moisture by roots, together with the losses due to interception of precipitation by the shoots, usually more than offsets the effects of vegetation in retarding evaporation from the soil. Thus the soil tends to have more moisture in forest openings than under adjacent timber (Table 6) (162). This is one of the factors that militates against seedling survival in dense forests. When a forest is thinned streamflow increases (529). When it is replaced with grass that is more shallow rooted or has a long season of limited activity, streamflow is increased still more (689).

Each climate has a certain potential evapotranspiration set partly by wind, and primarily by the energy received as solar radiation, as this raises temperature and lowers relative humidity. So long as soil moisture is not limiting (which it frequently is) actual evapotranspiration equals potential evapotranspiration, and this is very similar for closed vegetation as diverse as grassland and forest, providing the surfaces of the canopies are relatively level, the foliage has similar reflectivity, and the vegetative activities of the plants are near maximum. Potential evapotranspiration can be measured directly by growing vegetation in large containers that are kept moist and are surrounded by a broad buffer belt of the same nature (369). It can be approximated rather well by calculations based on weather data, as indicated previously.

Precipitation Losses as Runoff

Aside from the losses of moisture by evaporation, another major factor reducing the usefulness of precipitated water to plants is the quantity that runs off the land directly and flows back into the seas as streams. The intensity of precipitation is one factor determining runoff, for the more rain falling per unit of time the more the infiltration capacity of the soil is exceeded. Also the character and condition of the soil and the nature of the topography are significant factors governing the proportion of water lost in this manner. Five major features of the earth's surface are important in this respect.

Moisture content of the soil.   The rate of infiltration of water into a soil is at the maximum when the soil is fairly dry, for after water is added the pore space becomes filled, the colloids swell, and the rate of entry of additional water declines to a low but uniform level. For this reason, some investigators have advocated making comparative infiltration tests 24 hours after the soils are thoroughly soaked, so that the moisture-content factor in infiltration will be fairly well standardized. This procedure would also overcome the problem presented by cracks which may develop as a heavy soil dries, and which enormously increase the rate of infiltration.

Soil temperature.   Freezing weather produces a layer of frozen soil at the surface which effectively seals off the subsoil, and thereafter all rainfall and water from melting snow run off the surface. This explains why in cold climates the soil may be quite dry in spring a short distance below the surface in spite of high precipitation. When the soil is not frozen, temperature has relatively little effect on infiltration.

Porosity.   Water may infiltrate very coarse-textured soils so readily that none is lost as runoff even in the heaviest downpours. By contrast, the surface layer of a bare clay soil soaks up the first moisture that falls, then tends to swell and become a dense waterproof layer which sheds the remainder of the water.

On level topography the most important factor favoring rapid infiltration is the protection afforded the soil surface against the beating of raindrops by vegetation, litter, and similar material. These layers break the force of the raindrops and the water is scattered as a fine spray, whereas on soil not so protected the drops stir up fine mineral particles, then wash them into the surface, thus clogging the pores and retarding further infiltration (207).

Surface layers of organic matter are as porous as the coarsest of sands. Likewise, humus renders the soil structure so porous that infiltration into the *A1* horizon is characteristically very rapid when this horizon is high in humus content.

These influences of vegetation in maintaining permeability are exceptionally pronounced in virgin forests, where there is almost no runoff even on the steepest slopes and under the heaviest downpours, provided that the soil is not frozen. In an effort to find a means of disposing of impure waste water in New Jersey it was discovered that if piped into an oak forest, 49 ha would easily take up 38 million liters per day—the equivalent of 12.7 m of rainfall a year! Grassland is usually much less efficient in preventing runoff, although it is equally effective in preventing erosion. However, under disturbed conditions these relationships between forest and grassland are reversed. Following disturbance, forest soils lose their higher absorptive capacity more readily than grassland soils, for the superficial position of organic matter in forest-soil profiles allows it to be more easily destroyed than the deeply distributed humus of grassland soils.

Without minimizing the proved value of forests in preventing direct runoff, it must be acknowledged that the value of this factor in preventing floods has sometimes been popularly overestimated. Forest cover promotes more even flow from a watershed and reduces silting, but at the same time it reduces the total yield of water (owing to interception and transpiration), and its beneficial influences decline rapidly as the water-storage capacity of the soil is reached. During protracted rains, or during hot weather following deep snow accumulation, vegetation has negligible influence in reducing floods and flood damage. Floods in the Mississippi Valley have increased neither in frequency nor intensity despite extensive deforestation.

Smoothness of slope.   The greater the degree of irregularity of a slope the more obstructions there are to hinder rapid runoff, and the more pocketlike depressions to hold water and allow it to infiltrate. A surface cover of stones not only retards runoff but reduces evaporation, hence conserves moisture (436). A number of land-management practices are designed to offset the effects of smoothness of slope. On arable lands it is advised to plow and cultivate along the contour, and to grow crops in strips which follow the contour. Rangelands have been measurably benefited by constructing series of terraces and corrugations along the contour, even where the slope is almost imperceptible.

Steepness of slope.   As a result of their inherently high rates of runoff, steep slopes are too much of an erosional hazard to permit anything but limited land use. The usual recommendation is not to cultivate or graze such slopes, but to convert them into woodlots or other permanent vegetation that practically prevents runoff and accelerated erosion.

Not only does direct runoff render precipitation less effective in augmenting the supply of soil moisture, but also the soil solutes and the soil particles as well are washed away. As a result of accelerated erosion tons of topsoil rich in humus, the product of many years of accumulation, may be lost from a single hectare of misused land in a year. For each ha in the Tennessee River drainage there is an annual loss of 14 tons of silt, 254 kg of K_2O, 118 kg of MgO, 102 kg of CaO, 29 kg of N, and 16 kg of P_2O_5 (242).

That part of precipitation which circumvents evaporation and direct runoff sinks into the ground. In regions of moderate to heavy precipitation or of coarse soils, however, it is only a part of this water that remains available to plants, for some of it percolates below the reach of absorbing roots within a few hours after it falls.

From the standpoint of the average vascular plant, the effectivity of precipitation as a source of soil moisture can be summed up in the following equation:

Effective precipitation = (Precipitation + Condensation) –
  (Evaporation + Interception + Runoff + Colloidal adsorption + Percolation)

States of Water in the Soil

Soil is a porous medium which admits water, allows some to pass through, and retains the remainder. Infiltration, state of existence, and movement of water in the soil, therefore, comprise three important and interrelated aspects of soil-moisture relationships. The first of these has already received some attention, and discussions of the second and third follow.

Gravitational water.   After a heavy rain, or an application of irrigation water, the surface layer of a soil is temporarily saturated. Owing to the forces of gravity and capillarity the water in this saturated zone descends rapidly through the dry layers, leaving a moist zone in its wake (Fig. 27). If there is sufficient water, this wet layer may penetrate deeply enough to reach the permanently saturated part of the subsoil which characteristically lies upon bedrock.

Any water that moves downward through a moist soil in response to gravity is called *gravitational water*. It is available to plants only when showers follow one another in rapid succession; otherwise it percolates below the reach of roots within a few days. The water in the deep and more or less permanently saturated zone is called *ground water,* and the upper surface of this layer is the *water table.* The true water table, as differentiated from *perched* or *hanging water tables,* may lie at any distance up to 650 m below the soil surface (20).

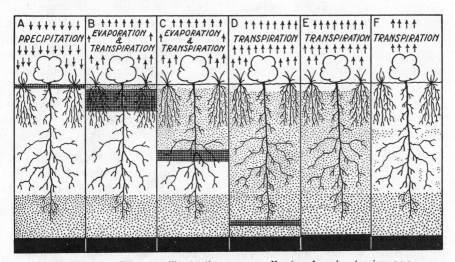

FIGURE 27. Diagram illustrating some effects of a short rainy season on a shrub and perennial grasses growing in an arid region. (*A*) When the rains begin the grasses are dormant because the soil is dry (unshaded) throughout the depth of their root systems, but the shrub is in active condition because its roots extend into the capillary fringe (heavy stippling) maintained by the ground water (black). (*B* and *C*) Free water (cross-hatched) percolates downward, leaving the soil moistened to the field capacity (light stippling). Water vaporizes from both plant and soil surfaces. (*D* and *E*) Free water approaches the water table, then becomes a part of ground water. Upper layer of soil which is subject to evaporation loss is desiccated to the extent that water loss is now chiefly transpirational. (*F*) Growth water above the capillary fringe is practically exhausted. Grasses again become dormant.

Capillary water. As gravitational water drains out of the upper layers of a soil it leaves behind much moisture in the form of films coating each particle and droplets suspended in the angles of the larger pores or completely filling the small pores. * This, the *capillary water,* is continuous from one particle to the next. It is held in the soil only in the sense that it does not respond to gravitational pull, and it is the source of almost all the water a plant extracts from the soil.

Hygroscopic water. Owing to the forces of evaporation from the soil surface and absorption by roots, the capillary water held by the soil is gradually depleted. As depletion progresses the force of attraction between the soil particles and the water held against their surfaces increases until finally the remaining water passes into a state where it is no longer in a liquid condition and thus ceases to be chemically or biologically active. Plants can absorb only a relatively small amount of this *hygroscopic water,* but it can be evaporated more or less completely, depending upon the severity of drouth. It exists in equilibrium with the vapor pressure of the atmosphere and therefore attains its maximum thickness (the *hygroscopic coefficient*) when the relative humidity is 100%.

The above three types of soil water are not sharply defined but form a continuous series from water which is not retained by the soil to water which is held with great force.

Combined water. Hygroscopic water cannot be entirely evaporated from a soil under ordinary atmospheric conditions because relative humidity never reaches zero, but it can be done by heating soil to constant weight in an oven at approximately 105°C. After the hygroscopic water is gone the only water that remains is in hydrated oxides of aluminum, iron, silicon, etc. Such *combined water* can be driven off only by resorting to much higher temperatures, but high temperatures also bring about irreversible changes in inorganic colloids, oxidation of humus, and decomposition of carbonates. These changes explain why determinations of the organic-matter content of soil by the loss-on-ignition method are not very accurate unless the combined water and carbonates are first removed.

Water vapor. That portion of the pore space in a soil that is not occupied by liquid water contains a soil atmosphere that always includes water vapor. So long as any capillary water is present this soil atmosphere is almost saturated with water vapor.

---

* Pores that are completely filled with water at this time are called *capillary pores.*

Movement of Soil Water

Gravitational and ground water.   The rate of downward percolation of gravitational water is determined by the number, size, and continuity of the noncapillary pores of the least permeable horizon, and by the degree to which the soil was already saturated.

As gravitational water reaches the water table or bedrock and becomes ground water, its further movement is in the direction of slope of these surfaces. Except in closed basins it finds an outlet back to the surface at lower elevations and thence flows toward the sea in stream channels. The water table is seldom level for any distance, the operation of several factors preventing it from attaining such an equilibrium.

The character of the vegetation has an important effect upon the height of the water table and related phenomena. The more massive the vegetation the more water it absorbs and transpires into the atmosphere, and this water is drawn either from the ground-water supply or from the soil before it penetrates to that level. Many studies have shown that the water table is characteristically depressed under groves of trees in otherwise open country. Also it has sometimes been observed that the removal of forest cover from shallow basins results in the appearance of swamps or ponds, showing that only vigorous absorption by the trees had previously kept the water table below the surface (843). Advantage can be taken of this phenomenon by planting trees in habitats where it is desired to reduce swampiness (132).

In view of the influence of trees upon the water table, together with the amount of rain and snow which they intercept, it is clear that the benefits derived from windbreaks in the way of preventing soil-blowing are gained only at a sacrifice of a portion of the water resources. Again, in hilly or mountainous country where vegetation is of most value as a protective covering for watersheds, theoretically the plant cover should be dominated by species that intercept and retain the least rain, which have the shallowest roots, and which are spaced as widely as possible without allowing enough direct runoff to cause accelerated erosion (411). It is true that the complete removal of vegetative cover gives the maximum total yield for a watershed, but the increase is due to a higher proportion of surface runoff which is concentrated in a shorter interval, and the resultant erosional losses, silting damage, and unevenness of water yield are so undesirable that every effort is ordinarily taken to prevent denudation.

The distribution of precipitation in space and time is a second major factor affecting the height of the water table. Unevenly distributed rainfall tends to. build up the water table first in one region then in another.

Fluctuations in precipitation from one season to another are reflected in corresponding fluctuations in the height of the water table. During the dry season the water table drops progressively, reaching its lowest level in the early part of the wet season before the water of the first rains has penetrated deeply. The highest altitude of the water table is attained at the close of the rainy season.

Fluctuations in climate cause ground-water fluctuations of still greater magnitude. This effect is most apparent where the water table is at such an elevation that depressions in the land surface contain ponds only during the wet phases of the cycles.

Finally, the water table is affected by topography, conforming to the contour but exhibiting less relief.

Capillary water.   The rate of movement of capillary water through the soil is influenced by some of the same factors that control the rate of infiltration, especially texture, structure, and temperature. In addition, its mobility varies directly with the thickness of the water films, for this determines the adhesive forces holding the water to the particles. Very thick films can move in any direction, but as they become moderately thin they become immobile. The amount of lateral movement of capillary water is never very great, and for this reason irrigation water must be applied close to the bases of the shoots if it is to benefit roots. For all practical purposes only the upward movement of capillary water is important, and movement in this direction is significant only within a meter or so of the water table.

In climates where the rainfall is scanty but the water table is not excessively deep, certain deep-rooted plants may be able to take advantage of water rising by capillarity from the water table upward (Fig. 27). Likewise most species of plants that are essentially confined to stream margins or lake shores are very dependent upon moisture rising from the water table. At about 85 cm above the water table, the rise of capillary water becomes too slow to be significant in replacing transpiration losses. The zone of capillary water maintained by vertical rise is called the *capillary fringe,* and plants depending upon its moisture are said to be *subirrigated.* In arid regions it is sometimes necessary to irrigate alfalfa only until its roots have penetrated to the capillary fringe, after which the field can be cropped repeatedly without resorting to further irrigation. Among the deep-rooted perennials native to such a region, the entire crop of seedlings which is produced each year may perish regularly during the dry seasons. The perpetuation of these species is made possible only by occasional years with above-average rainfall which permit the roots of the seedlings produced that year to establish connection with the permanent moisture supply of the subsoil before midsummer drouth becomes severe.

Plants of arid regions that depend upon maintaining their roots in the capillary fringe are called *phreatophytes* (e.g., many *Salix, Populus,* and the shrub in Fig. 27). They are lavish users of water, and with water resources becoming so critical, considerable interest has been shown in their elimination. However, this may increase streambank erosion and raise stream levels and so harm fish (118).

The moisture content of the soil in the capillary fringe is not uniform but decreases with height above the water table. In the lower portion the pore space is largely occupied by water and aeration is consequently very poor. Depending upon their aeration requirements, different species send their roots for varying distances into the capillary fringe. Very few can penetrate below the water table.

Ordinarily the water table is so deep that the capillary fringe ends well below the surface layer of soil that is strongly affected by direct evaporation, i.e., the upper 2–3 dm. Under experimental conditions, an undisturbed and subsequently unwatered soil with a deep water table was found to lose only 8.6 cm of water over a period of four years subsequent to a heavy irrigation (781). However, if the water table is high enough that the capillary fringe rises into this zone of surface desiccation, much moisture is lost by direct evaporation.

Water vapor. After capillary movement stops, small amounts of water are transferred through the soil by vaporization at points where the films are relatively thick, and condensation where they are thinner. In a somewhat similar manner, in winter, or at night, the relatively cold surface layer of soil increases in moisture content at the expense of the deeper and warmer horizons. The reverse takes place during the summer and during the day.

Soil-Moisture Constants

The actual water content of the soil is a variable, fluctuating as a net result of the action of many independent factors. On the other hand, such characteristics as the maximum amounts of capillary or hygroscopic water a particular soil can hold are determined chiefly by the kinds and sizes of the soil particles, and are therefore constants for each soil. With temperature remaining constant they are capable of reproduction within one or at most a few moisture per cent units.

Several types of constants can be determined by allowing a soil to establish an equilibrium with various forces which exert differing degrees of pull or tension upon the water it contains. Regardless of the type of constant, the amount of water held by the soil varies directly with the fineness of the state of division and with organic matter content.

Field capacity (capillary capacity). *Field capacity* is loosely defined as the percentage of moisture * when the drainage of a wetted soil diminishes sharply to a very slow rate. This condition has usually been attained by two to five days after a rain, provided the soils are pervious, uniform in texture and structure, the pores are not already clogged with water, temperatures are moderately high, and at least 8 dm of the profile are wetted by the rain. Since texture and structure usually change with depth, field capacity must be thought of in terms of a whole profile, and as varying somewhat with the season since the value is higher in cold weather due to the increase in viscosity of water. It is most accurately determined by flooding a deep soil, then covering it to prevent evapotranspiration, and after several days determining the water content of each horizon. The moisture in the soil at field capacity usually lies between 5% and 40%.

The chief significance of this constant lies in the fact that it indicates that point in the gradual thinning of capillary water at which the forces of capillarity (cohesion) are equal to the forces of adhesion, and water becomes essentially immobile. As long as there is water in excess of the field capacity the surface of the soil remains moist, because capillarity replenishes water lost by evaporation. When the moisture in a soil drops to or below this level, and thereby becomes immobile, roots must branch out and elongate to get more water than is contained in the soil previously adjacent to the root surfaces. It has been calculated that by the constant elongation of its multitude of root tips, a rye plant adds about 2 km of roots to its root system per day, thus making available all the water required by the plant, even though the water films are immobile (425). Because of this relationship roots grown in soil in which the moisture

---

* Moisture percentages are usually calculated on the basis of oven-dry (at 105°C) weight, which is taken as 100%. Thus if a sample weighs 60 g when moist and only 40 g when oven-dried, the moisture content is (20/40) × 100 = 50%. It is important to note that by this method of calculation the 60 g of moist soil is equivalent to 150%, and a statement that the 60-g sample has 50% moisture does not mean that there are 30 g of water and 30 of soil.

Problem: An analysis of a portion of a lot of air-dry soil showed the moisture content to be 4%. How much oven-dry soil is contained in the remainder, 950 g, of the lot? *Solution:* 104:100::950:x. Therefore x = 950 × 100/104 = 913.5 g.

Problem: How much water should be added to the 950-g sample to bring its moisture content up to 30%? For each 100 g dry soil, 30 g water is desired; therefore 9.135 × 30 = 274.1 g water needed. Then 913.5 + 274.1 = the total weight of soil plus water desired = 1187.6 g. Therefore 1187.6 − 950 = 237.6 = g water to be added.

frequently drops below the field capacity tend to be more branched than those grown in wetter soils.

Field capacity is also of great ecologic importance as a measure of the maximum storage capacity of the soil, and its determination can be put to good use in calculating the amount of irrigation water needed to wet a soil to the depth of root penetration. Depending upon texture, from 1–8 cm of water is required to raise the moisture content of 30 cm of dry soil to the field capacity. Once a layer of soils is wetted, the addition of more water only extends the depth of moisture and does not increase the moisture percentage of the first layer moistened.

In planning controlled experiments involving pot cultures it is important to take into consideration the fact that any moisture added in excess of the field capacity will either be lost as drainage or will produce a saturated layer in the bottoms of pots lacking drainage holes. Therefore irrigation water should not be added in excess of this amount. Also it is clear that soil moisture contents lower than the field capacity cannot be maintained. It is impossible to bring the moisture content of a dry soil mass up to a uniform moisture content less than the field capacity. In the absence of definite information as to the optimum soil-moisture requirements of a species, it is safe to assume that the field capacity approximates this value.

The weight method is at present the most satisfactory means of controlling soil moisture in accordance with the basic principles outlined above. Various types of automatic irrigators have been devised, but none maintains uniform, therefore known, moisture conditions.

Soils may contain water in excess of the field capacity for long periods if rainfall is frequent and texture or structure allow but a slow rate of percolation. Proximity to the water table brings about a similar condition.

Moisture equivalent.   The percentage of water which a saturated soil will hold against a centrifugal force equal to 1000 times gravity is called the *moisture equivalent.* A special type of centrifuge is used in making these determinations (786). This apparatus is provided with a set of sixteen 5 cm-square trays with screen bottoms. Air-dry samples after passing through a 2-mm screen are placed in the trays, the bottoms of which have been fitted previously with accurately cut squares of paper toweling. The trays are set in about 5 mm of water for several hours, then left to drain in a humid chamber for 24 hours before centrifuging.

It is important to have the thickness of the centrifuged samples as uniform as possible, for the percentage of water retained varies with the depth of the soil in the trays. A 1-cm-thick layer is accepted as standard, and this is approximated when a 15-mm layer of dry soil is used in each tray.

The weight of the trays and soils should be rather evenly distributed around the periphery of the centrifuge drum which holds the 16 trays. Opposite trays may be used for duplicate determinations, thereby making their weights easily equalized on a balance. If less than 16 trays are to be filled, the empty trays must be even in number and situated opposite each other in the machine.

The samples are centrifuged for 30–40 minutes after the centrifuge attains a speed of 2440 revolutions per minute. Although the machines are provided with a governor, the accuracy of control of speed is increased by constant observation of the tachometer and manipulation of the rheostat. After centrifuging, the soils are transferred to weighing dishes or cans and moisture analyses are made.

It is highly desirable to have made up a large and uniform quantity of loam of carefully checked moisture equivalent, and to include a pair of samples of this soil with each lot of unknowns. Then, when the percentages are calculated, errors due to slight variations in the speed of the centrifuge become apparent at once and adjustments can be made using coefficients based on the value the known sample should have exhibited.

Since the centrifugal force applied to the soil is approximately the same as the capillary tension of moisture at the field capacity, the mois-

Table 7

Relation of Moisture Equivalent to Textural Classes of Soils (543). It Will Be Observed That a Clay and a Fine Sandy Loam May Have the Same Moisture Equivalent. This Shows Why the Moisture Equivalent Is More Significant Than the Mechanical Analysis as a Criterion of the Ecologic Character of a Soil.

Textural class (U.S.D.A. system)	Moisture equivalent range
Coarse sand	2.2– 7.8%
Sand	3.0– 6.6
Fine sand	6.7–11.4
Very fine sand	9.7–12.9
Sandy loam	9.0–14.8
Fine sandy loam	9.8–25.3
Loam	15.4–29.2
Sandy clay	15.6–34.3
Clay loam	16.4–32.9
Silt loam	18.3–41.3
Silty clay loam	19.8–29.9
Clay	20.9–40.2

ture equivalent of a soil is very close to its field capacity, except for sands, in which the moisture equivalent is regularly a little lower.

The moisture equivalent varies regularly with texture, ranging from about 2% for sands to about 40% for heavy clays (Table 7), with considerably higher values for soils high in organic matter or volcanic ash. Since this constant is easily ascertained, the moisture equivalent has been widely used as a criterion of soil texture and such dependent variables as water storage capacity, fertility and aeration. It provides a single-figure expression of the net effectivity of all the solids, which eliminates the necessity of trying to integrate a series of values that are obtained through a mechanical analysis.

Permanent wilting percentage. If the roots of a plant are well established in a moist soil and no more water is added, the plant progressively reduces the amount of capillary water in that soil by absorption and transpiration. As capillary moisture approaches exhaustion the water-supplying power of the soil becomes insufficient to maintain the plant cells in a turgid condition, with the result that the plant wilts. The force with which soils hold water increases slowly until the tension reaches about –15–20 bars, but here there is an abrupt increase which brings about a correspondingly sharp reduction in water uptake by the plant and wilting takes place (214). Although transpiration declines and the water potential within the plant drops as this crisis is approached, the total soil-moisture stress (physical tension * + osmotic potential of soil solution) declines so abruptly within a narrow range of moisture content that the roots' increasing capacity to absorb water is rapidly outstripped.

A loss of turgidity in tender plants is first manifested by the wilting of the mature leaves, for old leaves appear to be less able to reduce their stomatal apertures and they have higher water potentials than young leaves. The earliest wilting may be evident only during the hottest part of the day, but finally it becomes so complete that the wilted plants will not revive even though the air about their foliage is brought to 100% relative humidity and transpiration is stopped. This condition of the plant is designated as *permanent wilting*. It differs from *temporary wilting* in

---

* The degree to which the free energy of water is reduced by its attraction to surfaces can be expressed as negative bars of potential. This is *matric potential*. Thermocouple psychrometer determinations of vapor pressure as applied to either plant tissue or soil measures the sum of osmotic and matric potentials. In plant tissue the osmotic potential is by far the major component; in nonsaline soils matric potential predominates. Since the thermocouple psychrometer can evaluate the water potential of soil, plant and atmosphere on the same basis, it provides a unique method of evaluating the water economy of plants. However, other methods of measuring water stress are also available.

that it is a result of soil drouth, whereas temporary wilting is a result of atmospheric drouth and is not so severe but that the plant will regain turgidity if relieved from high transpiration stress.

The percentage of water that remains in a soil when permanent wilting is attained is called the *permanent wilting percentage* or *wilting coefficient.* It varies between 1 and 15%, depending upon the texture of the soil, and includes all the hygroscopic water plus a portion of the inner capillary water. In routine determinations seedlings are usually grown in tumblers which must be buffered against wide temperature fluctuations by immersion in a water bath, since roots tend to be concentrated against the sides of the tumbler. When wilting is observed a bell jar is set over the tumbler without removing it from the water bath. The lower half of

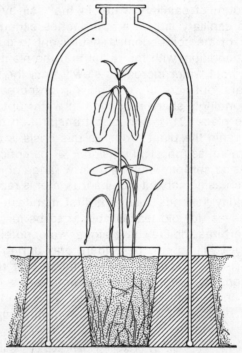

FIGURE 28. Method of determining the permanent wilting percentage of soils. Wheat and sunflower have been grown in glass tumblers of soil immersed in a water bath. A bell jar has been set over the plants to test for the permanence of wilting. Note the pendant tips of the grass leaves and the appearance of the lower pair of epicotyledonary leaves of the sunflower which were essentially horizontal and flat before wilting.

the tumbler of soil (roots removed) is used for making a moisture-content determination, after it is proved that the plants are permanently wilted (Fig. 28).

At an extremely slow rate plants can continue to reduce the water content of the soil below the permanent wilting percentage, for dormant and even dead roots can absorb water, the dead roots apparently transmitting water upward by wicklike action (92). Therefore the permanent wilting percentage is not a measure of absolutely unavailable water, although this distinction is more technical than practical. Since growth usually does not stop until the moisture content has been reduced to the permanent wilting percentage, water in excess of this value may be called *growth water* or *readily available water.* From an ecologic standpoint a soil lacking growth water is a "dry" soil regardless of its moisture percentage. When the water content drops below the permanent wilting percentage, growth water is negative in amount and normal functions cannot be resumed until water is added in excess of the deficit below the permanent wilting percentage. As long as the water content remains below this point all plants desiccate progressively until they die even though they continue to extract small amounts of water from the soil.

Because plant roots are most abundant and therefore absorb most actively near the soil surface, and water loss by direct evaporation is most rapid here, the desiccation of a soil begins at the surface and extends to deeper levels as a rainless period advances (Fig. 29). This is clearly shown by the appearance of plants in arid and semiarid regions. Mosses, with their rhizoids confined to the upper few millimeters of the soil, become brown and dormant first. Next the shallow-rooted plants ripen their seeds and become dormant if perennials * or die if annuals, and seedlings the roots of which have not been able to penetrate far also succumb. Indeed, where growth water is not present throughout the summer, the relative rapidity of root penetration is generally the most important factor governing the success of seedlings (300, 624). The plants latest to turn brown or shed their leaves are those most deeply rooted. Thus the sequence of cessation of activity is indicative of the progress of soil desiccation as well as the relative depth of root systems, except, of course, in those species which normally terminate their seasonal activity before the soil dries out. Even in arid regions, however, there are many plants that remain active during the dry seasons because their roots penetrate so deeply that they are in contact with permanently moist

* Summer dormancy due to drouth is called *aestivation.*

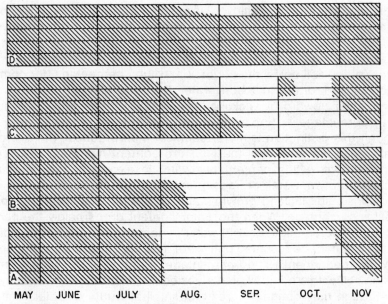

FIGURE 29. Seasonal distribution of growth water (shaded areas) in the upper 5 dm of the soil profile, at different altitudes in northern Idaho. Station A, steppe in the warm, dry climate of the basal plain at the foot of a mountain; B, *Pinus ponderosa* forest in the lower foothills; C, *Pseudotsuga menziesii-Abies grandis* forest on a mid-slope; D, *Thuja plicata* forest in a cool moist climate at high altitude. Desiccation is progressively later the higher the altitude, especially at depths below about 2.5 dm.

subsoil (Fig. 27). Plants can absorb enough water to replace transpiration losses if as little as one-quarter of the root system is in moist soil.

Fine-textured soils have high permanent wilting percentages; therefore after they dry out they require relatively large amounts of water if they are to be moistened to depths where roots are located. However, once they are moistened thoroughly the net growth-water storage capacity is so high (Table 8) that plants rooted in such soil continue active growth much later in the dry season than others situated in coarse soils. The more rocks and gravel contained in a volume of soil the lower its total water-holding capacity, for these large, nonabsorptive particles occupy space that would otherwise be filled with fine materials. In a region of low rainfall this condition is of considerable advantage to plants, for the soil between the rocks can be moistened deeply with relatively little rain.

With annuals and other tender plants the amount of water they can

Table 8
Interrelationships Among Certain Soil Moisture Constants (Percentages) of a Group of Soils (781).

Moisture constant	Clay	Loam	Silt loam	Sandy loam	Fine sand
Moisture equivalent	28.4	21.7	16.1	9.5	3.2
Permanent wilting percentage	13.4	10.3	7.5	2.9	1.0
Growth-water storage capacity	15.0	11.4	8.6	6.6	2.2

absorb beyond the permanent wilting percentage is almost insignificant in staving off death, which ensues within a few hours or days after this point is reached. Plants possessing succulent roots, stems, leaves, or fruits usually do not show evidence of wilting until long after soil moisture has dropped below the permanent wilting percentage, since a transfer of their water reserves from succulent to nonsucculent tissues will suffice to maintain turgidity for some time after growth water is exhausted. An extreme example of such moisture redistribution was provided by an experiment in which a branch of *Tradescantia* was severed and placed on a laboratory table, where it remained without water for two years. Although the branch transpired slowly all the time, it remained alive, and, by a transfer of water from old leaves to the tip of the branch, the length increased 150% and many new leaves were formed (584).

Many perennials are able to reduce their transpiration so strongly at the permanent wilting percentage that transpiration is scarcely greater than absorption, and thus death by desiccation is postponed for many days. Grasses such as *Bouteloua gracilis* and *Oryzopsis hymenoides* can endure months of soil drouth, and soil algae have been found living after half a century of storage under dry conditions. The duration of soil drouth is fully as important an ecologic character of a habitat as its attainment. Under experimental conditions no difficulty is involved in testing the ability of a plant to endure differing periods of soil drouth (176), although as explained earlier it is not possible to study the effect of different growth-water contents of the soil on plant growth. It is possible only to allow growth water to be exhausted to different levels before the field capacity is restored.

Certain species of plants have been shown capable of resisting permanent wilting until the moisture content of the soil is reduced to a point slightly lower than with other species. It might be expected that desert plants as a group would have developed an ability to absorb water so efficiently that in comparison with other ecologic types they would leave less water in the soil at permanent wilting. However, this is definitely not true. After making 1318 determinations of wilting percentages using a wide variety of ecologic types of plants, two investigators concluded that the variations in wilting percentages as determined by the use of different species and ecologic types of plants with the same soil are of such small magnitude that the wilting percentage is a characteristic, or constant, of the soil rather than of the plant (92). Cacti and nonsucculent desert shrubs (182) face a physiologic crisis at the same soil-moisture content as wheat and sunflower. The amount of residual moisture left in the soil is determined by an abrupt change in the physical forces controlling the availability of water in thin films rather than by insufficient suction force on the part of the plant.

From the ecologic standpoint the permanent wilting percentage is by far the most important of the soil-moisture constants. Its significance lies in the fact that when this percentage is subtracted from the actual water content of the soil, * both expressed as a percentage of oven-dry weight, the availability of water to all the plants rooted in that soil is known. Obviously a knowledge of the actual water content of the soil is almost wholly without biologic significance if the permanent wilting percentage is not known. This concept provides the common denominator by which the moisture relations of very diverse soils can be compared.

The concept of permanent wilting percentage is complicated by several factors, however. First, it is obvious that the more finely divided the

---

* The method of sampling a soil is equally as important as the analytic methods to be employed, for much care need be taken that the samples accurately represent the area and depth critical for the problem at hand. Soil samples for moisture or other types of analyses can be obtained with a trowel from the sides of a freshly dug pit, or by various types of augers and sampling tubes (446). With any of these methods samples of uniform cross section should be taken between the limits of depth chosen. Where the soil is stony a pit must be dug, but where free of stones samples on successive dates may be closely spaced if a tubular sampler is used and the holes are immediately refilled with soil at similar moisture content.

Where soil horizons are distinct, the vertical limits of the samples should coincide with the boundaries of the horizons; otherwise arbitrary depth limits are used, e.g., 0–10 cm, 10–20, etc.

It is very desirable that the samples be passed through a 2-mm screen at once to remove pebbles and roots, then placed in friction-top, rustproof cans. If these cans and their lids are stamped with serial numbers, and records of can weights are preserved, much time is saved in routine work. (Soil samples to be used for chemical analyses are usually placed in paper bags and allowed to become air-dry as soon as possible.)

Although the standard and most accurate method of determining soil moisture is by desiccation at 105°C (157, 490), it can be estimated by use of tensiometers (650), electrical conductance of fiberglass-gypsum blocks (78, 816), neutron scattering (778), or calculations based on climatic data (304, 534). None of the above methods integrates the combined effects of matric and osmotic potential.

root system the more uniformly the soil will be exhausted of its moisture, so that, in those plants the root systems of which show relatively little tendency to branch, the apparent permanent wilting percentage as determined by taking a large sample of soil is relatively high because of the small pockets of moist soil that remain scattered through the profile after the plant has wilted permanently (Fig. 27 F) (860). Small differences in the osmotic potential among species may also have some effect. However, the fact is highly significant that under experimental conditions species which give different wilting coefficient values when grown singly, wilt at the same time when grown together in the same pot! (123, 182). Grasses with their numerous, finely divided roots, are very desirable as materials for making wilting coefficient determinations.

Second, many plants do not show by their outward appearance when soil moisture is depleted to the critical point. Plants with succulent or firm-textured leaves may never wilt, although a physiologic crisis develops in these plants just as much as in those that show wilting plainly. Evidence of the attainment of the permanent wilting point in these plants may be obtained by observing the cessation of growth, abscission of leaves, reduction in transpiration, folding or rolling of leaves, etc. Because sunflower leaves or cotton cotyledons show wilting very plainly, an excellent method of making routine determinations of wilting percentages is to plant one broad-leaved and several grass seedlings in the same tumbler, the latter giving ideal root distribution and the former indicating very clearly the time of exhaustion of growth water (Fig. 28). Also, if a sunflower is planted in the same pot with a plant like pine which does not show wilting, it serves as an accurate indicator of the time when soil drouth becomes critical for both plants (176).

A third complicating factor is that, if the evaporative power of the air is suddenly increased during the period when soil moisture is being depleted to the permanent wilting percentage, plants will wilt before this point is reached. Thus drouth effects may be produced just as readily by increasing the transpiring power of the air as by a sudden decrease in moisture availability. However, if plants are not subjected to violent and unnatural conditions, the permanent wilting percentage is not affected by wide differences in the aerial environment (105).

The osmotic effects of salts in saline soils is additive to the increasing immobility of water films in drying soils, hence the wilting coefficient is. higher than texture alone would warrant. Halophytes are the only appropriate plant materials for determining the wilting point of saline soils (803).

Finally, owing to the increased viscosity of water the permanent wilting percentage is raised somewhat by a drop in temperature. Under natural conditions this increase is compensated by an increase in field capacity, but, when field moisture contents are evaluated in terms of laboratory determinations of the permanent wilting percentage, a certain amount of error is involved where the temperature conditions differ. Precise technique demands that the permanent wilting percentage be determined at the temperature of the soil at the time when the sample was taken for growth-water analysis. Chilling the soil increases water viscosity and reduces the absorptive capacity of roots to the extent of causing wilting. Therefore, as the soil approaches the permanent wilting percentage, tumblers of test plants should not be removed from their water bath.

The five points mentioned above show that the permanent wilting percentage is not as simple a constant as once thought, yet this fact is more than compensated for by the proved importance of this percentage in evaluating soil-moisture conditions for plant growth.

There is much disagreement in the literature on the extent to which all the water between field capacity and wilting coefficient is equally available. Much of the argument rests on the plant function measured (transpiration, fruit production, exudation from cut stumps, etc.) (299) and much rests on the conditions of the test (whether the roots are cramped in small containers or normally distributed, whether the soil mass was fully occupied by roots or not, etc.). Also the kind of plant is important. Potatoes require soil moisture near field capacity for maximum production whereas prune crops are not significantly reduced unless water drops to the wilting coefficient. Plant functions which govern establishment and successful development in nature may not be as critically affected by an approach to the wilting coefficient as are functions related to productivity, which becomes vital for the growth of sensitive crops such as potatoes (279, 727, 785, 797).

The literature of plant science includes numerous experiments concerning the effects of water content of the soil on plant growth, many of which were performed under "controlled" conditions using potted plants. A great number of these studies have undoubtedly yielded erroneous conclusions owing to a lack of understanding of the relations among soil, water, and plant, and they stand in need of validation by more critical techniques. Experiments designed to study other phenomena can easily give results which are strongly influenced by the technique of soil-moisture control and therefore do not provide a reliable answer to the questions prompting the studies. An experiment is performed under known and repeatable conditions only when moisture levels are adjusted in relation to the moisture equivalent (or field capacity) and the permanent wilting percentage as discussed above.

Interrelationships among soil-moisture constants. As the concept of soil-moisture constants emerged, early workers reasoned that it should be feasible to establish mathematical relationships among the constants so that if one were determined accurately others could be derived by formulae. They proposed a series of such formulae based on averages of numerous tests, apparently assuming that the variations that always had to be submerged by averaging were inconsequential or represented experimental error. This has proved wrong—the mathematical relation between two soil-moisture constants for one soil does not necessarily hold for another soil. For example, the moisture equivalent is roughly twice the wilting coefficient (Table 8), but the actual ratios in different soils have been found to range from 1.1 to 8.0! (784). Also, the water retained by a wet soil after it is subjected to −15 pressure is closely related to the wilting coefficient in many soils, but there are exceptions which render the approximation unsafe for critical ecologic work (489, 652).

One conversion factor safe to depend on in ecologic research makes minor adjustments of the moisture equivalent to obtain values that more closely approximate field capacity (606) and this manipulation is of use only in establishing the maximum weight of a unit of soil that is to be brought up to the field capacity periodically in an experiment in which soil moisture is to be controlled:

$$FC = 2.62 + (ME \times 0.865)$$

Field capacity may also be approximated by residual soil moisture after subjecting a wet soil to 0.3 bars pressure (151, 669).

Importance of Soil Moisture to Plants

Many studies have shown that soil moisture is an extremely important aspect of plant environment and that responses to variations in it are quite diverse. The following brief references by no means constitute a complete catalog of the numerous ways in which this factor has been found to influence plant growth.

Soil moisture begins to have an effect upon plants even before germination, for the seeds of many plants (e.g., *Salix* and *Populus*) must make contact with moist soil within a few days after maturation or perish. Such plants are consequently abundant only where soil drouth is slight at most. The seeds of other species which become dehydrated while dormant may not germinate readily unless soil moisture exceeds the field capacity for a time (201). This is due to the fact that the seeds can exhaust growth water only from the points where they make contact with soil particles, and, since capillary water is immobile at and below the field capacity, they may not obtain sufficient moisture from these points to extrude their radicles. Furthermore, certain seeds of desert plants seem to contain water-soluble compounds that inhibit germination until excess moisture has leached these materials away, and this mechanism insures dormancy until drouth has been broken (830).

As a soil dries the movement of nutrient ions toward the root in the mass flow of water decreases, and even diffusion is hampered by the increasingly more tortuous pathways around gas-filled pores. Consequently nutrient deficiency may become more important than the reduced water availability for rapidly growing plants in relatively infertile soils.

Roots elongate slowly through wet soil, but as the water content drops below the field capacity elongation speeds up considerably. Since the rate of shoot growth is not correspondingly accelerated, the root/shoot ratio increases as soil moisture decreases. Roots can extend into dry soil if another part of the root system has access to growth water (376).

Because most land plants cannot send their roots down into saturated soil, a high water table enforces shallow rooting (225). If a water table during one season is so high that it enforces shallow rooting but at another season the soil dries out below this level, only plants capable of aestivation can survive on the habitat even though the soil dries but little below the surface (Fig. 30) (99).

Successful irrigation demands the application of no more water than is needed to moisten the soil to the depth of the root systems and repeating this operation shortly before the growth-water supply becomes limiting. Insufficient watering moistens only the upper layers and causes roots to become concentrated there; excess watering produces a similar result through oxygen deficiency in the lower horizons.

Shallow-rooted plants are subject to wind throw, are easily injured by cultivation, are sensitive to soil drouth and frost, and cannot benefit from the nutrients contained in deeper soil layers. Because the root system is crowded and dwarfed, the shoot system is likewise small (225).

The vigor of shoots is also affected by the relative amount of water in unsaturated soils. Certain species of grass, for example, adopt the bunch habit and reproduce more vigorously by seeds when grown under a low moisture supply but become strictly rhizomatous and reproduce mainly vegetatively when the soil is quite moist (258). In nature, the drier the soil the earlier a plant matures. Moisture stress induces flowering in certain plants (14, 212), in others it is detrimental to the reproductive process.

In discussing the permanent wilting percentage it was emphasized that this value represents rather uniformly the minimal soil moisture content for maintaining nonsucculent plants in an active condition. Subjecting plants to permanent wilting increases the intensity of transpiration, and reduces the sizes of cells and organs, the first exposure to such a condition having the most pronounced effect. Slight drouth during the growing season also increases the plant's resistance to subsequent severe drouth and to frost (697), and in fruit trees it tends to increase the number of flower buds produced.

In habitats where the upper soil horizons regularly desiccate, the growing seasons of those plants the roots of which do not extend to the permanently moist subsoil are confined to the rainy season or a brief period following this season. Activity on the part of plants which have established contact with deep-lying moisture is not necessarily correlated with the rainy season (Fig. 27).

In comparison with minimum requirements, the optimal soil moisture point is less precisely fixed among plants. For many species essentially optimum growth is obtained at almost any moisture level between the permanent wilting percentage and the field capacity. However, in irrigation practice it is hazardous to approach the permanent wilting percent-

FIGURE 30. Foreground is the flat bottom of a playa which is too severe a habitat for plants because it is alternately flooded and desiccated during the wet and dry seasons, respectively. Muroc Lake, California.

age too closely, for a sudden change to drier weather may reduce soil moisture to the danger point before water supplies can be applied and become effective. In humid climates raising water content to the field capacity involves some hazard to crop production in that unforeseen rainstorms may cause temporary waterlogging.

The maximum moisture content tolerated differs widely among species. Here, however, insufficient aeration becomes the true limiting factor, for plants can hardly be injured by too much water. Exceptions to this statement are provided chiefly by the rupture of fleshy roots or stems in carrots, beets, potatoes, etc., when an unusual supply of water is suddenly made available after a moderately dry period.

Soil moisture has a marked influence upon the pathogenicity of certain soil-borne disease organisms. If such organisms become important parasites only within certain ranges of soil moisture, the easiest control measure may be the careful regulation of the water supply (249).

Several methods are available for the experimental study of the importance of soil moisture in the field. Irrigation is the obvious means of testing the effect of increase. Reduction of soil moisture in lowlands can often be accomplished by artificial drainage. On uplands the soil surface can be covered during some or all rainstorms with plastic sheeting or roofing tilted so as to drain away some or all the water falling under the canopy (496). Collars can be fitted around solid-stemmed plants to intercept and remove stemflow as well.

The Water-Balance Problem of Land Plants

Primitive plants are believed to have originated in the seas where such phenomena as transpiration, wilting, and drouth are inherently absent. In this aqueous medium protoplasm evolved to a high state of complexity while attuned to a water-saturated condition. Subsequently, when plant life began to encroach upon the land, an entirely new hazard to its existence was encountered, namely, the problem of keeping the water content of the protoplasm above a certain high minimum necessary for the maintenance of its vital functions. In the terrestrial environment plants must endure an almost continual loss of water to the atmosphere; therefore landward migration became possible only as efficient adaptations to the new conditions were developed.

Adaptations permitting the maintenance of a satisfactory balance between the loss and absorption of water evolved in two directions. On the one hand there were developed nearly impervious coverings of cutinized or suberized tissues which greatly mitigated the transpiration hazard. On the other hand roots and rhizoids, structures highly efficient in extracting moisture from the soil, were evolved. Water flows freely to aquatic plants as they need it, but plants rooted in drained soil must have a well-branched, fast-growing root system to obtain the immobile growth water.

Cutin and suberin are very effective in retarding the loss of water from a plant surface; however, they have the serious disadvantage of precluding the exchange of gases between living protoplasm and the atmosphere—a function which is absolutely necessary for both photosynthesis and normal respiration. This difficulty was largely offset by the pulling apart of certain cells, so as to leave intercellular gaps (the stomata and lenticels) in the superficial tissues which are provided with the waxy coverings named above. Such a solution to the problem is a distinct compromise, however, for, even though all but a very small fraction of the exposed surfaces is protected by essentially waterproof coverings, the stomata and lenticels which are distributed over these surfaces permit the loss of great quantities of water vapor. Furthermore, since the air outside the leaf contains only 0.03% CO_2 and 21% O_2 while the air in the mesophyll is nearly saturated with water vapor, transpiration is much more rapid than CO_2 or O_2 intake. Approximately 90–95% of the water transpired by leaves is lost through the stomata, showing that the cuticle is relatively efficient. Only in rare instances does cuticular transpiration equal or exceed stomatal.

Cuticular transpiration is not subject to physiologic regulation. On the other hand stomatal transpiration is physiologically regulated by the action of the guard cells in controlling the size of the stomata. Even though most of the water transpired by leaves is lost through the stomata, which have an ability to regulate this diffusion, their primary function is not to be thought of as one permitting water loss, any more than the function of a door in an office building is to let heat escape from the building in winter. Just as a loss of heat cannot be avoided if the door must be kept open to permit continuous streams of persons to enter and leave the building during business hours, similarly a loss of water vapor from the leaf is inevitable when the stomata are open to permit the exchange of CO_2 and O_2.

Up to a certain point transpiration may be beneficial. It increases the rate of rise of nutrients to the upper part of the plant, and makes possible the transpirational cooling of leaves. Possibly of more importance is the fact that it cools leaves at times when a matter of a very few degrees may mean the difference between efficient or inefficient functioning or even between life and death, for leaves exposed to the sun become much hotter than air when transpiration is stopped artificially or slowed down by wilting. The temperature of a leaf undergoing rapid transpiration may remain as much as 15°C below the temperature of the surrounding air.

Transpiration beyond the amount useful in cooling leaves is potentially detrimental to the plant. When vapor loss is so vigorous that leaves lose their high turgor pressure, normal functions of the protoplasm are impeded. Growth occurs only when absorption exceeds transpiration, for an excess of water is required to swell the vacuoles of newly divided protoplasts. Finally, excessive transpiration may desiccate protoplasm below the minimum water content which permits it to remain alive. This minimum is rather high in active tissues. The protoplasm of most leaves, for example, is killed when the water content is reduced below 30–50%.

The rate of transpiration from a leaf changes with variations in: (a) the evaporative power of the air as determined by saturation deficit and wind; (b) the direction and magnitude of the difference between the temperatures of the leaf surface and the air (a net result of respiration, transpiration, insolation, etc.), as this difference affects vapor-pressure gradients; (c) the degree of saturation of the tissues with water, as this affects the degree of stomatal opening and the ability of the protoplasmic colloids to give up water; (d) the response of the guard cells to light, which opens stomata; and (e) the action of light in increasing the permeability of protoplasm. So important are factors b–e that in actuality there is seldom much correlation between the daily marches of transpiration and evaporation, as noted earlier (Fig. 18). However, it will be observed that nearly all these influences are directly or indirectly controlled by solar radiation. This explains the striking differences between nocturnal and diurnal transpiration in most plants, and the close relationship between transpiration and determinations of total radiant energy (304).

Owing to the fact that the average vascular plant absorbs water through its roots at the same time that it loses water to the air through its shoots, the ratio between these two processes determines the state of hydration of the tissues. This ratio between water income and water loss is called the *water balance of the plant*. The significance of water to plants cannot safely be reckoned in terms of either water intake or transpiration, but only in terms of water balance, and this status is best quantified as water potential.

Inasmuch as absorption and transpiration are partly controlled by the environment and partly by the plant, there are both internal and external aspects of the water balance. The external aspects consist of (a) the amount of water available to the absorbing organs, and (b) the intensity of transpiration-promoting factors. As means of expressing this water balance of the environment nothing better has been devised than the methods of evaluating precipitation effectivity which have been discussed.

The internal aspects of the water balance depend largely upon those structural and functional characteristics of the plant bodies which tend to offset, or to aggravate, the natural limitations of environment.

Many studies have shown that the absorbing and conducting systems of terrestrial plants are relatively inefficient in supplying water to meet the demands of transpiration, even at times when growth water is plentiful. Between the epidermis and the xylem of an absorbing root lie several tissues composed of living cells which seem capable of transmitting water at only a very slow rate. Furthermore, the xylem of many plants is relatively inefficient in conducting water, once it penetrates to this tissue. In conifers, especially, the conductive elements consist entirely of small tracheids which would be quite incapable of conducting water rapidly enough to meet the demands of leaves with a high transpiration rate. The heavy cutinization, sunken stomata, etc., in this group, many members of which grow in moist or wet habitats, has been pointed out as being advantageous because of their relative inefficiency of conduction (810), and the relatively low transpiration rates of these plants in summer bear out this viewpoint (820).

As a result of the inefficient conduction of water from the absorbing to the transpiring surfaces, most plants, even if rooted in moist soil, lose more water during the daylight hours than they can absorb and conduct to the leaves during this period (422). * In the average plant this results in a tension in the xylem, a slackening of elongation, a decrease in the area and thickness of leaf blades, a shrinkage of the stem, a decrease in water content which may amount to 40% of the fresh weight, and in certain plants a closing of the flowers (303). Should the water deficit attain considerable magnitude, turgor may decrease until delicate leaves and stems wilt. These diurnal phenomena are most pronounced in very dry habitats; they are not apparent in submersed aquatics.

At night the transpiration rate declines sometimes to no more than 3–5% of the diurnal rate, and, since water uptake remains constant, this permits a reverse in the direction of the water-balance trend so that the deficit due to transpiration is gradually changed to a surplus which may even result in guttation.

As a protracted drouth develops the period of temporary daytime

---

* When the weight method is used in studying transpiration it must be kept in mind that a decrease in the weight of a potted plant may be much greater than the amount of moisture that it removes from the soil.

wilting lengthens and recovery is progressively later in the night, until a point is reached when complete recovery is not achieved even by sunrise and the plant remains permanently wilted. In annuals and woody plants this commonly leads to death in usually less than two weeks, but many perennial herbs can aestivate and survive for a long time in desiccated condition (182, 802).

The structure of plants appears to be more strongly influenced by water-balance conditions under which they are grown than by any other single factor of environment. Physiologic processes are also affected by moisture, but in comparison with other factors the influence of moisture is not especially pronounced in this respect. In comparison with plants grown under optimum moisture conditions, those grown under an unfavorable water balance have the following characteristics:

Morphologic Features
1. Reduced size of shoot (i.e., *nanism*).
2. Increased size of root system.
3. Smaller cells in the leaves, which in turn results in:
   a. Smaller and thicker blades or blade segments.
   b. Stomata smaller and closer together.
   c. Smaller vein islets.
   d. More hairs per unit area, if the surfaces are pubescent.
4. Thicker cuticle and cell walls, with more lipids on the transpiring surfaces (134).
5. Better-developed palisade mesophyll.
6. Weaker development of sponge mesophyll.
7. Less sinuous epidermal walls.
8. Smaller intercellular spaces.
9. Smaller xylem cells, but greater proportion of heavily lignified tissues.
   (Organs with these structural characteristics are said to be *xeromorphic,* although nutritional stress has the same effect.)

Physiologic Features
10. More rapid rate of transpiration per unit area (related to items 1 and 3b above), although the net transpiration per plant may be reduced.
11. More rapid rate of photosynthesis per unit area (related to 3b and 5 above).
12. Lower starch to sugar ratio (382, 459).
13. Lower osmotic potential (related to 10, 11, and 12 above).
14. Lowered protoplasmic viscosity (382).
15. Increased protoplasmic permeability (382).

16. Greater resistance to wilting (related to 4 and 13 above).
17. Earlier flowering and fruiting (related to 11 above).
18. Increase the percentage of bound water * per unit dry weight of tissue (839).
19. Greater longevity.

Two plants grown under distinctly different conditions of external water balance normally differ in all these respects. In addition, the leaves closest to the base of the stem have first opportunity to benefit from water rising from the roots, so that, with a plant growing in an open sunny habitat, the farther from the stem base a leaf diverges, the lower its water supply and the more xeroplastic its structure. Morphologic and physiologic modifications associated with height of divergence are likewise influenced to some extent by the fact that the upper parts of the plant develop in a stratum of atmosphere where they are affected more strongly by wind and insolation.

Xeromorphy has been attributed largely to low turgor pressure in developing cells, allowing the walls to mature before the protoplasts are fully expanded. In woody plants, leaves developing from buds on wood formed under dry conditions during the previous season are xeroplastic even though moisture is abundant during their expansion. This is due to the fact that the developing leaves obtain their water principally through the relatively inefficient xylem of the preceding season's growth. On the other hand, since "xeromorphy" can be induced by growing plants in an infertile medium, possibly the reduced ability of a drying soil to yield nutrients may be of major importance (483).

Susceptibility to an unfavorable water balance varies continuously throughout the life cycle of the average plant. Seeds are often capable of enduring extreme drouth and, indeed, may require this condition to maintain their viability. The seedlings of cereals, at least, retain a high degree of drouth resistance until they are several days old, but after their first leaves unfold they become very sensitive to desiccation.

When plants approach maturity they lose some of the sensitivity to drouth which characterizes the seedling stages. A rather interesting difference in response to drouth during the main growing season is exhibited by maize and sorghum, which are similar in morphology and closely related taxonomically. Drouth in early summer causes sorghum to suspend development, with no serious subsequent ill effects, until

---

* The water contained in plant tissue may be classified as *free water* and *bound water*. Bound water is that fraction which is absorbed by the colloids so strongly that it does not act as a liquid. It cannot function as a solvent, it cannot be frozen, nor can it be lost by transpiration.

the crisis is past, but in maize, activity is not arrested so that pollination and other functions are seriously affected (521).

Man can do much to improve the water balance of crop plants by:

A. Improving environment by (149):
   1. Increasing the supply of water through:
      a. Irrigation.
      b. Reducing runoff.
   2. Reducing the rate of evapotranspiration by:
      a. Mulching with straw, paper, etc.
      b. Providing windbreaks.
      c. Reduction of leaf areas by moderate winter pruning.
      d. Spraying foliage with emulsions that reduce transpiration (183).
      e. Elimination of weeds.
      f. Thin rates of planting.
B. Increasing drouth resistance of plants by:
   1. Breeding drouth-resistant strains.
   2. Stimulation of drouth-hardiness (693) by:
      a. Keeping the supply of N to a minimum consistent with satis-factory nutrition, since excess N causes tender shoots.
      b. Extending the interval between irrigations to the maximum con-sistent with satisfactory growth, in order to enforce deep rooting and bring about protoplasmic changes which increase drouth resistance.

Quantity of Water Used by Plants

Water losses from the soil into the air are increased greatly by the activity of a plant cover. The aggregate leaf surface may be 11 times greater than the soil surface of an area, and plants can withdraw water from considerable depths, whereas surface evaporation affects only the upper 2–3 dm. Certain plants have been observed to transpire in one hour the equivalent of all the water contained in the tissues. In the course of a growing season even a plant with a very slow transpiration rate absorbs a quantity of water many times its weight, most of which is lost to the atmosphere.

When moisture is determined as a percentage of the soil's dry weight, this is a qualitative expression. To quantify it one must ascertain the bulk density of each horizon to the depth of rooting. The rainfall equivalent of soil moisture, in cm, is:

$$\frac{\% \text{ moisture content} \times \text{bulk density} \times \text{thickness of horizon in cm}}{100}$$

This calculation is hardly useful unless the evapotranspiration rate is known.

The advantage of expressing soil moisture as a percentage of dry weight lies in the ease of relating this value to the field capacity and permanent wilting percentage. This procedure is justified since for eco-logic purposes other than crop production, the difference in availability of water between field capacity and permanent wilting percentage is negligible.

The rate of water loss from soil due to plant activity is controlled by the character of both plants and environment. It is at a maximum under optimum levels of soil moisture and under moderately high transpiration stress. Thus, subirrigated alfalfa in dry climates transpires considerably more than the annual rainfall (836). Different species and types of plants, of course, make widely different demands on soil moisture even under the same environmental conditions. For example, a maize plant may transpire 2 liters of water a day and an oak may transpire 570 liters during the same interval. The annual use of water by forests has been estimated as equivalent to 7 cm of rainfall in boreal and 300 cm in tropical forests. During a hot, dry day in the period of active growth, transpiration intemperate latitudes may rise to about 6 mm a day, providing soil moisture is plentiful. The use of water by streamside phreatophytes is so great that where water conservation is very important it is the best practice to pipe water out of the channels near the source before it can be consumed by the miles of dense streamside vegetation by which it would otherwise flow. Both stream levels and water tables that are near the surface undergo a depression during the day as a result of rapid transpiration; at night a reverse trend sets in. The fact that this influence is truly the result of transpiration loss is proved by the observa-tions that (a) the fluctuations begin with the appearance of new foliage on deciduous vegetation and cease with the arrival of cold weather in autumn, (b) diurnal fluctuations in water tables do not occur under fallow fields, and (c) a shower or drop in temperature sufficient only to raise the relative humidity is followed immediately by a rise in the water table (836).

A knowledge of the water needs of different species is an invaluable aid in the selection of suitable forest and field crops for regions of limited rainfall (633). Thus, in semiarid regions, observations at seeding time as to the depth to which the soil has been wetted provide a basis for judging what kind of crop and rate of planting is likely to be most successful (241).

Under limited rainfall the continued growth of alfalfa on an area results in increasingly deeper root penetration and a depletion of moisture reserves to such a depth that they can be restored only by fallowing for a series of consecutive years. Therefore, the selection of crops to include in a rotation plan must take into account differences in the amounts of water extracted by different crops (32), possibly allowing the land to lie fallow the season following planting with a heavy user of water.

The liters of water used during the growing season, as related to the grams of dry matter produced, can be used as a criterion of the degree of efficiency in use of soil moisture. This value varies with species and with the environmental conditions. Some plants can produce 1 g of dry material with as little as 0.2 l water, whereas others would use ten times as much water.

For a given species the transpiration efficiency is decreased by any adverse environmental condition, especially by: (a) relative infertility of the soil, (b) extremely dry or wet soil, (c) high evaporative power of the air, or (d) the invasion of a vascular plant by parasitic fungi. In wet climates the efficiency may be several times that obtained by the same species under dry climates. Therefore, since transpiration efficiency is so strongly influenced by environmental factors, an absolute value for a species or variety can be expressed only as an average of numerous trials under widely different conditions. The results of one season's observations, however, are significant when different species grown under the same environmental conditions are compared.

All other factors being equal, those species of forest, forage, and field crops that make the best growth with the least amount of water are our most valuable plants in regions where the water supply is limited. However, it is important to bear in mind that the transpiration efficiency is not related to drouth resistance or to the relative moistness of the habitats to which the plant is native. Species that are closely related ecologically may have widely different transpiration efficiencies. Thus *Agropyron smithii* has an efficiency of only 0.96 while the associated *Bouteloua gracilis* rates 2.96 (93). Depth of rooting and season of activity in relation to the rainfall pattern account for many of the differences in efficiency.

Classification of Plants
Based on Water Relationships

Ecologists have long been interested in the classification of plants according to their water relationships. For a time much attention was directed toward finding a common physiologic or morphologic basis for defining groups, but it gradually became apparent that this objective is not realistic, that alternative adaptations of structural and functional nature may have equivalent value. Hence emphasis has shifted to a primarily environmental basis of definition.

In 1895 E. Warming proposed three classes, *hydrophytes, xerophytes,* and *mesophytes,* referring to plants of wet, dry, and moist habitats, respectively. * In common with all other ecologic classifications, the groups delimited on this basis include species of very diverse taxonomic affinities. In common with most biologic classifications, the limits between the groups are ill defined.

Hydrophytes

Hydrophytes include aquatics which normally grow in water, and swamp, marsh and bog plants which inhabit soils containing a quantity of water that would prove supraoptimal for the average plant. In all these habitats water per se is not detrimental, but the extreme slowness with which oxygen dissolves and diffuses into water or saturated soil produces a complex of critical conditions with which plants can cope only if specialized. Unlike other ecologic groups, the higher plants that are hydrophytes regularly extend their roots into saturated soil or water. Although the physiology of hydrophytes with respect to aeration will be discussed in a later chapter, their principal morphologic characteristics may well be taken up at this point.

One of the most outstanding structural peculiarities shared by most hydrophytes is the sponginess of their tissues (159). Owing to the disintegration of groups of cells, or the separation of cells so as to create enlarged intercellular spaces, cavities (*lacunae*) are developed which become or remain filled with gases. Most mesophytes have a continuous system of intercellular air spaces, but hydrophytes have especially large cavities. Parenchyma containing cavities is called *lacunar tissue;* periderm containing intercellular spaces is called *aerenchyma* (Fig. 31).

In the fruits of *Nuphar* and lotus large air cavities render the structures buoyant and aid in dissemination. Special bladdery swellings in the marine brown algae *Fucus, Nereocystis,* etc., serve to keep the shoots erect in the water so that the chlorenchyma is advantageously exposed to light. Buoyancy in the water hyacinth (*Eichhornia*) results from spongy

---

* The Greek roots *hydro, xero,* and *meso,* meaning wet, dry, and intermediate, respectively, are used in several series of ecologic terms, each of which has definitely limited application. *Hydric, xeric,* and *mesic* are commonly encountered in ecologic literature, but are really superfluous in the English language since they supplant ordinary words of exactly the same meaning: wet, dry, and moist. Clearly, these adjectives, if used, should be applied only to habitats. *Hydrophytic, xerophytic,* and *mesophytic* are properly used only in connection with plants but obviously should not be used in such a redundant combination as "xerophytic plant"; xerophyte is sufficient.

enlargements of the petioles, and in other floating aquatics aerenchyma in stems or roots perform the same function (686).

In completely submerged plants (e.g., *Elodea*) the air cavities can only accumulate the by-products of photosynthesis (O_2) during the day, and this is used up in respiration at night during which time CO_2 accumulates. Thus the lacunae serve as gas-storage compartments which are alternately filled with the by-products of complementary metabolic processes. The cavities in emergent or floating-leaved aquatics generally form a continuous system of air passageways by means of which submerged organs can exchange gases with the air via stomata in the emersed organs (155). These passageways are kept open by perforated diaphragms that cross at intervals (705), and thus the plants are essentially immune to the direct effects of low oxygen concentration about their submerged parts. At times when the shoots carry on photosynthesis, downward movement of oxygen must be quite rapid.

In addition to sponginess, plant organs that normally grow below the surface of the water are very different from those that rise above the water in that they usually lack a cuticle or periderm, and any stomata that form are nonfunctional. The nonfunctional stomata constitute an important type of evidence supporting the belief that vascular aquatics

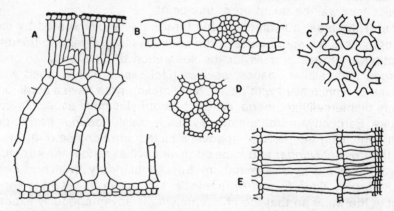

FIGURE 31. Some structural features of hydrophytes. (A) Section through the floating leaf blade of *Nymphaea* showing cuticle and stomata confined to the upper surface, and large lacunae. (B) Section through a delicate submersed leaf of *Elodea* showing lack of cuticle and stomata, and a thin lamina in which all cells have direct contact with the water. (C) Lacunar parenchyma in the stem of *Juncus*. (D) Lacunar parenchyma in the stem of *Hydrocotyle*. (E) Aerenchyma with two rows of normal periderm, in the stem of *Jussiaea*. (E from Coulter, Barnes, and Cowles (159), *A Textbook of Botany,* by permission of American Book Co.)

have been derived in evolution from terrestrial ancestors. As a result of the lack of cutin and suberin, submerged organs should be capable of absorbing water and nutrients directly, rather than getting them indirectly by means of root absorption. However, it has been found that a "transpiration stream" operates feebly in most vascular aquatics, and there is experimental evidence that root absorption is essential for optimal growth. Characteristically the supporting and water-conducting tissues of submersed vascular plants are also atrophied. This lack of mechanical tissue is amply compensated for by the buoyancy due to lacunae.

The roots of hydrophytes are usually shorter and less branched than those of mesophytes or xerophytes. Furthermore, the rootlets are often devoid of root hairs, and rootcaps are typically elongated and sheathlike, and are called *root pockets.*

Hydrophytes may be subdivided into five morphoecologic groups as follows:

Floating Hydrophytes. Floating hydrophytes are plants that are in contact with water and air but not with soil. Examples: duckweeds (*Lemna minor, Wolffia, Spirodela*), water hyacinth (*Eichhornia*), and *Salvinia.*

Suspended Hydrophytes. Plants such as phytoplankton, *Sargassum,* and the duckweed *Lemna trisulca,* are in contact with water alone. In this and in the preceding group the depth of water beneath the plants is of minor importance, and the members commonly travel great distances by water currents. This group differs from the first in the important respect that the plants are absolutely free from transpiration stress, and yet inhabit the best-lighted and best-aerated strata of bodies of water. However, at best, gas exchange is always a problem with suspended and submersed plants.

Submerged Anchored Hydrophytes. In this category are those plants that grow entirely under water and are attached to the substratum. Examples: *Elodea,* tapegrass (*Vallisneria*), eelgrass (*Zostera*), stonewort (*Chara*), hornwort (*Ceratophyllum*), many pondweeds (*Potamogeton*), and most macroscopic algae.

Floating-Leaved Anchored Hydrophytes. These include water lily (*Nymphaea*), spatterdock (*Nuphar*), water shield (*Brassenia*), certain pondweeds, and certain burreeds (*Sparganium*). In these as well as other plants with their leaves lying on the water surface, mechanisms that render the leaf surface difficult to wet are common. Thus in *Nymphaea* a waxy surface causes water drops to roll off quickly, and in *Salvinia* hairs hold water drops above the leaf surface. This adaptation is of obvious benefit to plants whose stomata are confined to the upper surfaces of the blades (Fig. 31).

Emergent Anchored Hydrophytes. These, the so-called "amphibious plants," grow in shallow water and extend their shoots well above the surface. Examples: rice (*Oryza sativa*), bullrush (*Scirpus*), cattail (*Typha*), cord grass (*Spartina*), water willow (*Justicia*), many sedges (*Carex*), bald cypress (*Taxodium*), and mangroves (*Rhizophora, Avicennia*). Included also are a number of species such as *Proserpinaca palustris,* which are well known for the striking dimorphism exhibited by leaves that diverge from the stem below the water level as compared to those that diverge above the water. Phreatophytes are transitional between this group and mesophytes.

Xerophytes

The term xerophyte has been given many interpretations. According to a loose qualitative usage, xerophytes are plants of relatively dry habitats; therefore almost all regions contain species that are xerophytes by comparison with others. When xerophytism is defined in this relative sense, the xerophytes of one region may be much more mesophytic than the mesophytes of another!

Attempts to define xerophytes on the basis of their morphology or their transpiration rates have ended in absolute failure. A truly ecologic definition approaching a fixed quantitative basis, which will be followed in the subsequent discussion, is that xerophytes are plants which grow on substrata that usually become depleted of growth water to a depth of at least 2 dm during a normal season. In arid regions all plants not confined to the margins of streams or lakes are considered xerophytes, whereas in regions of high rainfall the class would be represented only by shallow-rooted plants of sandy soils, by plants of dry ridgetops, by algae, mosses, and lichens which grow on tree bark or rock surfaces, etc.

By some means, morphologic, physiologic, or both, xerophytes must escape or endure recurrent drouth. Although the means by which this is accomplished are manifold, three general types of xerophytes, having little or nothing in common physiologically, morphologically, or taxonomically, can be recognized.

Ephemeral Annuals. Arid regions usually support an abundant flora of small annuals which can complete their life cycles in very short time. At the onset of the brief rainy seasons, which may come only at intervals of several years, these plants germinate, then quickly grow to maturity, flower, and set seed before the soil dries out to the depth of their diminutive root systems. They can withstand strong atmospheric drouth but not soil drouth. By passing through the dry season in the form of seeds they evade its effects. They are not xeromorphic; their principal morphologic adaptations are their small size, and large shoots in relation to roots. Their chief physiologic adaptation is an ability to complete their life cycles in a very short period.

Since these annuals really avoid rather than withstand critically dry seasons, some do not look upon them as true xerophytes, yet it is significant that by no means all plants with the annual habit can grow on deserts, and most desert ephemerals do not occur outside of deserts. Also, the percentage of annuals in the flora increases regularly with increasing aridity.

Succulents (389). Succulence results from the proliferation of cells in parenchymatous regions, accompanied by an enlargement of vacuoles and a reduction in the size of intercellular spaces. This morphologic character enables succulent plant organs to accumulate considerable quantities of water during brief rainy seasons, depending upon the extensibility of the individual cells and tissues. When moisture is abundant the succulent organs of xerophytes swell rapidly, then during the ensuing drouth they gradually shrivel as they become depleted of water. Thus water-storage tissue may be looked upon as a means by which the plant partly overcomes the natural unevenness of precipitation.

Succulence, to be effective, must be accompanied by low transpiration rates during the dry season, and in contrast with other major groups of xerophytes succulent plants (Figs. 32, 33) generally are conservative users of water. Every plant collector is familiar with the remarkable tenacity with which most succulent plants retain their moisture in a plant press. Sedums collected in flower commonly set seed before drying, even with repeated changing of driers. A stem of *Ibervillea sonorae,* which was stored dry in a museum, formed new growth every summer for eight consecutive summers, decreasing in weight only from 7.5 to 3.5 kg! (506).

Succulence may occur in roots (*Ceiba parvifolia*), stems (cacti, cactoid *Euphorbiaceae, Stapelia*), or leaves (*Agave, Aloe, Gasteria, Haworthia, Mesembryanthemum*). The reciprocal ratio of surface to weight of an organ may be used as a criterion of degree of succulence among these plants.

Of all the succulent desert plants the cacti have been most thoroughly studied. Aside from their succulence, cacti are also unique among other types of desert plants for their shallow root systems. Rains insufficient to moisten more than a few decimeters of the soil profile are of direct benefit only to those plants with the bulk of their absorbing roots in this layer. More deeply rooted species benefit from such rains only through the temporary rise in humidity that accompanies them. In some cacti, at least, the fine rootlets are drouth-deciduous, reappearing shortly after a supply of growth water has become available, then withering again soon after it is exhausted.

Studies of cacti and other glykophytic succulents have also shown that, in direct contrast to most plants, the stomata are closed during the day when transpiration stress is greatest, and open at night. Because of this, their relative transpiration is lowest during the day (699). The CO_2 used in photosynthesis during the day when stomata are closed is derived from organic acids accumulated during the night (759).

Since succulents appear to avoid drouth by means of their water reserves, some ecologists would exclude them from the category of true xerophytes. However, this does not seem practical, because succulents are best represented in desert vegetation. It seems best to regard succulence, like the ephemeral character of desert annuals, simply as a unique mode of adaptation found in certain xerophytes.

FIGURE 32. Typical xerophytes of the desert 170 km north of Mexico City. The ground is barren except for tiny ephemeral annuals. At lower left and lower right are succulent-leaved *Agave*. At upper left and in distance are succulent-stemmed cacti. Also visible are nonsucculent shrubs.

FIGURE 33. A cactoid African euphorbia, one of the many plants in this family which bear a striking resemblance of the vegetative parts to the American cacti in Fig. 32. The rod is marked in dm.

Nonsucculent Perennials. Many nonsucculent plants can endure periods of permanent wilting. In most if not all woody plants this endurance is limited to a short period, but in many grasses and forbs * it is a matter of months or years, so that all degrees of intergradation exist between very tolerant and absolutely intolerant species. The morphologic and physiologic features that enable nonsucculent perennials

---

* A *forb* is any herb which is neither a grass nor a sedge.

to withstand drouth are innumerable, but the principal ones are as follows.

Rapid and deep penetration of taproots. According to the environmental definition of a xerophyte which was adopted earlier, the categoɪy includes those plants that grow where the soil dries out deeper than a few decimeters. Any plant whose rate of root penetration is sufficient so that the seedling taproot can keep ahead of progressive desiccation from the surface downward, and is thus never subjected to soil drouth, is potentially capable of growing in habitats where only the subsoil remains moist the year around. This requirement excludes a vast number of species from such places, and among others there is considerable variation in ability to cope with differing degrees (depths) of soil drouth.

Many plants in this category can tolerate no more than very brief periods of permanent wilting. Even *Pinus edulis* (piñon) seedlings can tolerate no more than about 12 days of permanent wilting (176), and this small tree is characteristic of desert borders. If a physiologic definition of xerophytism were acceptable, such plants as *Pinus edulis, P. ponderosa* and *Artemisia tridentata* would be excluded, and even on the basis of the environmental definition which is being followed they represent the least xerophytic category.

The depth of penetration of plants in this category is often surprisingly great. *Prosopis juliflora* (mesquite) roots may extend to at least 53.3 m (619), and even *Medicago sativa* (alfalfa) roots have been found 39.3 m below the surface (536). The moisture supply at such depths may be too meager to maintain active growth except when supplemented by surface moisture that becomes available to lateral roots during and immediately after rain storms, but it prevents or at least shortens the period when the plants are subjected to completely dry soil throughout the extent of their root systems.

Extensive root systems. The essential identity of the permanent wilting percentage as determined with different species of plants shows that they all have approximately equal abilities to absorb moisture from the soil in contact with their roots. However, the extensiveness, degree of branching, and number and length of root hairs differ markedly among plants, * and these have an important bearing on the relative efficiency of the root system, for a small percentage of growth water contained in a large mass of soil yields equally as much moisture as a larger percentage contained in a smaller mass.

The extensiveness of root systems in proportion to shoot systems,

---

* The extent of root systems can be determined directly (596, 685), or by analyzing shoots for Li or radioisotopes that have been placed in the soil at definite points (627, 674), or by following the course of depletion of soil moisture (782).

which characterizes nonsucculent desert perennials, seems to be an important adaptation, not only because it increases the absolute absorptive capacity but also because it involves the exposure of only a relatively small proportion of the plant to the atmosphere and results in a wide spacing of the plants (Fig. 34).

FIGURE 34. Creosote bush (*Larrea divaricata*), showing natural tendency for wide spacing of desert shrubs.

Low osmotic potential and endurance of desiccation. Aside from annuals and succulents, the osmotic potentials of plants vary with their water supply (314, 459). The osmotic potential of desert plants typically lies between −15 and −30 bars, whereas in meophytes it lies above −15. This characteristic does not enable xerophytes to extract more water from the soil, nor is it believed to play a significant role in reducing transpiration. Possibly the low osmotic potential of plants in dry regions is no more than a development necessitated by the high solute content of the unleached soils. In this event the character would have the same value as in halophytes.

The theory has been advanced that an abundance of solutes in the protoplasts is of survival value in delaying if not preventing irreversible changes in protoplasmic colloids which might otherwise take place under extreme desiccation. The leaves of most mesophytes contain 100–300% water, based on dry weight, and, although the water content exhibits a rhythmic fluctuation during each 24-hour period, the leaves can tolerate relatively little dehydration without injury. On the other hand, desert perennials can withstand considerable dehydration, the leaves of the *Larrea tridentata* (creosote bush) being able to tolerate desiccation to 50% of their dry weight (664). Such plants lose water rather rapidly up to a point where an abrupt slowing indicates that only bound water remains (582).

The resurrection fern (*Selaginella lepidophylla*) and many mosses, fungi, lichens, algae, etc., possess this physiologic ability in an extreme degree and can remain alive in an air-dry condition for long periods. There is no apparent relationship between anatomy and endurance of protoplasmic desiccation. However, the gelatinous coating over the cells of bacteria and blue-green algae plays an important role here as shown by the fact that yeast cells can be desiccated to an equal degree without injury provided they are first covered with films of albuminous material.

Ability to reduce transpiration to an extremely low level during permanent wilting. Although some desert plants have relatively low transpiration rates, * the fact has been well established that most of the nonsucculent perennials transpire more freely than mesophytes when growth water is equally available to them. Control over water loss by these two

---

* The comparison of transpiration rates among plants involves many difficult problems. The bases for direct comparisons may be equal surface areas (g/dm²/hr), volumes, fresh weight, dry weight, or water contents.

Comparisons based on equal areas of transpiring surface must take into account wide differences in total transpiring surface, in size, form, or position of leaves on the plant, and must not allow results to be influenced by the fact that a leaf blade decreases in size during dehydration. If comparisons are based on fresh weight or water content the hour-to-hour changes in degree of hydration result in an inconstant standard of comparison. If they are based on either fresh or dry weight the presence or absence of succulent tissues which are not directly related to transpiration complicates comparisons between species differing in this respect. If comparisons are based on volumes the interpretations of results are difficult on account of wide differences in the amounts of intercellular space. As a result of all these complicating factors the relationships among species in regard to their transpiration rates usually differ considerably with different bases of comparison.

Relative transpiration is the rate of transpiration per unit area from the plant surface divided by the rate of evaporation from a free water surface exposed to the same conditions. Theoretically, the lower this ratio, the greater the resistance of the plant to the evaporative power of the prevailing atmospheric conditions. However, it will be recalled that different types of exaporimeters yield widely different values for the same evaporative conditions, and, furthermore, plants and evaporimeters respond very differently to changes in wind and light. Because of these circumstances the concept of relative transpiration has little real value unless applied comparatively under conditions where wind movement is prevented, light is uniform, and the same type of evaporimeter is used throughout.

ecologic classes becomes strikingly different, however, when the state of permanent wilting is attained.

The characteristics of structure and function which appear to be responsible for such a strong degree of transpiration reduction in xerophytes are varied. Broad-leaved shrubs may put out a special set of leaves during the dry season, which reduce the transpiring surface a third below a set that is photosynthetically more efficient and is put out during the less extreme season (583). Still others produce delicate leaves at the onset of a rainy period, then drop them to stand leafless during the following drouth. *Fouquieria* and crown-of-thorns (*Euphorbia splendens*) exemplify this habit. Their leaves are so simply constructed

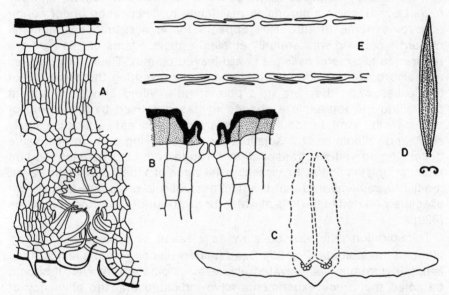

FIGURE 35.   Some structural features of xerophytes. (A) Section of *Nerium* leaf showing thick cuticles, and stomata confined to the lining of pits which contain trichomes. (B) Section of succulent leaf of *Agave* showing thick cuticle (solid) and outer cellulose wall of epidermis (stippled), and individually sunken stomata. (C) Outline of section through a leaf of *Poa* showing method of folding when a water deficit develops. (D) Under side of leaf of *Cercocarpus* showing stomatal surface protected by pubescence and by the permanently revolute margins of the blade. The latter feature is better shown in the cross-section silhouette. (E) Outline of section through a leaf of *Eleagnus* showing the peltate trichomes that thinly shade the upper leaf surface and create a dead air space next the stomata-bearing lower surface. (Part A from Wm. H. Brown *The Plant Kingdom* by permission of Ginn and Co.)

that the maximum photosynthetic value seems to have been attained with the minimum of structural materials. The leaf-shedding habit is very efficient as a means of transpiration reduction, for bare branches covered with corky or resinous layers transpire very little. This is shown by differences in the winter transpiration of deciduous and evergreen coniferous trees. Although the conifers strongly reduce their transpiration during periods of stress, it still exceeds that of bare deciduous trees by two to four times (820).

Many xerophytes that retain their leaves have special structural adaptations (Fig. 35) that decrease transpiration, although they are effective chiefly during periods of permanent wilting. Plants with evergreen leaves that are heavily cutinized (e.g., *Quercus dumosa* and piñon), or waxy (e.g., *Larrea*), show a surprising resistance to desiccation under conditions of extreme drouth. This appears to be a result of a stomatal closure * coupled with a highly efficient cuticle. Plants of this type are referred to as *sclerophylls* (i.e., hard-leaved plants). Their thick cuticles may also be of considerable advantage in preventing the breakage of the leaves while they are in a permanently wilted condition, and in preventing the forceable exchange of gases caused by a bending of blades by the wind. In addition, a heavily cutinized leaf surface is usually shiny and reflects much insolation, therefore having a theoretical value of tending to reduce transpiration through reducing heat absorption under strong insolation. In assessing the value of a thick cuticle, however, it must be remembered that the thickness of this protective layer is not absolutely related to the loss of water through it under average conditions (620).

Transpiration may also be slow as a result of protection variously afforded the stomata (Fig. 35). Dead hairs projecting from a leaf surface keep air currents well elevated above the stomata; however, it should be noted that some experiments have indicated that the efficiency of pubescence in transpiration reduction may be negligible (674). Stomata sunken in pits below the level of the epidermal surface are likewise well below the level of atmospheric turbulence, and plants with that characteristic have generally been found capable of strong reduction in transpiration. In still other species the stomata may temporarily become plugged with wax.

---

* It has been demonstrated that as drouth develops the stomata of mesophytes close before those of xerophytes. In mesophytes closure brings about an appreciable reduction in rate of water loss, but at the same time photosynthesis must diminish for lack of CO_2. On the other hand, xerophytes are often able to continue photosynthesis until a strong water deficit develops, and then when the stomata close transpiration is reduced to a distinctly lower level. Thus the leaves are at once more effective in synthesis and in moisture conservation.

When the effects of drouth are first felt the leaves of certain xerophytes change form or position so that the amount of direct radiation received per unit area is reduced and often part of the transpiring surface becomes shielded from direct contact with the air. The leaflets of many desert legumes fold upward together in such a manner that approximately half the leaf surface is protected. Grass leaves may have longitudinal furrows in their upper surfaces which permit them to roll or fold lengthwise (Fig. 35). The leaves of *Polytrichum* fold longitudinally in much the same manner.

In many species of erect shrubby *Arctostaphylos* the leaf blades are permanently oriented in a vertical position so that they are not subjected to full insolation. The grayish or light green color of these shrubs as well as of most desert plants is thought to be of value in reflecting light rays that would otherwise be absorbed and converted into heat, thus promoting rapid transpiration (Fig. 36).

Reduction in size of leaf blades. The fact has long been well known that all leaf blades, pinnae, and pinnules of desert plants are small; i.e.,

FIGURE 36. A xeromorphic species of *Arctostaphylos*. The white-scurfy vertically oriented leaf blades intercept a minimum of the strong solar radiation to which they are exposed in the dry summers of California.

FIGURE 37. Branch of creosote bush (*Larrea divaricata*), a desert shrub with evergreen, resin-coated, microphyllous foliage. The leaves are about 10 mm long.

the floras are *microphyllous*. Usually the area of these units is less than 1 square centimeter (Fig. 37) (750).

Theoretically transpiration can be reduced either by reducing the rate of water loss per unit area or by decreasing the aggregate area of the transpiring surface. However, careful measurements have shown that in microphyllous plants the small size of the leaves may be compensated for by greater numbers, with the result that nothing has been gained in the way of reducing the total transpiring surface (691). The important advantage of small leaves in dry environments is that they do not become overheated and therefore do not transpire excessively, when exposed to strong insolation. The larger the blade the thicker the enveloping film of still air that envelops it and retards the outward diffusion of water molecules (794).

In certain plants of the desert the trend toward blade reduction has progressed so far that the leaves are vestigial (e.g., *Ephedra,* Fig. 38) and in still others the blades have been entirely lost (e.g., the Australian wattles, *Acacia* spp.). The photosynthetic function in these plants is entirely relegated to petioles or to stems, both of which organs are well supplied with veins in proportion to the amount of parenchymatous tissue.

Reduction in size of cells.   Xerophytes are characterized by relatively small cells and small vacuoles. Just what role cell size plays in drouth resistance is obscure, but it is quite possible that in tissue composed of small cells, a given degree of dehydration has less tendency to pull the protoplast away from the cell wall and rupture plasmodesmata. However obscure the explanation may be, the correlation has practical significance. In certain crop species, but not in all that have been studied,

FIGURE 38.   Joint fir (*Ephedra nevadensis*), a leafless switch plant of the desert.

relatively drouth resistant strains can be selected as unerringly on the basis of cell size as by actual field trials. *

In concluding this discussion of xerophytes it should be pointed out once more that there are no definite anatomic or physiologic characteristics common to all members of this class. Each species has solved its water-balance problem by its own peculiar combination of adaptive characters. One plant possessed of certain of these features usually seems equally as much at home on the desert as another growing beside it but possessed of entirely different adaptive features.

Mesophytes

Mesophytes, the third and last of the ecologic classes of plants based on water relations, include those species which, on the one hand, cannot inhabit water or wet soil, nor, on the other, can survive on habitats where growth water is significantly depleted. The special structural and physiologic adaptations found in hydrophytes and xerophytes are lacking except in certain species of mesophytes derived in evolution from one of the other groups.

Role of Water in Plant Reproduction

Fertilization

The sperms of plants lower than the *Spermatophyta,* with few exceptions, must swim to reach the egg. This presents no problem for aquatic species, but terrestrial cryptograms usually must await times when precipitation or condensation provides continuous films of water over plant surfaces and soil to make fertilization possible. The significance of the pollen tube of the *Spermatophyta* and the trichogyne of the *Ascomycetae* lies in the fact that these organs make the plants independent of direct environmental moisture in the process of fertilization.

In mosses saucer-shaped structures lined with antheridia favor the scattering of sperms through the kinetic energy of falling raindrops.

Pollination (233)

Water carries pollen for aquatics such as *Elodea, Ceratophyllum, Naias, Potamogeton,* and *Vallisneria.* Of these plants *Vallisneria* has an especially interesting pollination sequence. The staminate flower breaks off the inflorescence and rises to the water surface where it floats about.

* Drouth resistance also tends to vary directly with salt tolerance and cold resistance.

The scape of the pistillate flower elongates until the stigma reaches the surface of the water where it remains until contacted by pollen from a staminate flower. After fertilization the scape curls up, drawing the young fruit under water where it completes its development.

Among terrestrial species rains during the pollinating season may have a very detrimental effect. Pollinating insects are inactive during such weather, and wind-borne pollen is kept washed out of the air.

Dissemination

Dissemination is a term for the transportation of detached reproductive structures, the *disseminules* or *diaspores,* away from the parent plant. Such structures are morphologically varied, consisting of spores, seeds, whole fruits, parts of fruits, groups of fruits, whole inflorescences, bublets, vegetative fragments of the parent plant, etc. Dissemination by water is sometimes designated as *hydrochory,* and a water-disseminated plant as a *hydrochore.*

Because primitive plant life appears to have been aquatic, water may be considered the oldest medium of dissemination. Streams and ocean currents carry the spores of hydrophytic cryptogams such as algae and water molds, the floating fruits of *Cocos, Nuphar,* and *Xanthium,* the floating seedlings of *Rhizophora* (Fig. 13), fragments of the vegetative bodies of *Elodea* and algae, and entire floating plants of duckweed (*Lemnaceae*) and *Salvinia.*

Disseminules of the birdnest fungi (*Nidulariaceae*) and the gemmae of *Marchantia* lie loosely in saucer-shaped structures and are dispersed by splashing raindrops. Such mechanisms have been called "splash cups."

A few land plants such as *Salvia lyrata* and *Kalanchoe tubiflora* have disseminules attached loosely to horizontal structures which, when struck by a raindrop, act as a springboard to hurl the disseminule (97).

Water dissemination is common only among hydrophytes, for the seeds of mesophytes and xerophytes are easily killed by lack of aeration when wetted. For example, countless numbers of the nuts of the Seychelles Islands palm (*Lodoicea sechellarum*) have been found washed up on the shores of India and the Malay Archipelago, yet the species has never gained a foothold outside the Seychelles. In contrast the fruits of the coconut palm (*Cocos nucifera*) can endure long periods of submersion in salt water and for this reason appear to have floated completely around the world by means of ocean currents. It is significant that most seed plants on remote islands, like the coconut palm, have floating disseminules that are not injured by salt water. Zoologists consider the presence of certain small mammals and ants on remote islands as strong evidence that driftwood rafts permit transoceanic dispersal, and the possibility is at least as great that plants too have spread in this manner.

The history of *Elodea canadensis* after its introduction into Europe from North America provides an excellent illustration of the dissemination of fresh-water hydrophytes. Within half a century it had become a common weed of ditches over wide areas in Europe. Since only the pistillate plant of this dioecious species was introduced at first, there is no question but that the early distribution was by means of vegetative fragments. A parallel example is provided by *Elodea densa,* originally from Brazil and Argentina, which has spread widely in North America although the pistillate plants have not yet been introduced.

To plants distributed by water, even narrow strips of land serve as barriers preventing migration from one body of water to another. For example, since the Suez Canal was dug, six species of marine algae have spread northward from the Red Sea into the Mediterranean.

Man's Influence on
the Water Factor

The distribution and quality of the earth's finite water supply has been greatly modified by man, especially in the last century or two. Although the implications for human welfare are uppermost in our minds, only matters pertaining to the water available to plants are relevant here.

Algae and vascular aquatics have been strongly affected by pollution, i.e., undesirable alterations of pre-*Homo* environment. Previous mention has been made of poisonous solutes draining into streams from mine dumps, and to these can be added toxic sprays leaching from lands where they have been used in pest control, or added directly to water in attempts to reduce stands of *Eichhornia* or other water plants. Mills manufacturing paper pulp produce waste sulfite liquor which they have been permitted to dump into streams. Each irrigation district must use a surplus of water to avoid salinization, so the waste water contains more salt than the incoming supply. In consequence, a river becomes progressively more saline due to the addition of salts and reduction of water to dilute them, as it flows through successive irrigation districts. Since all of these solutes tend to attenuate aquatic plant life, and plant life tends to keep waters well oxygenated, animals that obtain their oxygen supply from water are severely stressed.

Excess fertilizers applied to agricultural land, chiefly ammonium, find their way into streams and have an opposite effect—that of stimulating excess algal growth which in turn has adverse effects on plants attached to the bottom by reducing their illumination. Since soap is no longer fashionable, phosphorus, a basic constituent of detergents, adds another major nutrient stimulating algal growth, with human and animal wastes supplying a complete spectrum of fertility elements. As organic wastes increase green algae (e.g., *Ulothrix*) disappear and are replaced by blue-greens (especially *Oscillatoria*). Near the source of such wastes oxygen deficiency due to active decay is a major factor determining the nature of a highly specialized but impoverished flora; farther downstream, after the organic residues have been largely decomposed, fertility is excessive and very different algae become superabundant. Such eutrophication of streams could be prevented in an ecologically sound manner by returning human and animal wastes to the croplands where they originated, but in ponds and lakes effective ways of reducing this type of pollution have yet to be demonstrated.

Most of these alterations of the quality of freshwaters are a consequence of the obviously outmoded concept of a stream serving as a sewer to carry noxious materials away, and conceiving a body of still water as having an infinite capacity for diluting wastes until these are innocuous. Pollution has become so prevalent that most domestic water supplies have to be chlorinated to make them safe for human consumption. This poses quite a problem in controlled laboratory and glasshouse experiments where the chlorine content of the domestic water supply happens to be injurious to the organisms whose ecology is to be studied.

Irrigation commonly has another important influence on plant life. In arid eastern Washington where there was no water table a few decades ago, heavy applications of irrigation water have saturated the soil overlying basaltic bedrock, so that in depressions the xerophytic *Artemisia tridentata* and its associates have been killed and *Typha latifolia* marsh is now developing.

Along sea coasts excessive wastage of well water has been lowering water tables so that salt water, formerly held back by the slow outward flow of subsoil water from land to sea, has been encroaching on the land as shown by its increasing contamination of well water. This makes possible the landward encroachment of the coastal belt of halophytes.

Erosion from cropland increases the turbidity and silting of running waters, followed by sedimentation at the inlets of lakes and reservoirs. If plants fixed to the bottom are not eliminated by shading, the blanket of sediment has a pronounced effect on the kinds of any bottom-flora that may be tolerant of its accretion.

In many places gulleys resulting from accelerated erosion, or artificial drainage to permit the cultivation of local wet spots, have reduced water tables over wide area.

Other influences of man on the water balance of plants derive from his influence on climate. Although attempts at increasing rainfall, diminishing hail, modifying hurricanes, etc., have been largely unsuccessful, ecologists everywhere are apprehensive of the possible adverse effects of weather modification, should this objective be realized. The consequences of shifts in balance among species, including pests and diseases, cannot be predicted, and from the results of most other major manipulations of water relations we can be certain that changes made in favor of one objective will have unexpected and undesirable side effects. As a simple case, a "rain-maker" hired by grain growers seeded clouds with iodine crystals and then claimed credit for the shower that followed. This shower undoubtedly benefited wheat crops on the uplands, but he was at once sued by cherry growers who claimed that the rain spread over their orchards along the river terraces and caused much economic loss by splitting their ripening cherries!

Inadvertently man has already altered rainfall patterns somewhat. Air that is warmed in an industrial-residential area rises vigorously and tends to increase rainfall locally. By increasing the dust and smoke content of the air, the tiny particles necessary to initiate droplet formation are becoming more and more plentiful. Excessive particulate matter can cause so many droplets to form that none can grow large enough to fall, so cloud and fog are increased.

At ground level the pavement and roofs of city areas quickly shunt so much of the precipitation into streams that, following storms, water levels rise sharply and the chance of local flooding is enhanced. City environment is distinctly drier than surrounding areas as a result of rapid runoff from paved surfaces, less transpiring vegetation, and warmer temperatures.

Frequently it has been claimed that man has created deserts. The truth or falsity of this hinges upon one's definition of "desert." Certainly man has created wastelands which were previously covered with vegetation, as around smelters or in badly overgrazed areas, but we have no evidence that any such areas are a consequence of degrading the environmental water balance until only the vegetation natural to deserts can inhabit the area. This would demand a reduction in rainfall, or great increase in heat without a corresponding rise in rainfall, and we have no evidence that either of these has been brought about by men so far.

3

The Temperature Factor

The effect of temperature on organisms is very obvious when we compare arctic with tropical vegetation, or the active growth and reproduction of summer with the dormant state of winter in extra-tropical regions. Yet such tremendous differences in the kinds of organisms and their seasonal relations are brought about by only a narrow range of the known temperature scale. There is relatively little biologic activity below 0°C or above 50°C, a range bounded at the lower end by the immobilization of water, and at the upper end by the heat destruction of vital proteins.

Some Physical
Terms and Concepts

Heat is a form of kinetic energy which can be transformed into other kinds of energy or can be transmitted from a relatively warm body to a colder one. This transfer of heat is constantly going on, and the direction and rates of transfer comprise one of the most important aspects of organic environment. The means of transfer are three: radiation, convection, and conduction.

Hot bodies such as the sun or an object that has been heated by the sun give off rays of various wavelengths which travel in straight lines, a phenomenon called *radiation.* Atmospheric gases intercept very little of this radiation, but when it impinges upon the surface of solids or liquids their particles are set in rapid vibration, resulting in a heated condition. The earth's surface quickly reradiates much of the heat it receives from the sun, but, since the wavelength distribution has been altered in the direction of longer waves, reradiation has more of a heating effect upon the atmosphere.

The great vibratory activity of molecules at the surface of a soil that has been heated by radiation (i.e., an *insolated* surface) is transmitted by repeated collisions to the molecules of the gases immediately above, as well as to the molecules of the underlying soil particles. This second means of transmission of heat energy is called *conduction.*

When the lowest layer of the atmosphere is warmed by radiation, reradiation, and conduction it expands, decreases in density, and is then replaced by means of numerous small overturns by cooler and denser masses of gas lying above. Frequently heat energy is also transported by horizontal currents moving warm air into a cool region. Any such transfer of heat by means of currents in gases or liquids is called *convection.*

The unit of measurement of heat energy is the *gram calorie,* which is defined as the quantity of energy which will raise the temperature of 1 g of water from 14.5 to 15.5°C. Insolation may be expressed as gram calories cm^{-2} (i.e., Langleys) per hour.

The term temperature is used to refer to a particular level or degree of molecular activity. Temperature is a qualitative term but heat is quantitative, for fewer gram calories are required to raise the temperature of a small body of water through the same number of degrees as would be required for a large body.

Temporal Variations in Temperature

Although some heat originates by compression deep within the interior of the earth and is conducted to the surface, and another small increment originates where organic matter is decomposing, by far the greatest part of the enrgy available at the earth's surface where life is concentrated is derived directly from the sun by radiation. The localization of this heat source, together with the movements of the earth in respect to it, create great variations in the temperature phase of environment.

The amount of heat received from the sun fluctuates owing to the momentary passing of clouds, to the movement of sunflecks across the forest floor, and to daily, seasonal, annual, and geologic phenomena. Climatic records and fossil deposits show the existence of periods during which temperatures slowly increased and precipitation decreased until a warm-dry maximum was attained, at which time the reverse trend set in until a cool-wet maximum was reached, the process repeating itself endlessly but with so much irregularity that the word "cycle" is hardly justified.

As the sun rises in the morning the earth's surface begins to gain more heat than it loses by reradiation so that its temperature rises progressively and rapidly. After several hours a relatively high surface temperature is attained and radiation gains are approximately equaled by losses due to reradiation and conduction. This equilibrium is maintained until insolation begins to weaken during the afternoon. After the sun sets, the earth's warmed surface continues to give up its accumulation of heat to the atmosphere by radiation, and since it receives no more of this energy from the sun its temperature declines steadily during the night. This nocturnal loss of heat is accelerated by the cooling effect of evaporation from the soil, so that soil temperatures characteristically drop below air temperatures, with the minimal surface temperature occurring just before sunrise. Because the daily maxima are higher and the nightly minima are lower, the surface temperature of exposed soil fluctuates more widely each 24-hour period than does air temperature (Fig. 39).

The rate at which a soil surface heats up is considerably more rapid than the rate of cooling. On this account the temperature of the soil averages higher than that of the air immediately above at all seasons. In temperate latitudes this relationship holds to a depth of about 15 cm.

Except for the very surface layer, the attainment of maximal and minimal soil temperatures is distinctly later than these points in the daily air temperature curves. The greatest range as well as the least lag in

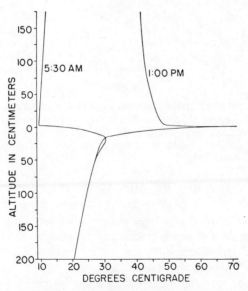

FIGURE 39. Vertical gradients in soil and air temperatures at approximately the hottest and the coolest periods of the day. Tucson, Arizona, June 21. [After Sinclair (709).]

soil temperatures is at the very surface. At a depth of 15 cm the lag usually amounts to about 4 hours; but at a depth of a meter, which is about the lower limit of diurnal fluctuations, the lag may be as great as 80 hours. Seasonal fluctuations (Fig. 40) are felt still deeper, and the lag at 3 m may amount to 5 months. Both the degree and depth to which daily variation in soil temperature extends are greatly reduced in winter, especially when the soil is frozen (402). Obviously any expression of soil temperature is without much significance unless the time of observation and depth at which it is made are specified.

The widest and most violent temperature fluctuations are probably those of intertidal areas. On these habitats the temperature drops many degrees within a matter of seconds each time waves return to cover a sun-heated beach.

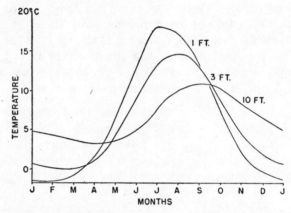

FIGURE 40. Annual march of mean monthly temperatures at three levels in the soil at Bozeman, Montana. This shows (a) reversed vertical gradients in summer and winter, (b) an increase in lag with depth, and (c) a decrease of range in variation with depth. [After Fitton and Brooks (246).]

At intermediate latitudes, the "temperate zone," temperatures differ considerably between night and day, and between summer and winter as well. But this situation changes progressively toward the equator or toward the pole. At the equator the temperatures of the warmest and coldest months commonly differ by only about 2°C, although diurnal fluctuations are at least five times as great. At the poles the reverse condition obtains: monthly means may differ as much as 36°C between summer and winter, whereas diurnal variation is quite small. Mean monthly temperatures therefore conceal much that is of ecologic importance, and annual means are still more deceptive.

Spatial Variations in Temperature

Color and Composition of Surfaces

The color of a soil surface affects the amount of radiation it can absorb and in turn governs the amount of heat that is stored and reradiated back into the atmosphere. White reflects all radiation; black absorbs it completely. When a bare, light-colored soil is subjected to insolation the reflection is so strong that the lower air strata become very hot but the soil remains fairly cool. On the other hand a dark soil surface, such as that of a burned area, absorbs more insolation and thus becomes relatively hot. Differences exceeding 20°C are not uncommon for adjacent dark- and light-colored surfaces (Fig. 41), and this has strong influence on promptness of germination (487).

The composition of the surface that receives and dissipates the daily increment of insolation is important in microclimates. On bare areas the surface alone is involved, but as a plant cover develops the top of the canopy becomes the active surface, and since it is irregular, the exchange surface is much thicker, but still the warmest and coldest points are elevated above the ground.

In shallow water solar radiation absorbed by the bottom is a major determinant of the water temperature. Peat surfaces have low conductivity so the rooting layer beneath remains cold, whereas they reradiate heat at night so efficiently that cranberry plants are often injured by frost. But if a dressing of light-colored sand (a better conductor and

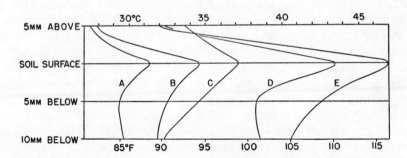

FIGURE 41. Vertical gradients in temperatures at and near the soil surface as affected by color, shading, and moisture content of the soil. Measurements made with a thermocouple in early afternoon of a clear day. (A) Gray, dry, thinly shaded surface. (B) Gray, wet, exposed surface. (C) White, dry, exposed surface. (D) Gray, dry, exposed surface. (E) Black, dry, exposed surface.

a poor radiator) is applied, the frost-free season may be almost doubled (301).

Porosity and Water Content of Soil

Coarse or well-aggregated soils respond to insolation more quickly than heavy or poorly aggregated soils, largely because of differences in drainage and water content.

The wetter the soil the slower its temperature changes (Fig. 41), because the specific heat of water is about five times as great as that of mineral particles, i.e., five times as much heat is required to elevate the temperature of water contained in pore space as would be required to heat an equivalent volume of soil minerals. This explains how a summer shower can cause a sudden cooling of the soil, and why boggy and swampy soils are always cooler during the summer than the adjoining uplands. The coolness of wet soil is especially pronounced in places where late-persisting snowbanks supply cold water to an area downslope as they melt in early summer.

From the above it should be apparent that sandy soils thaw earliest in spring, loams later, with muck and peat last. Although these differences are most pronounced in spring, they may also be important in winter, for frost damage to roots is usually greatest in coarse and well-drained soils.

Plant Cover

Where there is free air movement there is no difference in air temperatures in sun and shade (702), but in the absence of wind, conditions are quite different. During the day hot air formed over unshaded ground rises vertically and thus has but little influence upon the temperature of the air under adjacent shade. During the night, however, the cold air formed over open ground spreads out readily under adjacent cover. For this reason it is necessary to locate thermometers sometimes as much as 100 m back from the edge of the forest in order to measure temperature conditions that are truly characteristic of the forest.

Even a thin shade strongly reduces the heating of soil by solar radiation, and under full shade soil-surface temperatures remain cooler than air temperatures even during the hottest part of the day. In the absence of appreciable wind, the cool soil of shaded areas absorbs heat from the air more rapidly than heat can be transmitted by convection or conduction from unshaded areas. Furthermore, the greater humidity of air under vegetation increases the amount of heat needed to raise its temperature appreciably. For these two reasons forests generally depress maximal air temperatures as well as maximal soil temperatures.

At night the rate of loss of heat energy by reradiation is retarded by plant cover, with the result that nocturnal temperatures of both soil and air within vegetation characteristically do not drop as low as those of adjacent openings. Thus, on account of the opposed influences of vegetation during day and night, temperature fluctuates less widely under plant cover than where the soil is bare (Table 6). When the day and night influences are equal, mean temperatures do not differ within and without a forest, but sometimes the effect of vegetation in lowering the maximum is definitely more pronounced than in elevating the minimum, so that mean temperatures may be lower and the duration of frost longer under forest cover (504).

In local forest openings, and where plants are too widely spaced to shade much of the ground but are effective in keeping wind from blowing the heated air away, daily maximal temperatures may be raised by a plant cover (186, 715). Also the nocturnal minima may be lower on a vegetated than on a bare area if there is enough breeze to mix the air over the bare area but not enough to stir up the cold air which settles among the plants.

In cool and cold climates, forests reduce the depth of soil freezing but retain snow and hence delay warming in spring. Clearing forests where the subsoil is permanently frozen allows deeper thawing in summer, and the thawed soil often settles unevenly.

The temperature under plant cover decreases up mountain slopes on account of the lessening ability of the rarefied atmosphere to hold heat. Since the same atmospheric condition allows stronger insolation at high altitude, it follows that the disparity in temperature between shaded and sunny situations increases upslope.

Snow Cover

A layer of snow falling on frozen ground acts as an insulating blanket and may prevent thawing for a long time, or falling on unfrozen ground it may prevent freezing during protracted periods of low temperature (199). Since snowfalls usually precede very cold weather, the latter relationship usually prevails. There is always much less fluctuation of temperature within and below a blanket of snow than in the air above. Mulches as well as natural organic layers on the soil surface have similar thermal properties (246).

The reflection of radiation from a white snow surface raises considerably the temperature of the air immediately adjacent to it. Because the heating is only temporary, it may prove detrimental to plant organs situated at this level, especially since reradiation from snow at night causes a great drop in temperature.

Slope Exposure and Latitude (83, 252)

By projecting a beam of light through a hole in one piece of cardboard so that the light strikes another piece held at varying angles, it can easily be demonstrated that radiation is most concentrated and effective on surfaces forming a right angle with the path of the rays. In the field striking differences in environment due to the direction of slope exposure (i.e., aspect) can be demonstrated (83, 252). High in the mountains the effect of slope exposure may become so extreme that the minimal temperature of soil on south slopes is higher than the maximum of north slopes!

A slope of as little as 5° toward the pole reduces soil temperature approximately as much as 185 km of latitude in the same direction. Directly or indirectly as a result of this slope-insolation relationship, plants adapted to warm, dry lowlands extend highest in the mountains on those slopes and ridges oriented to get the maximum insolation, and plants adapted to the cool, moist environment of high altitudes reach their lowest limits in protected ravines and on poleward slopes. Plants characteristic of intermediate elevations in mountains characteristically occupy exactly reversed habitats along their upper and lower altitudinal limits.

On a much larger scale the same principle applies to the entire surface of the earth, since temperature tends to decrease from the equator to the poles where the sun is always low in the sky. This latitudinal effect is magnified by the screening action of the atmosphere. An average of about 2 g calories of energy per square centimeter per minute is received by the earth from the sun. When the sun is directly overhead the atmospheric blanket cuts out about 22% of this radiant energy, but with the sun at an angle of 5° in the sky the rays must pass for approximately 11 times as great distance through the atmosphere, and 99% of the energy is intercepted.

Vertical Gradients Near and Below the Soil Surface (263)

Previous mention has been made of the fact that soil-surface temperatures characteristically rise higher during the day and drop lower at night than the temperatures of the air above or the layers of soil beneath. The decrease in temperature upward from the soil surface is accounted for by the decreasing effectivity of reradiation and conduction, and to greater turbulence and hence mixing of the air (263). These vertical gradients are not particularly pronounced in dense shade or at night, but under full insolation they become very steep near the soil surface and remarkable differences can be measured at points separated by only a few millimeters (Figs. 39, 41). Differences among habitats are normally more pronounced in regard to the daily maximal than to the minimal soil temperatures.

Because average air temperatures decrease more than soil temperatures on ascending a mountain slope, and because soil temperatures are the higher, the disparity in temperature of these two contiguous bodies increases with altitude.

When interpreting standard meteorologic data the ecologically important fact must be taken into account that the temperatures recorded by an instrument exposed at a height of about 1.5 m * are several degrees lower during critically hot periods, as well as higher during critically cool periods, than the temperatures to which low-growing plants are subjected. English climatologists, in recognition of this fact, have established a special term *ground frost* for occasions when an unprotected thermometer lying on closely cropped grass registers –1°C or lower. The temperature at this level gives a much closer approximation of conditions to which seedlings and other frost-sensitive plants are exposed than do records obtained in standard shelters.

The vertical temperature gradient through the soil profile is exactly opposed in winter and summer. During winter a more rapid loss of heat from surface layers results in a downward gradient of increasing temperature, whereas in summer the subsoil warms up much later than the surface (Fig. 40).

Over about one-fifth of the earth's land area, including about half of Canada, the soil is perennially below the freezing point of water, a condition called *permafrost.* Only the thin surface layer which thaws each summer is available for root systems (615).

**Vertical Gradients in the Upper Atmosphere and
Temperature Inversion**

In general the temperature of the air lowers with increasing elevation above the land surface. Because of this gradient, or *lapse rate,* mountain slopes rise diagonally across progressively cooler strata of air. On an average the temperature in mountainous regions decreases about 0.5° C per 100 m.

Within the tropics increasing elevation reduces the warmth and duration of the daily period when temperatures rise above the threshold for plant activity, but the temperature characteristics of this favorable seg-

---

* At this height the nature of the ground surface ceases to materially affect temperature and relative humidity, thus making measurements useful in the geographic comparison of regional climates (or *macroclimates*). Obviously it is the microclimate immediately enveloping a plant organ that is of most immediate concern in autecology.

ment of the day remain essentially the same throughout the year. At high latitudes an increase in elevation not only reduces the heat and length of the warm period each day, but also shortens the season when any days favor activity.

The loss of heat from the earth's surface by reradiation at night is often so vigorous that quite early the soil becomes cooler than the air immediately above. This makes it possible for the lower layers of air to give up their heat to the soil by conduction, with the result that they become cooler than upper layers which have become heated by reradiation (Fig. 39). Whenever the air just above the ground becomes cooler in this way than the air at slightly higher levels, the condition is an *inversion* of the normal vertical temperature gradient. Inversion is prompted by: (*a*) long nights which allow more complete loss of heat by reradiation; (*b*) clear skies, because cloud and fog radiate almost as much heat downward as the soil radiates upward; (*c*) cold, dry air, which absorbs very little of the energy permitting it to escape the earth rapidly; (*d*) calm air, because turbulent air destroys a cool layer by mixing as fast as it settles out; and (*e*) snow cover, which in itself does not heat up much during the day and absorbs much heat from the contiguous air at night.

In mountainous topography the phenomenon of temperature inversion is magnified by the fact that the cold air of the upper slopes, having greater density than warm air, drains down ravines at night and slides under the mass of warm air which has accumulated in the valley during the day. Owing to this *cold air drainage* the valley floor at night is occupied by a slow-flowing layer of air which increases in depth as the night progresses and is cooler than the layer immediately above it that makes contact with the surrounding slopes and ridges. During the day a reversed stream of warm air may often be detected moving up valleys and ravines from the lowlands.

Under conditions of temperature inversion the vertical rise in air temperature does not continue indefinitely, but when a certain level (usually less than 300 m) is attained the gradient reverses and begins to decline from there upward at the usual lapse rate. The upper layer of air in the inverted zone therefore is a warm stratum suspended between bodies of cooler air. The frequency, intensity, and depth of inversion vary with both weather and topographic conditions.

In mountainous regions the zone where the warmest of these strata intersects the slopes on either side of the valley is called the *thermal belt* (Fig. 42). The steeper the slope the more pronounced the thermal belt unless the valley is very narrow. In locating orchards and vineyards in mountainous country, close attention should be paid to the elevation of this belt, for the latest frosts in spring and the earliest frosts in autumn occur on the valley floors below the thermal belt (379). So steep is the gradient and so sharply drawn the temperature strata in the cold layer of air on the valley floor that the lower portion of a tree and certain tender crops may suffer from frost while the branches above a certain level completely escape damage. Where cold air drainage is pronounced it may postpone the leafing out of trees by several weeks (621).

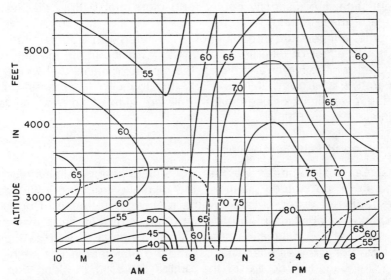

FIGURE 42. Isogram showing air temperatures for a day in August, on a south-facing slope in the southern Selkirks. The highest as well as the lowest temperatures are attained on the valley floor at 2300 feet (700 m). In late morning and afternoon the vertical stratification is normal, but at night the altitude of the maximal (indicated by broken line) rises far above the valley floor, forming an inversion layer about 1000 feet (335 m) deep. The thermal belt lies between approximately 3000 feet (915 m) and 4000 feet (1220 m). [After Hayes (325).]

Basins in topography, or the area immediately above a constriction in a valley, are often referred to as *frost pockets* because cold air drainage results in an accumulation of cold air in these situations to the extent that it exerts a profound influence upon the plant life there (363). Sometimes a bench along a slope has a similar disadvantage because cold air accumulates on it before descending further (161). Where cold air draining into a valley is filtered through a forest that has a certain capacity to store heat, frost hazard on the valley floor may be increased by deforestation (112).

Mountain and Valley Influences

In sunny climates wide mountain valleys often build up a large body of hot air each day due to reflection from the sides and feeble winds supplementing heat from the sky. Consequently their local climate is distinctly hotter (and drier) than the average for the elevation. It is to be noted that this is exactly the reverse of the topographic influence of narrow valleys which, owing to protection from insolation, and to cold air drainage, are distinctly cooler (and more moist) than the average climate of their altitude.

Another temperature phenomenon associated with mountain topography is the effect of differences in the size of mountain masses upon temperature. The larger and higher the mountain mass, the warmer the temperature at a given altitude. Thus the upper limits of forest growth and the lower limits of permanent snow are highest on relatively large mountain masses (605)

Distribution of Land and Water

The three most important factors governing geographic variation in temperature are latitude, altitude, and distance from the influence of large bodies of water. Commonly the relative importance of proximity to oceans is underrated, and the importance of latitude is overrated. For example, it is colder in January in some parts of Kansas than at Dutch Harbor in the Aleutian Islands, and the summer maxima are higher along the Canadian border of the United States than along the Gulf Coast.

The differing responses of land and water to insolation are responsible for such great deviations from the average latitude-temperature relationship. Because water reflects relatively more insolation, loses heat readily by evaporation, has such a high specific heat, and because vertical mixing distributes the heat, the temperature of a body of water changes more slowly than that of a body of land. The land becomes hotter during the summer because all heat absorbed stays in the surface horizon, and it cools to a lower level during winter because the heat is quickly lost from this thin horizon (Fig. 43). Bodies of water are therefore temperature stabilizers with respect to adjacent bodies of land. In addition to the specific-heat factor, the greater humidity associated with water acts as a shield against some of the heat radiated from the sun and as a blanket retarding reradiation from the earth. These influences are strongest on small islands situated in oceans, but they are distinct along the windward margins of continents and may be weakly evident in the vicinity of inland lakes (41). Orcharding in the vicinity of the Great

Lakes, for example, is greatly favored by the temperature-regulating influence of these bodies of water. In their vicinity, on the east and south sides, the annual minima are raised about 8°C and the frost-free season is about a month longer (793).

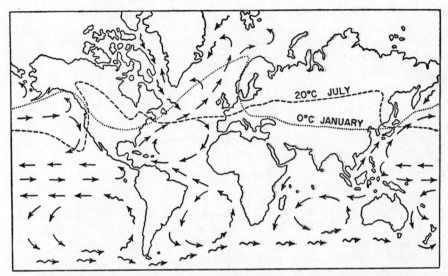

FIGURE 43. Direction of major ocean currents as indicated by arrows, the wavy arrows indicating cold currents. The isotherms of 20°C for July and 0°C for January in the northern hemisphere show that large continental areas heat up more rapidly in summer, and cool off more rapidly in winter, than oceanic areas.

From the above considerations it might be anticipated that the greatest annual ranges in temperature occur in the interiors of large continents. Geographers refer to the climate of areas where temperature extremes are wide because of a lack of oceanic influence as *continental* climates. Conversely, where temperature fluctuations are narrow the climate is *oceanic* (or *marine,* or *insular*). * Aside from the fact that temperature extremes are repressed by water masses, the temperature of bodies of water lags behind seasonal changes in solar radiation intensity, so that oceanic climates are likewise characterized by annual maxima and

---

* Oceanic climates also differ from continental climates in that they tend to have more cloud and rain. Owing to negligible annual variation in temperature everywhere in the tropics, a distinction between oceanic and continental climates is useful mainly in extratropical regions.

minima which come much later after the solstices than in continental climates (Fig. 44).

Again it is important to emphasize that the mean annual temperature (like mean annual rainfall) is an almost worthless statistic in ecology, for a severe continental climate can have the same mean as a mild oceanic climate (Fig. 44).

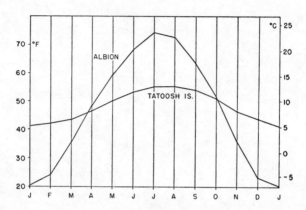

FIGURE 44. Annual march of mean monthly air temperatures at a station with oceanic climate (Tatoosh Island, Washington) and a station with continental climate (Albion, Nebraska). Although mean annual temperatures are almost identical, the continental climate has much the wider annual range. The attainment of the summer maximum is distinctly earlier in the continental climate.

Ocean Currents and Streams

Ocean currents which flow poleward carry water from warm parts of the oceans to high latitudes where the heat is transmitted from the water to the overlying air. Cold currents moving in an equatorial direction have the opposite effect. The climates of islands and of continental margins are strongly influenced by these streams. Thus the coast of British Columbia and Alaska is warmer than latitude alone would warrant on account of the Japan Current which approaches this region from the southwest (Fig. 43). On the opposite side of the continent at the same latitude, the Labrador-Newfoundland region is cooled below the latitudinal average by cold water currents which approach from the northeast. Arctic species extend far southward along coastlines thus cooled, and temperate species follow warm coastlines into high latitudes.

Cool streams flowing from alpine regions may descend so rapidly and heat up so slowly that they drag their cooling effect far down valleys, although aquatic and streamside plants are the chief species affected. Cold irrigation water also has significant effects in lowering soil temperatures.

Direction of Air Movement

The direction of movement of air masses often plays a critical part in temperature relations. Air masses moving from a polar region are colder than those approaching from the tropics. Winds blowing from a mountain mass are generally cooler than those from a reverse direction. Where east-west chains of mountains occur, the regions on the equatorial sides tend to be much the warmer because of the protection afforded against polar air masses. This effect of mountains as climatic barriers is heightened by the fact that half the total water vapor of the atmosphere is below a height of 2000 m, and this vapor has a powerful effect on radiation. All large mountain ranges separate areas with very different temperature characteristics.

Winds blowing steadily off an ocean may carry oceanic influence far inland along the path of major wind movement. Thus along the major storm track which starts eastward across North America in the latitude of the Canada-United States border (Fig. 20), oceanic influence upon temperature extremes is detectable as far east as the divide of the Rockies.

Vertical Temperature Gradients in Bodies of Still Water

Temperature is infinitely more uniform in aquatic than in terrestrial habitats, and the larger the body of water the less the fluctuations. Even on a bright day insolation can raise the surface temperature of the ocean only about 1°C.

In shallow bodies of water where aquatic vegetation is not dense enough to interfere with water currents set up by wind friction, the entire body of water circulates and consequently remains uniform in temperature. But in deep ponds lacking a strong current, only the upper few meters of water set in circulation by wind friction accumulates heat during the summer. At this season, from the surface downward there is a vertical gradient of slowly decreasing temperature extending a few meters, then a narrow zone with rapidly diminishing temperature, and below this is a third, cold zone, in which there is almost no further decrease in temperature. During the winter the warm uppermost layer radiates its heat slowly and becomes cooler than the bottom layer, creating a temperature condition exactly the reverse of that in summer. The narrow intermediate zone of rapid temperature transition, below which there is practically no seasonal fluctuation, is called the *thermocline.* The water below the thermocline is called the *hypolimnion,* and the portion above the *epilimnion.* The hypolimnion remains at about 4°C (the temperature at which fresh water is at its maximum density) the year around in temperate latitudes.

Importance of Temperature to Plants

Plant Temperatures

The temperature of plant organs tends to follow that of the immediate environment closely. This is especially true of root temperatures, which are almost identical with soil temperatures, except where a rising transpiration stream pulls cool water through roots in the warm surface layer of soil. Shoot temperatures may differ several degrees from air temperatures, depending on the absorption of solar radiation which elevates tissue temperature or upon the vigor of reradiation and evaporation which tend to cool tissues below air temperatures. The thin leaves of some plants remain cooler (up to 15°C) than air even when strongly insolated, whereas large thick leaves may become 30°C warmer than air. With massive organs of plants, and organs covered with hair or cork, the temperature changes of the inner tissues lag behind those of the environment, the lag often being sufficient to prevent injury from short periods of extreme temperatures. In mocroorganisms and a few vascular plants, respiration may be so vigorous under favorable circumstances as to cause marked elevation of plant temperatures above environmental temperatures. Rapid respiration in the inflorescence of certain *Araceae* raises temperatures around the spadix by many degrees, and so may confer frost protection for sensitive sex organs.

Temperature and Transpiration

Reference was made earlier to the fact that transpiration increases directly with the magnitude of the difference in temperature between the leaf surface and the adjacent air. Temperature also changes the ratio of cuticular to stomatal transpiration: the higher the temperature, the greater the cuticular component. Thus at a temperature of 49°C the nocturnal rate of transpiration in *Helianthus annuus* was observed to rise to 91% of the diurnal rate, even though the stomata remained closed at night.

Cardinal Temperatures for Physiologic Processes

Cardinal temperatures * differ for the same function in different plants. For example, the minimal temperature for growth in melons, sorghums, and the date palm lie between 15 and 18°C, and the corresponding value for many plants of high latitudes lie between −2 and 8°C (Fig. 45) (622). The maximal temperature for growth in boreal shade plants is not many degrees above freezing, but a species of *Opuntia* was observed to make growth when its tissue temperature was 56.5°C and the air temperature was 58°C (507). Certain arctic marine algae and the snow

FIGURE 45. Shoots of *Valeriana sitchensis* emerging through late-persisting snow in spring. By absorbing light rays that penetrate the snow, and reradiating the energy as heat, the shoot melted the snow as it elongated. Low temperatures apparently had kept the plant etiolated. Glacier Park, Montana.

---

* *Cardinal temperatures* are the *minimum* below which a function is not detectable, the *maximum* above which it is not detectable, and the *optimum* at which the function progresses at maximum velocity.

algae complete their life cycles in habitats where the temperature never rises significantly above 0°C, whereas hotspring algae may live in water uniformly as hot as 93°C. Because of evolutionary adaptation it is generally true that the warmer the native habitat of a species the higher up the temperature scale its cardinal temperatures lie.

Different functions of the same plant may have different cardinal temperatures. In many if not most plants the optimal temperature for photosynthesis is distinctly lower than the optimum for respiration. In the white potato the rate of photosynthesis rises to a sharp maximum at about 20°C, but respiration at this temperature is only 12% of its maximum rate. With an increase to approximately 48°C respiration reaches its optimum, but the photosynthetic rate has declined to zero by this time. Because both growth and reproduction depend upon a more rapid rate of accumulation than of oxidation of organic compounds, plants are at a disadvantage whenever the temperature rises above the optimum for photosynthesis. This is believed to be the explanation for the facts that peaches, apples, white potatoes, etc., do not accumulate normal food reserves when planted below certain altitudinal or latitudinal limits, and that anthocyanins, which are usually associated with abundant sugars, fail to develop at high temperatures. This photosynthesis-respiration relationship may prove to be of considerable importance in setting the lower altitudinal and latitudinal limits of many plants. Quite possibly the more rapid rate of carbohydrate accumulation observed in certain crop plants as temperatures decline at the end of summer is in part at least a result of temperature control of the photosynthesis-respiration relationship.

Various organs of the same plant may have different cardinal temperatures for the same function. Roots, the temperature of which follows that of the soil, appear to have lower threshoild (i.e., minimal) temperatures for growth than do shoots (567). In many plants of temperate regions the roots continue growing as long as the soil is not frozen, although in general the roots of most plants in temperate climates become relatively inactive for at least a part of the winter. Because of definite and often different temperature requirements of roots and shoots, the relations between soil and air temperatures have a pronounced effect upon the welfare of certain species.

Cardinal temperatures vary also with the age of the plant, with its physiologic condition, with the duration of particular temperature levels, and with variations in other environmental factors. Strictly speaking, then, a cardinal temperature is a range rather than a fixed point on the temperature scale, and with plants growing in natural environments it follows that the optimal temperature conditions for the successful completion of the life cycle embraces a range set by particular maxima and minima which may limit development only at one or two points during the cycle. Thus the temperature requirement of different functions at each stage of development must lie within the variations in temperature which prevail during the season corresponding to that stage of development. At each phase of development there is, for the organism as a whole, an optimum range which is most conducive to the harmonious interaction of all physiologic processes.

Thermoperiodism (829)

The rising and setting of the sun each day set in motion complex rhythmic variations in several important environmental factors. As the sun rises, relative humidity decreases while light and temperature increase, and the direction of these changes reverses sometime after noon. Many plants have become so adjusted to this regular diurnal sequence of events that they require lowered temperatures at night and will not exhibit normal behavior when grown in artificially uniform environments. The response of plants to rhythmic diurnal fluctuations in temperature is called *thermoperiodism*. Its physiologic basis seems to rest upon the fact that separate and complementary processes, e.g., growth and photosynthesis, go on under day and night conditions, and these processes have different cardinal temperatures. For example, high day temperatures favor rapid photosynthesis and low night temperatures reduce respiration to a low level, so the photosynthate produced during the day is conserved. Thus it would appear that within certain limits the greater the disparity between day and night temperatures the more efficient the temperature relations with reference to the energy economy of the plant. As an example, photosynthesis in the soybean at 28.2°C fixes 15 times as much CO_2 as nocturnal respiration under 11.8°C releases, but when plants are grown under a uniform temperature of 28.2°C the ratio falls to 9:1 (211). The lower ratio is due to both a reduction in photosynthesis and an increase in nocturnal respiration, the latter factor supporting the theory outlined above.

This theory obviously does not hold for those plants the growth of which is more closely correlated with diurnal than with nocturnal temperature levels (657), or vice versa. In *Sequoiadendron* and *Pisum sativum* (garden pea) growth is more closely correlated with day temperatures, whereas in *Pinus sabiniana* and *Lycopersicum esculentum* (tomato) night temperatures are the more critical (Fig. 46). In still other plants just the maintenance of a diurnal temperature difference is more important than the absolute levels, and in *Salvia splendens* it makes no difference whether it is the high or the low level that is associated with night! The bulk of the experimental data, however, show that vegetative growth (828), flowering (278), fruiting, and germination (313, 754) are normal only under alternating temperatures.

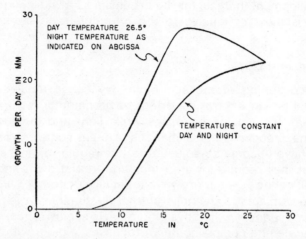

FIGURE 46. Comparison of stem growth of tomato plants under constant and under alternating temperatures. [After Went (828).]

Temperature Efficiency

Because heat energy increases the kinetic activity of molecules, the higher the temperature the more rapid is the rate of chemical reaction and, consequently, the physiologic processes. The ratio of a rate of reaction or function at a given temperature to its rate at a temperature 10°C lower is called the *temperature coefficient* and is designated by the symbol Q_{10}. In ordinary chemical reactions the temperature coefficient usually varies between 2 and 3, depending on the particular reaction, but in a single complex organismal function the coefficient varies from zero above and below the maximum and minimum temperatures, respectively, to approximately 2 in the region of the optimum temperature (Fig. 47). The temperature coefficient must be expressed in relation to a particular range on the temperature scale; moreover, it changes with variation in other environmental conditions as well as with age of the plant. Because of these limitations the concept finds very little use in biology.

Ecologists have long been interested in evaluating temperature data in terms of efficiency in allowing plants to grow and reproduce vigorously, but the problem is very complex and only unrefined methods yielding generalized conclusions have as yet achieved any measure of success.

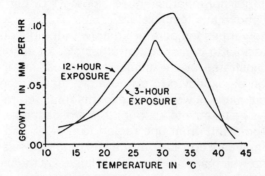

FIGURE 47. Rate of shoot elongation of *Zea mays* seedlings in weak light during 3- and 12-hour exposures to constant temperatures. [After Lehenbauer (454).]

On the basis of data for the growth of maize seedlings under 12-hour exposures to uniform temperature (Fig. 47), Livingston (466) prepared a table of temperature-efficiency values which he suggested be used by summing the individual values corresponding to mean daily temperatures into a seasonal total. As Livingston pointed out, the investigator who uses these values should bear in mind the serious limitations that they are known to be physiologically reliable for only the elongation of immature shoots of one species under the particular intensity and duration of day and night temperatures to which the experimental plants were exposed. Subsequent studies which have revealed the importance of thermoperiodism indicate that constant-temperature conditions yield very little information of real ecologic value, and that there is some doubt as to whether it will ever be possible to express mathematically a close relationship between growth and temperature (828).

With temperature data as with precipitation, the monthly or at least the seasonal data should be kept distinct, for seasons have different physiologic significance and the time when heat is most available is important. In spring new growth is the predominant function, but during summer the restoration of food reserves becomes the major vegetative process. Superimposed on these are the heat requirements for reproductive activity, first for the initiation of the flower primordia, then for the development of the flower, and finally for the maturation of fruit and seed.

Reproductive processes are the first to suffer from insufficient heat

(180), and good advantage can sometimes be taken of this fact in selecting regions for the growth of crops according to whether fruiting or vegetative organs are the desired products. Thus, most strains of lettuce produce heads but not inflorescences below 21°C, and inflorescences without heads above this temperature (753).

By correlating various temperature data with the distribution of plants it has been possible to demonstrate rather close relationships in some instances, as illustrated by the timberline discussion above. However, this method can be expected to do no more than suggest causal relationships and supplement work of an experimental nature. The important fact must always be taken into account that *most* aspects of temperature exhibit similar gradients from one region to another, so that what may appear to be significant correlations may be but expressions of concomitant phenomena which are not causally related. By the method of correlation coefficients it has been shown that temperature statistics believed to have great ecologic significance exhibit geographic trends that almost exactly parallel other statistics of very questionable significance (412).

In addition to the suggestive information derived from correlations between plant distribution and climatic data, there is good evidence from other sources that the upward altitudinal limits of many plants on mountain slopes are determined by temperature efficiency. Most montane plants extend highest on slopes facing the equator, and, when low-altitude trees were transplanted to high altitudes in the southern Rocky Mountains, their decadence and death after a few years has been interpreted as being due to insufficient heat (601). At the poleward limits of species distribution, many plants are confined to warm ridges, south-facing slopes, or sandy soils.

For the most part plant adaptations to low-temperature efficiencies involve obscure protoplasmic characters which result in low cardinal temperatures. In evergreen conifers native to temperate regions the threshold value for photosynthesis, for example, is commonly as low as $-3°C$. Many low-growing plants of cold climates have such low cardinal temperatures that they begin growth under the snow several weeks before it disappears and thereby are effectively equipped with photosynthetic apparatus when the most favorable growing conditions arrive (Fig. 45). It has been pointed out that the predominately prostrate forms of plants of arctic and alpine regions, and the rosette form of many winter annuals and perennials with basal leaves, may be of considerable advantage in allowing these plants to exploit the warm stratum of air which lies next to the earth's surface, and thus carry on photosynthesis at times when taller shoots would be too cool for this function. Finally, the evergreen habit allows plants to benefit more than one season from the photosynthetic capacity of a particular set of leaves. At high altitudes and latitudes where temperatures are low and seasons short, spruces and firs often retain their needles 15 to 20 years. Such plants synthesize small quantities of carbohydrate during brief periods of favorable temperature during the winter.

The effect of low temperature in retarding respiration and so conserving food reserves is illustrated by the practice of storing dry seeds in cold rooms when prolonged viability is desired. Seeds of *Lupinus arcticus* obtained from permafrost more than 10,000 years old were found to have unimpaired germinability!

Temperature and Phenology

Annuals characteristically complete their life cycles with no resting period between germination and seed-setting. Some perennials likewise develop continuously, environment permitting, but most of them undergo rest periods in which active growth and flowering cease even though the environment continues to be favorable. In temperate regions activity often stops in early summer, and the rest requirement for certain spring-flowering perennials is in some instances so nearly satisfied by the end of summer that, if the advent of winter weather is delayed, these plants have a brief though futile period of flowering in autumn (284). Although root growth of at least some perennial plants in temperate regions is almost uninterrupted by winter, the shoots have rest periods which normally cannot be broken without exposure to low temperatures. Their low-temperature requirements are usually met before the end of winter weather so that they can resume activity as soon as their threshold temperatures are exceeded, provided no other condition such as suitable daylength or moisture is lacking. It has often been observed that the beginning of activity in these plants is relatively late or early, depending on the arrival of suitable temperatures in spring, * and that activity always begins later at progressively higher latitudes and altitudes. Phenologic † events average 4 days later with each degree of latitude in a polar direction, 5 degrees of longitude in an eastward direction (in North America), and 13 m of altitude in an upward direction (355). These averages are subject to so many deviations due to local topography and other factors that they have little practical value. Such observations,

---

* Rainfall always varies inversely with temperature, but the temperature variable is the one most important in phenology. High rainfall, cloudiness, low temperature, and late development are nearly always correlated.

† *Phenology* embraces all studies of the relationships between climatic factors and periodic phenomena in organisms.

however, led to an early theory that a stage of development depends upon the receipt of a certain quantity of heat units. In 1735 de Reaumur conceived the idea that if one assumes a threshold value for plant development, and sums the daily temperatures in excess of this, starting in spring, a certain sum would be correlated with each stage of plant development as the season advanced. There is just enough evidence in support of this concept that it still attracts some workers. However, the limitations are serious. The air temperatures used are scarcely a match for tissue temperatures in any part of the plant. Thresholds, variously assumed to lie at some point between 0–7°C, differ from one species to another, for different processes, and for different stages of development. There is an inherent assumption that each degree of change along the temperature scale has the same physiologic significance, whereas just above the threshold each degree confers small benefits in comparison with more optimal levels, and above the optimum benefits decline until further increase is detrimental (Fig. 47). The length of day differs among months, and this governs the duration (Fig. 47) of high temperatures without affecting the mean as usually calculated. *
Is the mean more significant than the maximum or the minimum?

A refinement of the temperature summation approach meets some of the objections lodged against it by providing an estimate of the day-degree values for each combination of maximum and minimum temperatures, and for different presumptive thresholds (464). Another worker weighted day-degrees by day length (574). A still more refined approach led to an evaluation of the overall thermal efficiency of each degree on the temperature scale with reference to the development of apricot fruits, and so accounted for the negative effects of hot days. This method made it possible to predict within one day the date when apricots would be ready for harvesting. However, the equations proved useful for only this type of fruit and for only the location where the study was made! (101, 102.)

If phenologic maps are to be made for the delimitation of biologically equivalent climates (124), microclimates must be standardized, for the rate of development of a shrub close to the equatorward side of a

---

* Most of the climatologic data available to biologists have been collected by meteorologists, and for the most part neither the types of data nor the methods of summarizing them have been selected with reference to their utility in biology. For example, daily maximal and minimal temperatures appear to be more important to plant growth than the daily means which are generally calculated for summaries. Again, *median* of precipitation values are more significant than *means,* although they are never calculated. As for solar radiation, evaporation, and relative humidity, the basic data are taken at so few stations that only broad generalizations can result from their analysis.

building, another isolated on flat topography and another under a tree canopy may differ considerably, and only the plant standing in the open is appropriate to use if different places are to be compared in the same year. Also, since genetic variation results in phenologic variation with a species, precise work involves growing vegetatively propagated material distributed among the study sites, a technique which has been followed in an international network of phenologic stations scattered over Europe.

Stimulating Effects of Low Temperature

The twigs of *Rosa, Rhus, Sambucus,* etc., continue growth until frost kills the tips and enforces a rest period. Inactivity imposed and maintained only by environmental limitations is *quiescence.* In contrast, many plants native to cool and cold climates must each year undergo a rest period that is not enforced primarily by low temperature. After growing vigorously for a time they become *dormant* even though external conditions remain favorable for growth (202). Ordinarily this dormancy is broken only by temperatures below about 5 to 8°C, the effect of short periods of exposure below this level being cumulative, yet susceptible of being nullified by subsequent high temperatures. If a plant is moved too far in a poleward direction, its chilling requirement is satisfied early in winter so that growth is resumed before the period of frost hazard is past. If moved too far in an equatorial direction the chilling requirement may not be met by the end of winter, the dormant buds will not open promptly with the arrival of warm weather, and the development of leaves, flowers, and fruits is repressed (670). Peaches require 400 or more hours of temperature at or below 7°C, the length of time varying with the variety. Blueberries need 800 hours at this level, and apples require even more. Winter stimulation accordingly has been found to set the lower altitudinal and latitudinal limits of profitable culture of these plants (174, 196). The possible role of this factor in setting the lower altitudinal and latitudinal limits of native plants has not yet been investigated.

The requirement of a rest period is by no means confined to plants of cool and cold climates, for many plants of the perpetually warm and moist tropics do not maintain a uniform rate of activity (352). Here the rest period may be attuned to the season with least rainfall, but in many plants it is definitely autogenic and unrelated to environmental vicissitudes. Since temperate-zone plants were probably derived from tropical ancestors, it has not been difficult to adapt dormancy to dire ecologic necessity.

Low temperatures are often necessary to stimulate the formation of flower buds. For example, *Calceolaria* and *Senecio cineraria* will not form flower buds if the temperature is kept above 15.5°C (625). In most herbaceous plants the temperature level necessary to stimulate the formation of flower primordia is lower than the level that favors rapid flower development, whereas very high temperatures repress flowering even to the extent of nullifying an initial thermoinduction (753).

In the tropics where temperatures vary but little throughout the year, small deviations from normal can have surprising influence in triggering flowering activity. Coffee, certain orchids and other plants may flower gregariously a week or so after a thunderstorm that ends a series of dry days, and this has been interpreted as a response to the shock of sudden coolness associated with the rain (647).

The seeds of many plants of cold regions require chilling under moist conditions for a period after apparent maturation if they are to germinate vigorously. The common practice of supplying these conditions artificially is spoken of as *stratification,* and temperatures varying from –1 to + 2°C for one to several months are used, depending on the species.

The physiologic explanation of these conditioning effects is unknown, but in some seeds low temperature may function to render the seed coats more permeable (110) though in other instances it may be necessary to alter the ratio of products formed by different physiologic processes. An interesting complexity has been discovered in certain species in which the radicle and epicotyl of the embryo require successive exposures to low temperature levels for development. As a result the seedlings do not emerge until the second year. The radicle develops the first summer in response to the stimulation received the first winter, but growth soon comes to a stop and cannot be resumed until after the epicotyl is stimulated by the low temperatures of the second winter (42). There may be some connection between the above discovery and the fact that fluctuating temperatures are necessary for the best germination of most seeds (29, 809).

Many seedlings and rosettes of cool and cold climate plants do not have a rest period the breaking of which requires low temperature, yet they require exposure to low temperature to come into flower promptly. For example, winter wheat sown in spring does not flower before the plants are killed by drouth or frost, but if the grain is soaked until it starts to germinate and then is subjected to a temperature just above freezing for a period, it can be sown in spring and a crop quickly produced. Since cold treatment is effective through promoting the formation of an essential metabolite, the stimulus derived from chilling is retained even if the seed is subsequently dried. Thus if seed wheat is moistened to 50% of the dry weight, then chilled at 2°C for about two weeks, it can be dried again and sown many weeks later. However, exposure to heat during this time may nullify the effects of chilling.

This type of physiologic conditioning in plants lacking dormancy is called *vernalization* (i.e., bringing into the spring state). Although generally impractical in routine agriculture, for special purposes it can be used to good advantage with many cool-climate plants. For example, it expedites plant breeding in glass houses, since early- and late-developing plants can be brought into flower simultaneously, also several successive generations can be grown in a relatively short time for progeny tests.

The same form of stimulation can also have undesired effects. Sometimes biennials complete their life cycles in one season as a result of rosette stimulation by late spring frost.

Cold Injury (459)

The migration of plants from their ancestral environment of the seas onto the land necessitated marked adaptations for enduring the wide variations in temperature that characterized the newer environment. Although there is no region on earth which is so cold or so hot but that some plants live there, plant adaptations are not so perfectly adjusted but that temperature extremes frequently cause injury or death. Moreover, the absolute extremes that result in the immediate death of protoplasm are much farther apart than those extremes that are detrimental by inhibiting growth and other functions.

When temperature drops below the minimum for growth a plant becomes dormant, even though respiration and sometimes photosynthesis slowly continue. Chlorosis likewise may result from such chilling (397). With further loss of heat a point is usually attained below which the protoplasm is fatally injured. Three main phenomena appear to be involved in killing by low temperature. (a) Proteins may be precipitated directly, especially in plants that are killed before temperatures drop to the freezing point of water. (b) At lower temperatures intercellular ice commonly forms, drawing water out of the protoplasts. This causes a dehydration * which, below a certain critical temperature or after a prolonged period, allows irreversible precipitations in protoplasm. Possibly mere deformation of the shrinking cell may be lethal. Also, when the ice crystals melt rapidly the cell walls may expand more rapidly than the protoplasts can swell, and thus may tear the two apart. (c) Rapid

---

* This accounts for the wilting so commonly observed, and for the injected areas which show up over a leaf blade when the intercellular ice first melts.

freezing causes ice to form within protoplasts. This ice formation is nearly always fatal, presumably because crystal growth disrupts protoplasmic organization.

The ability of plants to endure low-temperature extremes varies widely among species. Certain plants of tropical affinity (*Coleus,* rice, cotton, sudan grass, etc.) are injured by exposure to temperatures which are low but yet above the freezing point. Other plants are not injured until they are frozen; still others native to cold climates can endure periods when the tissues are frozen solidly and the temperature drops to −62°C. The freezing point of plant sap, because of its solute content, usually lies several degrees below 0°C, but certain plants, mostly cryptogams and seeds, cannot be frozen at any temperature (even −270°C), and these are immune to low-temperature injury.

A plant is not equally resistant to low temperatures at all stages of its life cycle. Seeds and spores are generally the most resistant stages. Among trees seedlings are commonly more sensitive to cold than older plants, but with grasses the relationship may be reversed.

All organs of the same plant are not necessarily equally resistant to low temperature at the same time. The ovule may be killed without apparent injury to the carpel; gynoecium is more sensitive than other flower parts; flowers tend to be more sensitive to cold than fruits or leaves; leaves and roots are more sensitive to the same degree of frost than stems. Young leaves are more resistant than old, and the leaf apex is often the only part of this organ to be injured. Trees are commonly injured severely by frost that harms herbs temporarily if at all. Because meristematic activity usually begins in the buds of woody plants, then spreads gradually down the cambium, the younger the wood the more likely it is to be injured by the late spring frosts. Still, the cambium of woody dicot stems often lives though living wood cells immediately within are killed. The injured cells in the xylem thereafter constitute a permanent record of the date of such injury, and the growth ring thus affected is often called a *frost ring* (188).

The frost-free season. In his efforts to extend the production of food and ornamental plants into colder regions, man has persistently introduced plants into regions cooler than those to which the plants are adapted. Consequently, frost occurring after growth has begun, or before crops have been harvested, commonly causes much loss among such important crop plants as tomatoes, cucurbits, potatoes, beans, maize, and sorghum. Deciduous fruit and nut trees often suffer damage to opening buds, flowers, or young fruits, frequently to the destruction of a crop (317). The growing of all such plants must consequently fall within the frost-free part of the year, and considerable importance is attached to methods of protecting them from any out-of-season frosts that may occur. Planting sites should be selected in relation to cold-air drainage or hedges should be grown to deflect cold air. Frost pockets are especially unfavorable habitats because their accumulation of hot air in daytime stimulates growth early in spring when the nocturnal frosts are still very severe.

On account of the significance of frost in agriculture, much importance has been attached to that climatic statistic the *frost-free season;* yet, when all facts are taken into consideration, the significance of this concept is limited even for tender plants. It has been pointed out that (*a*) a given degree of frost is not equally injurious to all species, (*b*) some plants are injured before temperatures drop to the frost level, (*c*) the distribution of frost is so patchy as a result of topographic influence that the data provided by one instrument are of very little general value, (*d*) plant temperatures are determined by other factors in addition to air temperatures (e.g., water content, evaporation rate, etc.), and (*e*) the frost-free season has a wide range of variation at any one station over a period of years (437).

The welfare of by far the majority of plants native to cool or cold regions bears no relationship to the frost-free season, so that this climatic statistic has no real value except in agriculture. Minor exceptions to this general rule are provided by the seedlings of certain trees, a small percentage of which may be killed during late spring frosts, and by the development of frost rings in wood which reduces the quality of the lumber. The frost-free season should never be called simply the growing season (412).

Injury due to winter minima. Except for *Loranthaceae* (798) and perhaps one species of cactus, over those parts of the earth's surface regularly having periods of subfreezing weather there is no evidence that the polar or upper altitudinal limits of distribution of plants are set by winter minima, although at the warm edge of this region frost killing is frequently observed (771). Nevertheless even in areas where plants are adapted to endure very cold weather, severe injury if not death commonly results from a sudden drop in temperature. Certain species of plants that are perennials or biennials where winters are mild become annuals in colder regions. In the northern Mississippi Valley the shoots of the vine *Menispermum canadense* are regularly killed back to the ground in winter, but in the southern part of this valley this plant has a thick woody stem that attains considerable diameter. Cold climates generally become limiting for species distribution more from the standpoint of inadequate summer heat than from winter damage to tissues.

Adaptive resistance to low-temperature injury. The degree of injury that plants suffer from low temperature depends (*a*) on the degree and duration of minimum temperatures, (*b*) the suddenness of change, (*c*) on previous physiologic conditioning by chilling, the level of mineral nutrition, moisture content of tissues, and length of day, and (*d*) on structural adaptations. Plant organs the surfaces of which are covered with a waxy bloom or with dense pubescence can endure freezing temperatures for a relatively long time without ice formation inside the tissues. Also, small cells are correlated with cold resistance.

Temporary protoplasmic adaptation affording a measure of immunity to low-temperature injury is called *hardening* (9). It can be induced artificially by most conditions that result in a sudden checking of growth, especially by chilling to within a few degrees of freezing for at least a few hours, or by drouth. The lower the temperature used in hardening the greater the resultant degree of cold resistance. Glasshouse-grown seedlings are generally hardened by placing them temporarily in a cold frame and watering them but lightly. In irrigated orchards the trees can be hardened against early autumn frost simply by withholding water; on unirrigated land cover crops sown late in summer constitute the best means of reducing soil moisture. These practices lower the injurious temperature level a few to many degrees, depending on the species and on the duration of the hardening period.

The perennials of cool and cold climates regularly develop hardiness in autumn, mainly in response to shortening days that produce a hormone in the buds or leaves which is then transmitted to other organs. The first cool nights are then effective in increasing the degree of hardiness by acting directly on all tissues, even though day temperatures may remain high enough to favor growth and other functions. In spring this hardiness is lost with the renewal of activity, and in summer condition the plants are easily killed by temperatures far above those they regularly endure in winter. The importance of this seasonal adaptation can easily be underestimated simply because it is ubiquitous in extratropical plants.

Several interrelated physiologic changes are known to take place during hardening. Protoplasm develops low structural viscosity and therefore is better able to accommodate deformation. The free-water content decreases so there is little water available for disruptive crystal growth. Soluble proteins, lipids and sugars increase, especially in aerial organs, and this lowers the freezing point of the sap (68). The importance of abundant carbohydrates is shown by the fact that defoliation or any other condition which prevents their accumulation renders plants more susceptible to cold injury.

Different species have developed a measure of immunity to low-temperature injury by adaptations leading in different directions, although the results often have the same value. Thus there appears to be no one standard criterion of hardiness that has universal application, although the following correlations have been found. So long as comparisons are confined to closely related plants, differences in the tissue water content in midwinter may be a fairly reliable criterion of hardiness (847). Among plum varieties it has been observed that the degree of hardiness is directly proportional to the constancy with which the tissue moisture content is maintained during periods of fluctuating temperature in winter (739). In general, the slower the growth rate, or the smaller the cells, the greater is the hardiness. Finally, small decreases in osmotic potential are associated with considerable margins of resistance to low temperature. It must be emphasized again that all these physicochemical and morphologic criteria of hardiness fail when unrelated plants are compared. Direct methods involving the freezing of intact shoots permit more reliable estimates of field responses (198).

Winter Drouth Injury

On account of the slowness with which the temperature of soil changes, the temperature of the air is alternately higher and lower than that of the soil. At those times during the cold season when air temperature is the lower, cold injury per se is the chief hazard to plants; but when the air becomes warmer than the soil, plants have great difficulty in replacing water lost by the shoots in transpiration. With a drop in temperature from 25 to 0°C, the viscosity of water is doubled. Therefore even when there is an abundance of growth water, low temperature greatly reduces the ability of the soil to supply water to the roots (Table 10). The optimum temperature for absorption by roots is generally around 30°C or higher, and soil temperatures in the root horizons are usually lower than this even in summer. The fact has long been known that plants can be wilted by cooling the soil about their roots, as in irrigating with cold water from mountain streams; the warmer the native climate of the plant, the higher the temperature evoking response (221). Experimentally it has been shown that cold soil induces the same structural modifications as drouth, thus demonstrating the correctness of the application of the term physiologic drouth to this phenomenon (763). Furthermore, as a rule, leaf tips and margins are affected first, whereas in freezing injury the leaf as a whole is killed.

Table 10

Effect of Temperature Upon the Rate of Absorption of Water from a Soil by an Unglazed Pottery Surface (421). Since the Moisture Content of the Soil Was 5% Below the Moisture Equivalent, These Data Represent Only the Transfer of Water from Particles in Direct Contact with the Pottery.

Temperature, °C	Water absorbed, mg cm^{-1}hr^{-1}
0.0	57.2
8.2	96.6
24.0	132.2
34.8	171.8

High protoplasmic viscosity in winter undoubtedly adds to the water-balance difficulty because water must diffuse through a series of protoplasts in getting from the root epidermis to the xylem, then from the xylem to the transpiring cells (307, 422). In addition to the effect of low temperature on the viscosity of water in the soil and the viscosity of root protoplasm, its influence in retarding the rate of root growth is important. With soil moisture at the field capacity the ability of the plant to continue absorption is directly related to the rate at which rootlets can penetrate new soil masses. This situation is particularly serious when the soil surface freezes after a dry summer, sealing the profile against full moisture recharge. Frozen shoots of course conduct no water, and each portion of such a shoot is completely dependent on its own water resources.

Low temperature does not have an equally retarding influence upon evaporation, however. High winds, and especially warm breezes that break a period of cold weather, have strong drying effects. These conditions bring on severe physiologic drouth in winter, with a result that the shoots are damaged or the plants killed outright. Evergreen plants are particularly susceptible to this indirect type of low-temperature injury which has been aptly called *parch blight* (814). Their foliage is commonly observed to turn brown in late winter or spring after desiccation. For these plants, snow cover during the cold season is of immense value, as witnessed by the frequency with which shoots are killed back only as far as the snow surface (262) and by the healthy condition of the foliage protected by snow (Frontispiece). Nurserymen sometimes apply plastic sprays to evergreens, which serve as antitranspirants to reduce parch blight.

Raunkiaer (639) developed a widely used classification of the life forms of plants based on the degree of protection afforded the perennating buds during the dormant season. Although life form is of obvious significance in relation to winter killing, it is by no means an adaptation to temperature conditions alone. The five principal categories of this classification are:

1. *Phanerophytes,* trees and tall shrubs.
2. *Chamaephytes,* low shrubs.
3. *Hemicryptophytes,* buds at the soil surface so that they are generally protected by snow or organic debris.
4. *Geophytes,* buds beneath the soil.
5. *Therophytes,* annuals with the embryonic bud protected by a seed coat.

Mechanical Injury Due to Low Temperature

Lesions. As stated earlier, the temperature of plant shoots tends to follow closely any temperature change of the air. When air temperature drops at night in winter, tree trunks lose their heat rapidly, and below approximately 0°C this results in intercellular ice formation and marked tissue shrinkage. Because the outer layers of bark and wood cool more rapidly than the inner part of the woody cylinder, the outer layers are subjected to considerable tension in a tangential direction. When the temperature drop is sudden this tension is often sufficient to cause the stem to crack open along one radius with a loud report. Such frost cracks (Fig. 48) close up again as the temperature rises, but cambial union is not likely to prevent subsequent reopening unless a series of mild winters allows several continuous layers of wood to form before stress again develops. In addition to the vertical cracks extending radially inward, a tangential lesion may develop in a particular growth ring in the zone of greatest tension, or the bark may separate from the wood. The crowns of perennial forbs, especially *Medicago sativa* (alfalfa), often suffer similar injury. In all cases lesions are important because they permit parasitic fungi to gain entrance into the living tissues (823).

Frost heaving. The freezing and thawing of soils is thought by some to be beneficial from the standpoint of formation and maintenance of aggregate structure, yet it tends to keep soil moisture drawn to the surface where it can be lost by evaporation, it promotes erosion during rainy weather, and it is often directly injurious to plants. When freezing is rapid the soil freezes as solidly as concrete before the water it contains can segregate, and without much change in volume. When the temperature drops slowly below the freezing point, a thin layer of the surface freezes, then water is drawn to the bottom of this layer where it forms

FIGURE 48.   (Left) Close-up of a frost crack in a dead, decorticated tree. Repeated opening on successive winters had caused the cambium to produce liplike ridges of wound tissue on either side of the long straight crack. (Right) Temporarily healed series of cracks along the same radius of a tree with spiral grain.

as a layer of vertically oriented crystals that continue to grow in length. Additional layers of such needle ice may accumulate from below on successive nights, building up the frozen mass of segregated water and lifting the first-frozen layer higher and higher (680). Very little soil is lifted in this process, but the stems frozen fast in the surface crust tend to be lifted (heaved) by the subsequent growth of crystals below. Alternating periods of such heaving and thawing may pull plants farther and farther out of the ground, the crowns sometimes rising several decimeters.

Because of the extremely sluggish movement of capillary water at and below the field capacity, frost heaving is possible only when soils have a higher water content, as immediately after a shower or the melting of snow. Furthermore, the phenomenon is pronounced only where radiation is not retarded by plant cover, snow, or litter and duff.

Heaving is most injurious to seedlings which have weak axes and therefore cannot resist much tensile force, but other plants, especially those with thick taproots that are poorly anchored by laterals, may suffer also. Winter annuals should be planted at the earliest possible date in order that they may be strong enough to endure heaving when it occurs. If the mechanical injury is insufficient to kill the plants by tearing the tissues apart, exposure of the roots to desiccation may do so. Heaving may cause each grass or sedge plant in a wet area to become lifted as a conspicuous mound (706).

Frost-churning of soils (358). In regions where soils freeze deeply in winter, and on habitats within those regions where percolation is impeded by high water tables, shallow bedrock, or permafrost, the layer that is subject to annual freezing and thawing is very unstable. Isodiametric areas, often closely spaced, hence hexagonal, bulge upward in winter to form low mounds. When these mounds begin to thaw in spring the surface becomes a film of soft mud which allows stones to slide a little centrifugally. More stones are brought to the surface by frost-lifting each year, so that eventually all stones in the isodiametric *frost boil* work their way to the surface and then to the margin where they accumulate. Close spacing of the frost boils thus produces a reticulated *stone net* (Figs. 49, 50). When the process occurs on sloping ground the nets may become so elongated across the contour as to be aptly described as *stone stripes*. The churning action of this process tears root systems apart and kills seedlings while still young, so that the frost boil is not permanently colonized until activity ceases. The cause of origin, longevity, and the forces that bring about senescence are unknown, but the phenomenon is common in arctic and high alpine regions, exhibiting many varieties of expression.

In addition to stone nets and stripes, frost churning commonly produces special microrelief. Unequal freezing may force up permanent *mounds.* Thawing soil often flows as a sheet and may form *turf-banked terraces* behind plant communities that impede flow. In arctic and alpine tundras frost is the most important agent of erosion, for it affects almost the whole area, moving silt as well as boulders, whereas streams are usually small and clear, hence can do little cutting.

FIGURE 49. Young frost boils in which the centrifugal migration of stones is not yet manifest, although vegetation has been almost eliminated from the areas uplifted in winter. Swedish Lapland.

FIGURE 50. Old frost boils that have become senescent and well vegetated. Cool, moist crevices among the segregated stones of the net, now that movement has ceased, provide habitats for certain mosses and lichens which were not formerly represented in the area. Swedish Lapland.

High-Temperature Injury

Aside from its role in desiccation, and in bringing about a disbalance between respiration and photosynthesis, high temperature per se can injure and kill protoplasm. When temperatures rise above the maximum for growth a plant enters a quiescent state, sometimes accompanied by chlorosis, and with further heating a lethal level is eventually attained. Declining physiologic activity above the optimum is widely held to be a consequence of inactivation of enzymes or other proteins, yet this does not seem adequate to account for the low optima of shade plants in cold climates and of arctic marine algae. Possibly there may develop serious imbalances resulting from differences in the Q_{10} of metabolic processes. Leaf functions become impaired at about 42°C. Lethal temperatures for active shoot tissues are generally in the range of 50–60°C.

The principal adaptational features which protect plants against high-temperature injury are: (a) smallness and thinness of leaf blades, coupled with a high rate of transpiration, which prevent leaves exposed to the sun from becoming more than about 5°C warmer than air, and may possibly account for the fact that they are seldom injured by heat; (b) a vertical orientation of leaf blades which always reduces the tissue temperatures at least 3 to 5°C below that of leaves turned at right angles to the sun's rays; (c) whitish color of surfaces which reflects rays that would otherwise be absorbed and become heat energy; (d) a covering of dead hairs or scurf which shades living cells; (e) a thick, corky bark which insulates the phloem and cambium; and (f) a low moisture content of the protoplasm, and a high content of osmotically active substances. Any development of resistance to frost or drouth usually involves an increase of resistance to heat injury (459).

Stem girdle. Stems are more frequently injured by high temperature than leaves or roots. The high level of heat obtaining just at the soil surface often scorches tender stems at that level, a phenomenon known as *stem girdle.* Seedlings and herbs may be killed by contact with hot soil. Injury is first manifested by a discolored band a few millimeters wide, followed by a shrinkage of the tissues thus discolored. In flax, the stem immediately above the injured zone swells to form a *heat canker.* Stem girdle apparently causes death by killing the conductive and cambial tissues, or by injury that allows pathogens to become established.

Injury to coniferous tree seedlings begins at soil temperatures of about 45°C, but some species can tolerate prolonged contact with soil at 70°C (176). In part, at least, the better survival of seedlings under shade or in the protective shelter of rocks, logs, shrubs, etc., is to be explained on this basis (420). Lithophytic lichens of arid climates must possess the greatest degree of resistance to this type of excessive heating. Like seeds, their superior tolerance is undoubtedly a result of their low degree of hydration during the dry season. On the other hand, some bacteria can endure boiling water, so that water content is not necessarily associated with heat tolerance.

Sunscald. Old woody stems and other plant organs are subject to another type of heat injury. During the dormant season bark temperatures on the sunny sides of trees may rise so high during the afternoon and drop so rapidly in the evening as to cause the death of the living tissues. Cambial temperature on the sunny side of a peach tree has been observed to rise as high as 30°C while the air temperature remained slightly below 0°(219). Owing to the movement of clouds, changes of 10°may occur in the insolated tissues within a few minutes. Under such conditions the injury (Fig. 51), commonly called *sunscald,* is more a result of the degree of fluctuation than the intensity of heat (368). Such practices as whitewashing or wrapping trunks, or pruning so as to leave a few lower branches for shade on the exposed sides of trees, all tend to reduce sunscald. White paint alone can reduce cambial temperatures by 40°C.

FIGURE 51. Sunscald injury on the southwest side of a young tree of *Pinus monticola*. Stems become immune to such injury after a thick layer of dead cork tissue has accumulated.

Fleshy fruits are often scalded when subjected to strong insolation after experiencing a period of abundant moisture. Also, the tips or margins of leaves may die back as a result of heating.

It is to be emphasized that sunscald is primarily a result of rapidity of temperature change, for it may take place when the plant is not exposed to either the absolute maximal or minimal values that the protoplasm can tolerate. The importance of sudden and violent temperature changes to protoplasm is illustrated further by the fact that, before the era of colchicine, such treatments were the most effective known methods of inducing polyploidy in plants.

Other types of heat injury. High temperatures are detrimental to dormant seeds, in part at least because the higher the temperature the more rapid the exhaustion of food reserves by respiration. For plants of temperate regions, dormant seed usually should be stored at 0 to 5°C (36, 76).

High temperature tends to stimulate the opening of stomata, even when leaves are wilted. This response is sluggish in drouth-resistant plants but prompt in others, and it is possible that drouth resistance is largely a matter of the relative ability of the plant to keep its stomata closed under the stimulus of high temperatures. In *Poa annua* temperature rising to 29°C in a glasshouse was enough to kill anthers, but did not impair the function of pistils.

The roots of plants in undisturbed soil are generally deep enough to escape injury due to excessive heat, but where plants are grown in containers, the direct insolation of the containers becomes an important factor because roots tend to be concentrated in a layer against the inside walls. Shielding such containers from sunlight, or immersing them in a water bath, will guard against this undesired effect.

Living tissue can be conditioned to withstand relatively high temperatures if given a period of exposure to near-lethal levels previously—a phenomenon quite the complement of frost hardening.

Heat Stimulation

Just as some plants require low temperature to initiate flower primoidea, others require high levels (153, 180). This phenomenon occurs chiefly in climates where the summers are cool, and since the average temperature of the summer season correlates best with the vigor of flowering, the phenomenon is probably a matter of temperature efficiency governing the accumulation of abundant food reserves.

Temperature and Plant Diseases (133)

The ability of a parasitic fungus to gain entrance into as well as to

develop within a host organism is often strongly conditioned by temperature. For example, at temperatures below 13°C the seedlings of most
strains of maize become very susceptible to disease, whereas flax becomes susceptible to *Fusarium* wilt only at temperatures above 14°C.
Host plants commonly extend into climates where temperature restricts
their parasites (798), and it is often possible to subject a diseased plant
to temperatures lethal only to its parasite.

The Time Factor in Temperature Relations

The rate of temperature change is more important than the degree
of change. Sudden changes tend to be more deleterious to plants than
slow changes of the same magnitude, apparently because the protoplasm
requires a certain amount of time to adjust itself to new temperature
levels (Fig. 52). If dormant woody plants are chilled slowly then warmed
slowly, they can withstand temperatures far below any levels attained
even at the poles of the earth (824). It will be recalled that the rate
of cooling, rather than the absolute minimum, determines whether or
not the plant is injured by the formation of intracellular ice or the development of stem lesions. Likewise, the temperature extremes that result
in sunscald are no greater than the extremes endured at other times
when the change from high to low temperature is more gradual. Tissues
of an insolated leaf may be killed in spots where drops of cool water
fall on them (224). Injury associated with freezing is increased either
when temperatures drop rapidly, or when the tissues are thawed rapidly.
Under natural conditions such injury may result from the fact that the

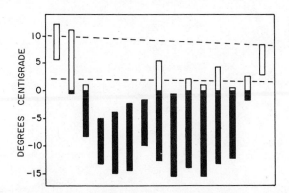

FIGURE 52. Diurnal ranges of temperature (vertical bars) during
November 6–20, 1955, at Centralia, Washington. In this period an
invasion of cold polar air sent even the daily maxima well below the
normal daily range (dashed lines) for this season, and native plants
as well as aliens were extensively damaged. Shading of the columns
indicates temperatures below the freezing point of water.

conditions bringing about rapid thawing at the same time cause a sudden increase in transpiration stress. Although the relative importance of temperature and concomitant variables is difficult to assess, it may be true that certain plants confined to protected habitats are excluded from others because of more rapid temperature fluctuations there.

Within certain limits rapid rates of temperature change may be of benefit to plants. Certain species will not flower unless the plant experiences a rapid drop in temperature, the same differential being without effect if accomplished over a longer period.

It has been demonstrated that, when a plant is transferred from one environment to another with a different temperature, the rate of physiologic processes quickly changes, but the new level of activity is not long maintained. Also, when a lot of seeds is divided and placed under conditions suitable for germination but at different temperature levels, a well-marked optimum is evident at first, but as time advances satisfactory germination is secured over a much broader range (218). Consequently germination percentages and cardinal temperatures must always be defined in terms of duration as well as level (Fig. 47).

Time also has an important bearing upon the degree of injury sustained at extreme temperatures. A plant may withstand an extreme of a given intensity for a short time, whereas the same temperature maintained for a longer period would prove fatal (Fig. 53). The giant cactus (*Cereus giganteus*) can tolerate no more than 18 hours of temperature below freezing, and this restricts its natural distribution (700). This phenomenon may be explained in part by the fact that, owing to undercooling, plant tissues may not freeze until critical temperatures have prevailed for some

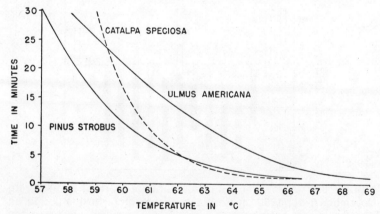

FIGURE 53. Time required to kill cortical parenchyma cells at different temperature levels. [After Lorenz (481).]

time. The length of time in the frozen state, within certain limits, seems to be of no great importance. In this connection the time factor can be expressed as the number of consecutive hours at or below a certain level, or as *hour-degrees* of freezing weather in which intensity as well as duration is taken into consideration. Sometimes brief interruption of a cold period may nullify its effects (721).

Temperature and Plant Geography

In the preceding discussions evidence has been presented that temperature limits the ranges of plant species directly or indirectly as follows:

1. Poleward or upper altitudinal limits are set by:
   a. Extremes so low as to kill the protoplasm.
   b. The relative efficiency of summer heat in respect to the accumulation of carbon compounds.
   c. The relative adequacy of summer temperature to stimulate reproduction.
   d. Parasites that become active only at low temperatures.
   e. Adverse combinations of day and night temperatures.
2. Equatorial or lower altitudinal limits are set by:
   a. Extremes so high as to kill the protoplasm.
   b. The photosynthetic-respiration relationship.
   c. The inadequacy of winter temperatures to stimulate germination, growth, and sexual reproduction.
   d. Parasites that become active only at high temperatures.
   e. Adverse combinations of day and night temperatures.

In addition to these major aspects of temperature, numerous special but indirect relationships may be significant. For example, the hypothesis has been suggested that forest extends into arctic tundra only along river valleys because of the snow accumulations there which protect the soil from freezing deeply at an early date in winter.

Actually many other factors, climatic, edaphic, or biotic, frequently prevent a plant from attaining its theoretical maximum range with respect to temperature requirements. Significant in this respect is the fact that the geographic area over which individual crops may be grown usually takes the form of a belt elongated in an east-west direction, but such a distribution pattern is much less pronounced in regard to the natural ranges of plant species. This indicates that cultural practices can accomplish much in the way of making other conditions suitable for plant growth, but temperature limitations are less easily overcome.

It has long been known that when equal areas are considered, the number of plant species increases in an equatorial direction from both poles. Although other factors such as relative stability of climate and land masses have undoubtedly played an important part in bringing about this condition, the north-south temperature gradient may have had something to do with the rate of evolution. Under experimental conditions, at least, the rate of mutation has been found to increase directly with temperature.

Temperature Measurement and Control

Our understanding of any environmental factor is fundamentally conditioned by the degree of refinement in the methods employed in measuring that factor. Not only must the number and distribution of measurements be adequate from a statistical standpoint, but also in the interpretation of data it is essential that the limitations of each instrument be understood. It is easy to accumulate great quantities of data, but it is not easy to determine with precision the temperature of most objects, surfaces, and media. There follows a brief discussion of the principal instruments used in measuring temperature in ecologic studies, together with suggestions concerning their application.

Simple Thermometers

Ordinary mercury-in-glass thermometers can be used for instantaneous determinations of temperature in places where there are no sharp temperature gradients. They can be inserted in water, in soil, and into thick plant organs. If they must be removed to be read, this must be done with alacrity, for the mercury column begins to change noticeably within a second or two. It should be noted if the thermometer is intended for total immersion or partial immersion.

These thermometers should never be used for soil temperatures except at some depth on account of the sharp temperature gradients near the surface (Figs. 40, 42). A pointed steel rod with a diameter slightly larger than the thermometer is useful in making a hole in the soil to the depth to which temperature is to be measured. If used in tubes inserted in the soil (152, 264), the bulbs can be insulated with a coating of paraffin to reduce their rates of temperature change when removed for reading.

Thermometers are generally calibrated quite carefully by the maker, but the frequency with which these instruments develop unexplainable inaccuracies serves as a caution against using any one of them indefinitely without checking its precision.

True air temperatures can be approximated most closely only if the sensitive element of the apparatus used is protected from direct sunlight and reflected light, as in a well-ventilated and double-roofed shelter large enough so that the sensing element is at least 3 dm from any side of the box. Alternatively, thermometers may be suspended horizontally below two flat sheets of aluminum, with another sheet below to intercept ground radiation. However, the points must not be overlooked that in ecologic work the tissue temperatures are the critical values, and plant organs are subject to strong radiational loss of heat at night as well as direct and indirect insolation during the day. Temperatures as obtained in an instrument shelter are buffered against radiation and insolation so that they never rise as high or drop as low as plant temperatures. For this reason some have proposed using more direct exposure of instruments with a view to getting a closer approximation of tissue temperatures.

Maximum-Minimum Thermometers

A common and widely used apparatus that makes automatic records of maximal and minimal temperatures consists of a pair of thermometers, one mercurial and one alcoholic. The bulb of the former is mounted a few millimeters higher than the stem; the latter is tilted the same amount in the opposite direction. The bore of the mercurial member is constricted just above the bulb to such an extent that, although rising temperature can still force mercury from the bulb into the bore, the extruded mercury cannot return. Thus this thermometer registers the maximal temperature until it is reset by centrifugal whirling, which forces the excess mercury in the bore back into the bulb. Since the mercury column is broken at the constriction, the thermometer must be gently turned to an upright position to read the correct value.

In the alcoholic member there is a short, thin rod of metal in the extremity of the liquid column. This rod, or index, is light enough so that a receding meniscus will draw it back toward the bulb as temperature drops, but there is no force to cause it to move back if the alcohol should subsequently expand. The position of the distal end of the index then indicates the minimum temperature, and the instrument is reset by tilting so that the metal index slides down the bore and rests against the meniscus once more.

Another (Six's) type of maximum-minimum thermometer has a U-tube with a central section of mercury, colorless creosote filling the remainder of the bore except for a bubble of air at one end to accommodate expansion and contraction. The mercury column moves forward and backward with each change of temperature, pushing an index at either end. Friction prevents the indices from moving back as the mercury recedes so that one index records the maximal and the other the minimal temperatures. A magnet is used to pull the indices back against the mercury column and reset the instrument. Costing approximately half as much as the pair of thermometers described above, and being slightly less fragile, Six's thermometer is generally the more satisfactory for field work.

Although usually employed in connection with air temperatures, maximum-minimum apparatus has also been used to study water and soil temperatures (264).

The "mean temperatures" reported by the U.S. Weather Bureau are in reality medians based upon daily readings of a pair of maximum-minimum thermometers. As approximations of true mean temperatures these leave much to be desired, for the median is usually higher than the mean (154).

Thermographs

A continuous record of temperature is frequently an essential part of experimentation, for, as previously discussed, the duration of different temperature levels is often of much significance. *Thermograms* obtained by instruments called *thermographs* provide the only adequate data for calculating temperature efficiency, and they are the perfect records of the degree of constancy secured in "constant"-temperature apparatus. This instrument is also invaluable in obtaining temperature records at stations that cannot be visited more often than once a week.

The most popular form of air thermograph at present has a sensitive unit consisting of a flattened, curved, metal tube filled with liquid. Changes in temperature alter the curvature of this tube. One end of it is fixed, and the other motivates, by lever mechanism, a pen that moves across a paper chart furnished by the instrument maker. The chart, marked off in degrees, hours, and days, is wrapped around a drum. The drum is motivated by a clock movement and makes one revolution in seven days. These instruments are advertised as being accurate to 0.28°C, but it is essential that a maximum-minimum thermometer be kept in the instrument shelter so that the degree of inaccuracy can be recorded and suitable corrections applied to the data taken from the thermogram. Instruments are on the market that make two records at once on the same chart, such as air temperature and soil temperature or air temperature and relative humidity. At each visit the chart must be changed, any correction data recorded on the thermogram, and the actual date written on the new chart as it is installed. The clock must be wound and then the drum rotated by hand until the pen is on the correct line. If necessary the pen should be refilled with ink (a glycerine solution of dye is satisfactory); then capillary connection must be established between the ink drop and the paper through the leaves of the pen. Whenever it is necessary to readjust the pen to indicate a more correct temperature, the instrument should first be allowed to attain a perfect equilibrium with the air.

The sensitive unit of the soil thermograph is a cylindrical tube about 27 mm in diameter and several decimeters in length. Because of its size it should be used only below about the 2-dm level. The large bulb, filled with toluol, is connected by means of a flexible cable to the recording mechanism, and, since both the cable and the recording mechanism are sensitive to temperature changes, accuracy is insured only when the cable is buried at approximately the same depth as the bulb, and the drum housing is placed in a covered box sunk in the earth (523, 770). The pen is adjusted after observation of the soil temperature at the level of the bulb some time after the bulb has been in place.

Thermocouples

Because the sensitive elements of most of the instruments described above are at least several millimeters in diameter, none is suited to measuring the temperature of small areas where the gradient is steep. However, a *thermocouple* is adequate for measuring the temperature of the cambium layer, the inside of a leaf, the very surface layer of soil

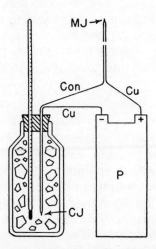

FIGURE 54. Diagram of a thermocouple made of copper (Cu) and constantan (Con) wires connected to a potentiometer (P). The cold junction (CJ) is in a thermos bottle of chipped ice. Below the measuring junction (MJ) there have been omitted long segments of the wires which make the apparatus more convenient to use.

grains, etc. A thermocouple is made by fusing together in a Bunsen flame the ends of two fine wires of dissimilar metals, usually copper and constantan (an alloy of copper and nickel). At the junction of these metals a difference of electrical potential is set up, and the magnitude of this difference is proportional to the temperature of the junction. Two such junctions are connected in series, one being kept at a known constant temperature (such as 0°C in a thermos bottle of distilled water containing crushed ice), and the other being placed where the temperature is to be measured (Fig. 54). The thermocouple is also peculiarly adapted to measuring the temperature of water at considerable depth, and a more substantial type of junction can be made which can be thrust into the soil at various depths (523).

A potentiometer is used to measure the differences in potential between the two junctions, and temperatures can be read off prepared tables.

Thermistors (510, 623, 629)

*Therm*ally sensitive res*istors* (*thermistors*) are instruments somewhat similar to thermocouples in that a potentiometer is used to measure the resistance offered to an electric current passing through an extremely small globule composed of a sintered mixture of metallic oxides. Advantages claimed for this instrument over the thermocouple are: (*a*) length of wires between measuring junction and potentiometer has no effect on readings, (*b*) a reference junction at known temperature is not needed, (*c*) greater sensitivity, and (*d*) lower cost of apparatus.

Electric apparatus is available for making continuous records of temperature at thermistor or thermocouple junctions located at different places nearby (445).

Sucrose Inversion

If a sterile, freshly prepared solution of sucrose is sealed with some enzyme in a small glass capsule, inversion to monosaccharides is subsequently proportionate to temperature efficiency. A polarimeter is used to compare the relative amounts of monosaccharides at the beginning and end of a time period. This method of temperature evaluation has the advantage of automatically integrating the different efficiencies of different levels on the temperature scale, but has the disadvantage that the Q_{10} of inversion is not variable as is the Q_{10} of plant processes. Furthermore, the method is comparative, and the results cannot be equated with the usual type of temperature data.

Melting-Point Indicators (708)

Wax shavings, or bits of other solids with narrow ranges of melting point, may be laid on insolated surfaces to indicate by changes in their shape whether or not temperatures rise to specific levels.

Biologic Methods

Despite the fact that the seasonal progress of plant development is influenced by a number of factors, phenology constitutes a method of evaluating the earliness or lateness of seasonal temperatures which has certain advantages over direct measurements of weather variables. From phenologic data maps can be drawn with lines (*isophenes*) connecting locations where plants are in the same stage of development at the same time. This biologic method has been found very useful in Europe but has not yet been applied to any extent elsewhere. In using the method species must be selected that are insensitive to daylength (see discussion in Chapter 4).

Temperature Control Indoors

Temperature control for seeds and heterotrophic organisms which do not need light is not particularly complicated provided that refrigerating and heating units with suitable thermostats are available. The same applies to studies where only soil temperature need be controlled (516).

Studies of the influence of air temperature on the shoots of green plants, however, demand units of larger size. Some source of light must be provided during part of each day, and the relative humidity must be controlled (89, 278, 760, 827). Relative humidity is regulated in accordance with the various temperature levels in such a manner that the saturation deficit is uniform throughout.

When it is desirable to alternate temperatures from a low level at night to a higher one during the day, a satisfactory practice is to move the plants back and forth between two chambers set at the desired temperatures. The high-temperature period in this case should never exceed 8 hours (313).

Temperature Control Outdoors

On a strictly local scale significant temporary modifications of temperature are possible. For the most part this has been practiced to benefit economic plants.

Soil temperatures can be raised with a covering of clear plastic which transmits much energy that is converted to sensible heat and trapped beneath, as in a glasshouse. In Alaska it has been found effective to make deep furrows in an east-west direction, blacken the south slope with coal dust and erect a vertical sheet of aluminum as a reflector along the summit of the ridge. Heat-demanding plants are grown on the warm southerly exposure, with cold-tolerant species on the opposite slope (51). Rows of brush or paper tilted at a 30° angle toward the equator and located about 30 cm poleward of a row of plants can increase temperatures materially on a calm day and give some frost protection at night. Paper cones ("hot caps") are often set over seedlings with the same effect. Covering the soil with flagstones is used to raise soil and air temperatures in some European vineyards (436).

Soil temperatures may be lowered by plowing furrows in a north-south direction so that when the sun is at zenith no surfaces are oriented to absorb the maximum heat. Straw mulches lower soil temperature and minimize its diurnal fluctuation, but they raise air temperatures during the day and lower them at night.

Modifications of air temperature on a somewhat larger scale are often attempted to prevent damage to orchards and other crops from light frosts which result from temperature inversion (866). For a time small smoky fires (smudges) were maintained all night at regularly spaced intervals over the crop area in the hope that much of the heat reradiated from the soil would be reflected downward again by the pall of smoke. Electric wind machines have been found much more effective than smudging, through mixing air and so diluting the cold layer next to the ground. One wind machine may protect up to 10 ha of orchard. A third method is to spray water over the crop as the temperature drops below freezing, which takes advantage of the heat already in the water plus that released as the water starts to freeze. This method is practical only where water is plentiful. In England approximately 18,000 1 ha^{-1} hr^{-1} are used for this purpose. However, spraying becomes hazardous if the water freezes and stems become so weighted with ice that they break.

Orchard heaters operated to provide heat rather than smoke, are effective for a mass incursion of cold air (322). Oil heaters in California have burned up to 227 million liters of oil in a year.

The temperature of a creek flowing through a forest may be increased as much as 6°C during warm summer months if the forest is removed by fire or logging (103). However, strips of shrubs or trees that may escape burning, or may be left by judicious logging practice, can effectively prevent this somewhat temporary but biologically significant stress on aquatic organisms.

Man's Influence on
the Temperature Factor

The local influences on temperature considered above are strictly ephemeral. But man has additional influences on the heat balance that are more widespread and more difficult to control.

As a city develops air temperatures within its area increase (440). Smoke particles and gases discharged into the air reduce reradiation of heat, wind dissipation of heat is minimized, and in cold climates the loss of heat from factories, buildings, engines and inhabitants all add materially to these effects. Roofs, pavement and walls absorb heat without being cooled by transpiration as are the vegetation surfaces of the countryside. Owing to all this added warmth, plants near the cold limits of their geographical ranges are much more likely to prosper in city environment.

The current trend toward an increasing use of nuclear fuel as an energy source creates important environmental problems, foremost among which is that of dissipating waste heat. Tremendous quantities of cool water taken from rivers or lakes have heretofore been simply dumped back in heated condition regardless of biologic consequences. Nuclear power interests would like permission to raise the temperature of river water 5°C in winter and 2°C in summer. The importance of such a change is set in perspective when it is recalled that during the last major glaciation average temperatures dropped only 5°C lower than at present, yet this caused plant ranges to extend equatorward up to as much as 500 km in extra-tropical regions. Aquatic organisms have even narrower ranges of temperature tolerance so that the dumping of hot water into a river usually kills the normal biota. If the supply of hot water were constant a new biota tolerant of the warmth would develop, but the discharges are not constant so the stream becomes inhabitable by very few species, mainly bacteria and blue-green algae. Warm water has limited capacity to hold dissolved O_2, and tends to spread out as a layer over cool water below. With the warm layer thus sealing off the cool water from an atmospheric source of O_2 replenishment, and at the same time increasing the rate of respiration, anaerobic conditions arise. Secondary changes in CO_2 pressure, solubility of toxic compounds, etc. then accentuate the abnormality of the thermally polluted stream.

To dissipate waste heat by spraying hot water into the air is wasteful of water and alters the microclimate as does an alternative use of a heat exchanger. Potential uses of hot water in combating frost damage to crops and lengthening the frost-free season are too limited in temporal or spatial need to be significant. Finally, this burgeoning release of heat into the biosphere could have sufficient influence on global climates to upset biological relations for plants and animals everywhere. Energy release is doubling every 10 years.

Another of man's influences on temperature that clearly involves the entire atmosphere of the earth is the indirect consequence of converting all the carbon in millions of years' accumulation of coal, oil and gas into CO_2. This gas, acting as a screen against the reradiation of heat from the earth's surface, is apparently responsible for raising the average temperature of the atmosphere approximately 0.56°C since the start of the industrial revolution (261). This effect too is increasing geometrically.

4

The Light Factor

The energy necessary to sustain life on earth is derived from sunlight directly by green plants, or indirectly by other organisms which, except for chemosynthetic bacteria, must eventually depend upon organic compounds synthesized by green plants. Chlorophyll, through its ability to absorb radiant energy from the sun and convert it into chemical energy contained in simple sugar molecules, provides the essential connecting link between nearly all living organisms and solar energy. In addition, light exerts many stimulating effects upon plants, especially upon the differentiation of tissues and organs. In fact, light is rivaled only by water in its influence upon the morphology and anatomy of plants. Less evident, but certainly none the less important, are the effects of light on the physiologic processes and chemical composition of plants.

Subdivisions of Radiant Energy and Units of Measurement

All radiant energy the earth receives from the sun is in the form of electromagnetic waves varying in length from about 5000 to 290 millimicrons. This series, the *solar spectrum,* can be conveniently subdivided on the basis of wavelength as follows.

Light

Wavelengths between approximately 750 and 400 millimicrons comprise that segment of the solar spectrum which passes through the earth's atmospheric envelope with the least diminution of energy. Capitalizing on this, plants have evolved special pigments (chlorophylls, carotenoids, flavins, etc.) to exploit this abundant energy

211

source, just as animals have exploited it by developing eyes. This part of the spectrum is called *light* or *luminous energy* because it alone can be seen with the eye (Fig. 55). Approximately 40–60% of the total energy of solar radiation reaching the earth's surface lies within this narrow range, depending on whether the climate is rainy or desertic. Green plants grow normally only when exposed to a combination of most of the light wavelengths (616, 693).

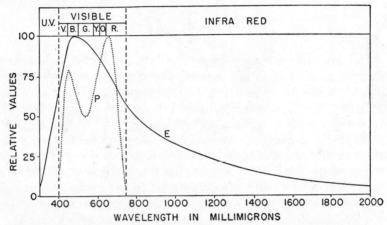

FIGURE 55. Distribution of energy (E) in the solar spectrum at the earth's surface, and the relative rates of photosynthesis (P) of wheat at different wavelengths of light of equal intensities. The inefficiency of green light is simply a consequence of reflection and transmission reducing utilization (812). [E after Fowle (250); P after Balegh and Biddulph (37).]

When sunlight is passed through a prism it is dispersed into a series of wavelengths exhibiting different colors as follows: red 750–626 millimicrons, orange 626–595, yellow 595–574, green 574–490, blue 490–435, violet 435–400. All these colors making up the *visible spectrum* affect photosynthesis, but yellow and green are utilized very little, the principal wavelengths absorbed lying in the violet-blue and orange-red regions (Fig. 55). Phototropism is governed chiefly by blue-violet wavelengths.

Infrared Radiation

In this category are included wavelengths longer than the longest to which the eye is sensitive—hence the name infrared, meaning "below red." Animals can detect the presence of this type of energy only by the sensation of heat which it produces, the longer the wavelength the greater being its heating effect.

Infrared rays coming from the sun are no longer than 3000 millimicrons, and this is referred to as the *near infrared* range. *All* solar radiation is eventually converted to wavelengths exceeding 3000 millimicrons (*far infrared*) then reradiated back into space. Far infrared is important biologically only for its heating effects, which were the subject of the preceding chapter. Near infrared is important to plants through its influence on hormones governing germination, responses to daylength, etc. (812.)

Ultraviolet (Actinic) Radiation

Wavelengths less than 390 millimicrons are too short to be seen (ultraviolet meaning "above violet"), but they are very active in certain chemical reactions. Plants do not require these wavelengths for normal growth and in general are not injuriously affected by them (116). Owing to the screening effect of ozone in the atmosphere they comprise but about 2% of radiation at the earth's surface. Furthermore the epidermis is essentially opaque to these rays. For these reasons ultraviolet radiation is not particularly important except to certain of the lower plants.

To a certain extent it and the shorter wavelengths of light as well, tend to promote the formation of anthocyanins, are responsible in part for phototropic phenomena, and by inactivating growth-promoting hormones check stem elongation.

For the sake of convenience, and because of its relatively minor importance, the effect of ultraviolet on plants will be considered along with light in this chapter.

Units of Measurement

Light can be expressed in terms of the gram-calorie unit of energy, if suitable filters are used to exclude infrared and ultraviolet. It can also be expressed in terms of intensity with reference to the illumination produced by a *standard candle.* Actually candles are no longer used, having been replaced by less variable standards, but the term *candle-power* has been retained for the units. The amount of light received at a distance of 1 m from a standard candle is called a *lux* (*L*) or *meter candle* (*M.C.*). The light intensity at 1 foot from a standard candle is called a *foot-candle* (*F.C.*), which equals 10.764 *L*. By common agreement of world scientists the lux has been accepted as the standard international unit for expressing light intensity. Light energy can be expressed as lux-seconds or lux-hours, for according to Bunsen and Roscoe's law a definite photochemical effect requires a definite amount of light energy regardless of its distribution in time.

Aside from the above type of absolute expression of light intensity, it is often convenient to express the amount of light received under water or vegetation as a percentage of the amount received where these screening influences are not operative.

The basis of comparison here (full sunlight) is not a fixed intensity but varies greatly with the season, the time of day, humidity, and other atmospheric factors (73). However, this limitation is more important to the physiologist than to the ecologist, for plant response to specific light intensities under laboratory conditions is not adequate to explain shade tolerance in the field. Percentage values made in the field automatically nullify the influence of a swarm of variables other than light energy which influence plants growing in the shade, and which are too complex to duplicate under laboratory conditions. It is the *balance* between production and use of metabolites, as influenced by *all* shading influences, that is critical for field survival.

Relative Importance of Quality of Light

As with the temperature factor, light varies in intensity and duration, but unlike temperature it also varies in quality, i.e., the proportions of red, yellow, blue, etc. For practical reasons, however, this complication remains chiefly of theoretical importance in plant ecology. It is not feasible to place much importance upon variations in wavelength under natural conditions for the following reasons. (*a*) Although it is true that variations in quality affect plant processes differently, each process is somewhat sensitive to all wavelengths of light, so that in ecologic work analyses of wavelength composition are difficult to interpret. (*b*) The influence of light quality upon plants differs so much from one species to another that, with the exception of those generalizations already mentioned, no other physiologic roles of different portions of the spectrum have yet been established. (*c*) With land plants growing under natural conditions the variations in light quality have never been demonstrated to be great enough to be critical as an environmental factor. For these three reasons the intensity and duration of light are the variables of chief ecologic importance.

Spatial Variations in Light Intensity

Effects of Atmosphere

Atmospheric gases, chiefly nitrogen and oxygen, absorb and disperse a small portion of the shorter wavelengths of light as it passes through the gaseous layer enveloping the earth. * The higher the elevation of a surface above sea level, the thinner the layer of air remaining above it, and the brighter the light. Mountain summits extending to an elevation of a kilometer are exposed to a maximum intensity of about 129,000 L whereas only 107,000 L are received at sea level. However, unless strongly influenced by cloud and fog, differences in light intensity due to elevation are not of sufficient magnitude to be of critical importance to plant life.

In contrast to N and O, all moisture contained in the air, visible as well as invisible vapors, exerts a powerful screening effect. For this reason the intensity of light is much greater in dry than in humid climates and is very low where cloud and fog are abundant as along the central and northern Pacific coast of North America or in tropical mountains. On a cloudy day light may be reduced to 4% of the normal intensity. Under these conditions a relatively high proportion of the longer light rays and infrared rays are absorbed by the moisture, and the shorter light rays and ultraviolet are scattered. Light scattered by gas molecules and water droplets becomes *diffuse light* or *sky light* as contrasted with *direct light*. On clear days diffuse light comprises but 10–15% of the total light, whereas on overcast days it may comprise up to 100% of the total.

As pointed out in connection with heat, the angle of the sun's rays with respect to the earth's surface at a given point determines the distance which the rays must pass through the atmospheric blanket to reach that spot and likewise determines the amount of surface over which a given amount of light is spread. Latitudinal variations in light intensity due to the height of the sun above the horizon are very important. In equatorial regions light is most intense and contains the highest proportion of direct sunlight. Progressing toward the poles the intensity decreases and the percentage of diffuse light increases.

Effects of Layers of Water

Submerged plants are subjected to weaker illumination than terrestrial ones, for part of the light is reflected back at the water surface, and of the remainder much is absorbed by the upper layers. The greenish or bluish color of bodies of water indicates that the principal wavelengths reflected are at the short end of the spectrum, especially between 420

---

* It is due to this absorption that ultraviolet wavelengths shorter than 290 millimicrons never reach the earth's surface.

and 550 millimicrons. When the surface of water is rough reflection is increased several times.

As light penetrates water the intensity decreases geometrically as depth increases arithmetically. Even in perfectly clear water only 50% of the light impinging upon the surface penetrates as far as 18 m, and at 120 m there may be barely sufficient light for feeble photosynthesis (321). More than half the earth's surface lies in perpetual darkness, beyond the reach of sunlight, under half a mile or more of ocean water.

Light penetration through snow often is sufficient to allow hardy plants to begin growth before the snow cover melts in spring. Photosynthesis may take place under as much as 40 cm of snow, although the condition of the snow may be such that much less light gets through (265).

Effects of Suspended Particles

Solid particles dispersed in the air (dust, smoke) or in water (clay, silt, plankton, bog colloids) have a great screening effect.

Turbidity in streams draining arid regions is often inevitable because of the sodium ions which cause colloidal dispersion. Streams from limestone regions tend to remain clear most of the time because of the flocculating action of calcium ions. Owing to accelerated erosion brought about by man, many streams which would otherwise be clear are so heavily laden with colloidal soil particles that submersed plants, and consequently the animal life dependent upon them, have disappeared.

In metropolitan areas smoke may cut off 90% of the light. Even more detrimental are the effects of smoke particles which settle out of the air and accumulate as films on plant surfaces, cutting down the amount of light available to the chlorenchyma. Plants with sticky or hairy surfaces suffer the most. Also evergreens are inherently more vulnerable to this form of injury than deciduous plants which use their leaves for only a few months, although exceptional evergreen plants are tolerant and deciduous plants sensitive to smoke films. In England evergreen conifers cannot grow where more than 19 tons of soot are deposited per square kilometer per year, and deposition in industrial areas may exceed ten times this amount. Even clean glass oriented at right angles to the light rays cuts out about 13% of sunlight, and when allowed to collect grime the interception may rise to several times this value.

Effects of Layers of Vegetation

Leaves transmit about 10% of the light impinging on them, so that most of the light that penetrates through foliage passes between the leaves as sunflecks or as sky light.

In evergreen coniferous forests the quality of light is scarcely affected by this screening action of the foliage, but in winter-deciduous forests the quality of light that has filtered through the leafy canopy differs materially from full sunlight: the proportion of red light is higher, and the proportion of blue and violet is lower (253). As far as is known these differences have no influence in determining the differences in undergrowth between these two kinds of forests, apparently the increase in proportion of long wavelengths offsetting the decrease in short wavelengths. Also most of the light useful to undergrowth plants is received as unmodified sunflecks.

In a complex plant community, the stature of any one plant in relation to that of its neighbors determines to a large extent the amount of light it receives. In a forest only the mature trees of the tallest species ever receive full insolation. Undershrubs receive subdued illumination; herbs and especially epigeous cryptogams grow in still weaker light. When in full leaf the canopy of very dense forests may reduce light to less than 1% of full sunlight. No autotrophic plants can live under such conditions, and the ground frequently remains bare until the death or injury of one or more of the trees improves lighting.

The reduction of light by a canopy of vegetation is very important ecologically, particularly after the intensity is reduced to about 20%, but, because other factors such as wind, relative humidity, soil moisture, and temperature vary concomitantly with reduction in light intensity (Table 6), it is extremely difficult to evaluate the influence of the light factor alone. It must therefore always be remembered that the word shade connotes a complex of factors.

Effects of Topography

The direction and slope of the land surface caused marked variations in the intensity and daily duration of insolation. In general the temperature aspect of this topographic factor is probably more important than the light aspect. However, on steep poleward slopes direct sunlight may be completely lacking at noon so that plants must rely heavily on sky light, which is only about 17% as intense as the light received by a surface level enough to get full direct lighting. To get the maximum possible sunlight a plant must grow where neither topographic nor other features of its surroundings are near enough or high enough to interfere with sky light from any direction.

Effects of Latitude

In a poleward direction the progressively shorter growing season and weaker light are more than compensated by the increasing length of day in summer (465). Since high temperatures reduce the effectivity of brighter light during the shorter days at low latitude, the assimilation accomplished during a summer day at high latitude exceeds that in a day at low latitude. Therefore, considerable agriculture is possible in arctic regions if plants such as cereals, berries and tomatoes, which need only a short growing season and either require or tolerate the long days, are grown. *Phleum pratense* (timothy) profits so much from the longer daily periods of photosynthesis that in a poleward direction its vegetative period is shortened and flowering is progressively earlier in relation to the beginning of its growing season (231).

Temporal Variations in Light

At dawn, at sunset, and in winter, light intensity is weak because the waves are traveling a long distance through the atmosphere and most of the light, especially the shorter wavelengths, is absorbed. When the sun is at the horizon the rays pass through approximately 20 times the thickness of the air they have to penetrate when the sun is overhead. For this reason the diurnal intensity is represented by a broad curve reaching a midday maximum at which time direct sunlight furnishes up to about 83% of the available energy, and tapering rapidly to weak light at either end where sky light becomes the sole constituent (Fig. 18). It is to be noted that effective daylength is not limited to the period between sunrise and sunset, for the sky light available before sunrise and after sunset is of considerable ecologic importance.

Winter sunlight in central North America is only one-tenth as bright as that of summer.

At the equator daylight prevails for about 12 hours out of every 24, in both summer and winter. Progressing toward the poles the length of day (i.e., the *photoperiod*) becomes increasingly longer than 12 hours in summer and increasingly shorter than 12 hours in winter. Even within the range of latitude encompassed by the United States daylength at the summer and winter solstices differs about 2 hours between the southern and northern tiers of states. Above the latitude of about 66° effective daylight in midsummer lasts through the 24-hour day, whereas in midwinter only faint indirect light is seen for a short period centering about noon.

Illumination in microclimates fluctuates hourly. Under a canopy of vegetation the movement of leaves by the wind, together with the varia-

tions in the movements of sunflecks and shadows across the groun\
results in rapid and wide variations in the amount of light energy receive\
at a given point. Thus, the light intensity at a leaf surface may rise abruptly
from 2% to 35% for a few minutes and then drop to its former low.
Because of the influences of wind, changing angle of the sun, differences
in time of day and of season, and the effects of weather, single measurements of light intensity under a vegetation canopy cannot be interpreted
very closely.

The turbidity of water, and hence its transmission of light, varies greatly
with the amount of wind action (125). Likewise the effect of plankton
on light penetration through water is very different depending upon
seasonal variations in light, temperature, and aeration. Wide fluctuations
in water transparency result from these superimposed influences.

Moonlight is bright enough (to 0.2 L) to satisfy light requirements of
certain seeds (408), to promote starch hydrolysis in leaves, to affect
leaf movements of legumes, and possibly to stimulate sexual activity
in certain marine algae (50, 688).

Importance of
Light to Plants

Light affects a number of plant functions, and, like temperature, the
positions of the cardinal values tend to vary with the particular function,
the kind of plant, the stage of the life cycle, and with variations in other
factors.

Photosynthesis

The basic pattern of the plant shoot is directed toward efficiency in
photosynthesis. The stem functions as a support enabling leaves to be
exposed advantageously to light, and the large surface of the thin photosynthetic organs favors the absorption of light energy. The structure
of the spongy mesophyll and the stomatal apparatus allow rapid gas
exchange. Even the fact that photosynthesis utilizes the visible wave-
lenths of radiation most heavily is significant, for this is the region of
the spectrum with the greatest energy values.

Despite the above adaptations, surprisingly little of the solar energy
available is captured in photosynthesis. Most leaves become *light saturated,* i.e., they get all the light they can use, with only about 20% of
full sunlight. Of this quantity only about 20% is then stored in the
sugar molecules produced. The resultant theoretical 4% efficiency is
in turn reduced by the near exhaustion of the CO_2 supply adjacent to

the leaf. After translocation, the conversion of sugar to myriads of other compounds, although not a part of the photosynthetic process, dissipates still another increment of energy, so that the dry matter content of the plant represents only 1–3% of the light supply that was available.

Leaves in the shade produce little photosynthate owing to the limited energy supply, but lacking the limitations imposed by light saturation and often having somewhat more than average supplies of CO_2, their relative efficiency may approach 20%.

Respiration is a never-ending process in every protoplast, by which carbon compounds are oxidized to liberate energy for the maintenance of vital activity. Whenever a plant is not carrying on photosynthesis its dry weight progressively decreases as a result of respiration. The amount of light required for photosynthesis to equal the respiratory use of carbon compounds, i.e., for CO_2 to be neither absorbed nor evolved, is called the *light compensation point*. This value is always higher than the absolute minimum for photosynthesis, ranging from about 27 to 4200 L in higher plants and varying somewhat throughout the year (463). In interpreting such values it is to be noted that they must be maintained throughout the 24-hour day if the plant is to maintain itself. In at least the early part of each day the plant is replacing the carbon compounds it respired during the night (362). In terms of a percentage of full sunlight, the light compensation point for tree seedlings usually lies between 2 and 30% (Table 11).

Table 11

Compensation Points of Some Tree Seedlings Based on 3-hour Tests of Potted Plants. Values Are Percentages of Full Winter Insolation in Maine as Measured with a Thermocouple (108).

Pinus ponderosa	30.6	Celtis occidentalis	11.5
P. sylvestris	28.7	Picea engelmannii	10.6
Thuja occidentalis	18.6	Pinus strobus	10.4
Larix laricina	17.6	Picea abies	8.7
Pseudotsuga menziesii	13.6	Tsuga canadensis	8.4
Pinus contorta var. latifolia	13.6	Fagus grandifolia	7.5
Quercus borealis	13.6	Acer saccharum	3.4

During protracted cloudy weather photosynthesis may lag behind respiration needs and food reserves decline to the extent that animals depending on forage starve (222).

Growth obviously demands synthesis in excess of respiration, so that the minimum requirements for this function are met only when light intensity exceeds or has exceeded the light compensation point. Thus the compensation point for seedlings of *Pinus strobus* is 1830 L, but twice this amount of energy is required to maintain growth. An increase in light causes a small increase in the respiratory rate as well as the photosynthetic rate, and therefore raises the light value of the compensation point. Still, the net effect of light increase at low intensities is highly beneficial, for the rate of photosynthesis increases so much more rapidly than the rate of respiration.

Although high temperatures in the tropics generally reduce plant production through their adverse effects on the photosynthetic-respiration relationship, many plants of hot climates have three special adaptations that tend to offset this limitation. (A) Light saturation approaches or exceeds full sunlight. (B) Their respiration does not increase in the light as in other plants. (C) As the CO_2 supply adjacent to their leaves dwindles through uptake, they are able to continue photosynthesis to much lower atmospheric concentrations (less than 10 ppm) than other plants, which suffer a net energy loss as the CO_2 concentration declines below about 50 ppm * (156, 553).

Unicellular green plants can survive in habitats where light is too weak for multicellular plants, simply for the reason that in the latter group each chlorenchyma cell must not only manufacture the food it requires for itself but must provide an excess for the nonphotosynthetic cells. The compensation points of simple algae are therefore lower than for differentiated plants (751), and they utilize a higher percentage of light energy (as much as 18%). This principle is also illustrated in climates where seasons warm enough for photosynthesis are very short and cool, for alpine and arctic timberlines represent places along climatic gradients where trees become unable to support a great mass of nonphotosynthetic cells, and so are reduced to shrub size.

The expression optimum light intensity used for plants growing in natural habitats does not imply that the intensity is really optimum for photosynthesis except at fleeting intervals. Much of the time and for most of the photosynthetic organs sunlight is either too weak or too intense for maximum assimilation. Optimum means only that under certain combinations of habitat factors the net effect of lighting conditions over a considerable period of time is more favorable for photosynthesis than it is under other combinations.

---

* Species with the special type of photosynthesis (the "C_4" pathway rather that the "C_3" pathway) permitting use of low concentrations of CO_2 can be recognized morphologically by having a sheath of enlarged parenchyma cells heavily charged with starch surrounding their vascular bundles.

Photosynthetic rates may be determined directly by the infrared gas analysis (80) or the alkali absorption techniques (711).

Heliophytes and Sciophytes

Plants may be classified ecologically according to their relative requirements of sunlight or shade. Those that grow best in full sunlight are called *heliophytes,* and those that grow best at lower light intensities are *sciophytes.* Among heliophytes there are some species which, though they grow best in the sun, can grow fairly well under shade. These are called *facultative sciophytes,* and those sun plants which cannot do so are *obligative heliophytes.* Sciophytes likewise can be divided into two groups, depending on their relative ability or inability to tolerate full sunlight.

With most submersed aquatics no difficulty is involved in assessing the importance of the light factor (679, 766), but in terrestrial habitats other factors, especially temperature and relative humidity, vary concomitantly with light intensity and it is very difficult to evaluate light effects alone. In fact, investigators have frequently concluded that the failure of seedlings under light intensities that appear to be critically low is really due to shade factors other than light. Also, in assessing the value of the light factor it must not be overlooked that photosynthesis is not the only function requiring light. In consequence an explanation of the relative difference in success of heliophytes and sciophytes in sun and shade is complex, resting on the net influence of a galaxy of concomitant variables operating on a series of interdependent plant functions.

In many species light requirements for photosynthesis may alone offer sufficient explanation of their superior growth in full sun. Inadequate light energy has a directly detrimental influence when respiratory requirements are not sufficiently exceeded, but the effects may also be indirect. For example, when insufficient light is available and photosynthesis is curtailed the photosynthate is nearly all retained by the shoots and the resultant dwarfing of root systems in turn jeopardizes the continued welfare of the plants. Also, in some instances soil drouth in the surface horizons is more severe in full sun than under shade, but the more rapid penetration of seedling taproots in the sun may more than offset the greater drouth there, with the result that only seedlings in the shade perish (Fig. 56).

Possibly some species grow best in sunny situations because they have high heat requirements. Others may escape destruction by fungi only under the low humidities that accompany bright light. Still others may require high light intensities to stimulate flowering, or to open the guard cells in order to obtain sufficient CO_2. In one study it was concluded that one of the most detrimental aspects of shade environment was the N deficiency brought about by a very slow rate of decay. The fact that certain plants have much higher N requirements than others may account for their exclusion from shaded habitats.

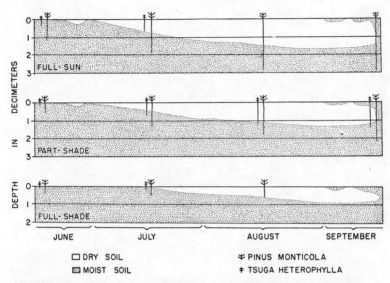

FIGURE 56. Interrelationships among shading, root penetration, and progressive drying of the soil in northern Idaho. [After Haig, et al. (300).] By studying the rate of root penetration of these two species in relation to the differing degrees of soil drouth illustrated in Fig. 29, it is readily seen why both are permanently excluded from vegetation zones represented by profiles A, B, and C in that diagram.

The explanation of the superior growth of sciophytes in shade is just as complicated. In the first place sciophytes must have low light requirements (70). The compensation point for heliophytes may be as high as 4200 L, but for shade plants it may be as low as 27 L. Deep-water algae and the algae and mosses that inhabit caves can grow under very weak light, often under intensities no greater than that of moonlight. One investigator found that the sciophytes he studied differed from the heliophytes in possessing superior abilities to increase their chlorophyll contents under low light intensities. Also it appears that others have concluded that sciophytes are plants producing their maxim leaf area in proportion to plant weight under shaded conditions, with heliophytes behaving the opposite (64). As a rule, heliophytes can carry on photosynthesis more rapidly under full insolation than can sciophytes, and the latter are more efficient than the former under low light intensities (Fig. 57). However, if sun plants are made to grow in the shade for a time, their compensation points lower and they lose their ability to profit by full sunlight.

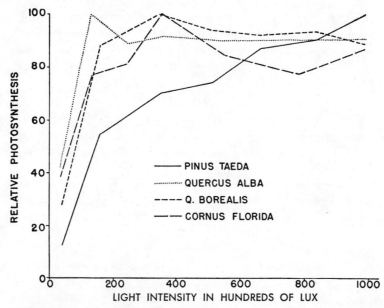

FIGURE 57. Relative photosynthesis of tree seedlings expressed as percentages of maximum observed rates. *Pinus,* an obligate helio-phyte, is relatively inefficient in weak light and assimilates most rapidly in bright light. Seedlings of all these species are dwarfed in the shade of a tree canopy, but this is regularly fatal only for *Pinus* (425). [After Kramer and Decker (425).]

Sciophytes may be at a disadvantage in full sunlight if they cannot manufacture chlorophyll at a rapid rate, for light continually decomposes chlorophyll, the plant remaining green only when it can maintain an equal rate of synthesis of these pigments. This effect of light is illustrated strikingly by *Selaginella serpens,* the intense green color of which fades noticeably during midday. Apparently the decrease in photosynthetic rate which is usually observed when a leaf is moved into bright light is due to rapid destruction of pigments, and the failure of sciophytes in sunny habitats may be due to this, at least in part.

Full sunlight is definitely supraoptimal in many instances where a physiologic explanation is obscure (520). In wheat it has been observed that bright light decreases the acidity of the cell sap and causes chlorosis which is believed due to pH interference with iron translocation (474). It has frequently been observed that when sciophytes are grown in full sunlight they develop an adverse internal water balance so quickly that the stomata close, thus causing photosynthesis almost to cease. The possible effect of the conversion of light waves into heat waves upon the photosynthesis-respiration balance may also be critical.

From the above it is clear that there is no consistent pattern of morphologic or physiologic adaptation differentiating heliophytes and sciophytes. They are critically defined only in terms of their usual habitat.

Adaptations Minimizing Injury from Bright Light

Several characteristics of plants have been interpreted as beneficial in reducing injury from supraoptimal light intensities. In leaves exposed to bright sunlight the disc-shaped chloroplasts tend to become oriented against vertical walls so that only one edge is exposed to direct lighting. Possibly of similar significance are the vertically oriented leaf blades of such plants of sunny habitats as prickly lettuce (*Lactuca serriola*) and manzanita (*Arctostaphylos* spp.). In general heliophyte leaf blades are not flat and are not oriented at right angles to the path of incident rays as are those of sciophytes.

The decrease in chlorophyll content that accompanies bright light has its beneficial aspects, for it results in less light being absorbed and more transmitted, and the excess absorbed light would be converted into heat, which has a detrimental effect on the internal water balance and on the photosynthesis-respiration balance.

It has been observed in many plants that light intensity is directly correlated with the formation of anthocyanins, and these pigments, which are located in the superficial layers of cells, act as a reflective screen retarding the penetration of light into underlying tissue (28). Red pigments reflect chiefly red rays, and, since long rays have the greater heating effect, their reflection greatly reduces danger from overheating. The temperature under red spots on fruit has been observed to be 22°C lower than under comparable green spots. The whitish surfaces of many heliophytes likewise provide as much protection against true light injury as against heat injury. The epidermis of shade plants transmits about 98% of the incident light, whereas the value for sun plants may be as low as 15%.

Motile algae are able to escape excessive illumination by vertical migration, and when the light is bright they are most abundant about a meter below a pond surface.

Light Relations in Plant Communities

In moist climates vegetation tends to be a complex series of superimposed layers, tall trees, low trees, shrubs, herbs, mosses, etc., and a high percentage of the flora are facultative or obligative sciophytes. In dry climates the reverse is true, although even here there are many low plants that grow only in the shade of larger ones (702).

Among tall plants light requirements are of most importance in the seedling stages, for by the time they are mature the foliage occupies an elevated position and receives good lighting. Because the seedlings of different trees have different light and shade requirements (Table 11), some succeed only in habitats where others fail. Heliophytes frequently become established in the full sunlight of the original habitat. By consulting Table 11 it is easy to find an explanation of the fact that when abandoned fields in New England revert to forest the aggressive *Pinus strobus* is able to dominate the initial forest stand, but as shade develops conditions become favorable to the sciophytic *Tsuga canadensis, Fagus grandifolia,* and *Acer saccharum,* and eventually shade completely prevents new seedlings of *Pinus* from starting on the area. Likewise, mixed stands of *Pinus ponderosa* and *Pseudotsuga menziesii* become pure stands of the more shade-tolerant tree within a few centuries. To return to the shade relationships of the New England trees, it should be noted that, although *Tsuga, Fagus,* and *Acer* have different light requirements, these differences are so small in relation to the heterogeneity of shade environment that all three commonly continue to inhabit an area together.

Most maples, beech, red oak, basswood, spruces, firs, yews, arborvitae, and hemlocks are obligative or facultative sciophytes, and most pines, soft maple, bur oak, willows, cottonwoods, aspens, tulip tree, birches, larches, and junipers are obligative or facultative heliophytes. Within these groups sun and shade requirements are of course far from uniform, and when other species are included there is a complete intergradation among them.

It must not be assumed that because shade-tolerant seedlings can live in dense shade they attain normal growth there. They are merely able to survive there for a long period as compared with others, thereby increasing their chances of benefiting from the death of an old tree that would leave a break in the forest canopy. Young trees have been known to persist in dense shade without making any diameter growth for as long as 46 years (767).

Differences in seedling reactions to shade are exceedingly important in silviculture. Forestry in most parts of the world depends on the natural reproduction which follows the removal of each tree crop, rather than on planting the desired species. Obviously the degree of completeness of logging is important because it determines the amount of sunlight available for the seedlings that will become established and provide the next timber crop. If shade-tolerant species are the more valuable, cutting must not as a rule be so heavy but that ample shade is left to encourage these species and discourage others. If light-demanding species are the more valuable, as they usually are, most of the timber should be felled at once, and sometimes it is desirable to burn the area lightly in addition in order to remove all shade. The problem of perpetuating light-demanding trees is a difficult one, for by the time a forest of these trees matures there is generally an understory of young shade-tolerant trees which, after the mature timber is harvested, tends to recover quickly from suppression and form a stand so dense that it completely prevents the reproduction of the desirable light-demanding species.

Underwater differences in cardinal light values are a very important factor governing the depth to which submerged plants can extend. Red algae as a group have lower light requirements than other algae. In the Puget Sound region this permits them to grow in the sea at depths of 25 m, whereas brown algae cannot synthesize carbohydrate below 15 m (766). According to the well-known theory of complementary chromatic adaptation, the red pigments associated with the chlorophyll of the Rhodophyceae enable them to make exceptionally efficient use of the weak light, consisting mostly of short wavelengths, which reaches them. Supporting this theory is experimental evidence that red algae absorb a greater percentage of blue light than do either green or brown algae. There is considerable question, however, whether the nongreen pigments have much bearing on the ability of freshwater aquatic plants to grow in deep water (210).

Attached algae and mosses grow as deep as 120 m in the exceptionally clear water of Crater Lake, Oregon, where they get less than 0.5% of sunlight. Vascular plants in fresh water usually extend no deeper than about 10 m even in the clearest lakes, and in shallower water the species form zones at different depths according to their light requirements. Only the lower limits of each of these zones seems to be set directly by the amount of light penetrating the water; at the upper limits competition and shading effects become increasingly more important.

Transpiration

The detrimental effects of high light intensities include their influence in promoting rapid transpiration. Light stimulates the guard cells to open, as well as increases the permeability of the plasma membranes. From earlier discussions it will be recalled that the stomata of most plants remain open all day and close at night, and data were presented which showed that transpiration increases rapidly at daybreak and slows to a very low level at sundown, if not earlier, owing to a tissue water-deficit (Fig. 18).

Leaves use only a small percentage of the available light energy in photosynthesis. Of the remainder approximately one-third is reflected back from the leaf or is transmitted through it, and about two-thirds is absorbed, changed into heat energy, then lost by radiation or used up in the vaporization of water. Since some of the light rays that penetrate tissues are always changed into long heat rays, it is apparent that light effects can never be completely divorced from heat effects, and, because heat influences transpiration and other physiologic processes, the investigation of light as an ecologic factor is very complicated. As suggested earlier, heat effects can in part be ruled out by using aquatic plants as experimental materials, but there is no assurance that the results of such studies are directly applicable to land plants.

Many specific instances have been described in which the influence of shade in reducing transpiration as well as direct loss of water from the soil is decidedly beneficial from the standpoint of seedling survival (696). On the other hand there are many places where the moisture factor is most critical in the shade. The shade-induced dwarfness of seedling root systems results in either greater mortality under shade despite the milder intensity of soil drouth there (Fig. 57), or in the dense plant cover producing the shade offering seedlings too much competition for moisture and nutrients. In general, moisture conditions are more favorable under light shade than under dense shade or full sunlight, but, as emphasized earlier, shade implies a number of concomitant environmental conditions, and their net effect cannot easily be predicted.

Growth Form and Physiologic Characteristics

The amount of light available to a plant as it develops exerts a profound influence on the structure and functions of the organs finally produced. Light influences on structure of the shoots are essentially negative, for the blue and violet rays are the most important wavelengths governing differentiation, and their action can best be described as stunting.

In comparison to plants grown in the shade, individuals developing under full sunlight usually exhibit the following characteristics:

Morphologic Features
1. Thicker stems with well-developed xylem and supporting tissues (603, 608, 693, 791).
2. Less leaf area per plant (65).
3. Shorter internodes (791).

4. More prolific branching.
5. More weakly developed endodermis, where present.
6. Smaller cells in leaf blades (in part a result of item 25 below), which usually results in:
   a. Usually smaller but thicker leaf blades or blade segments (113, 282, 310, 693, 791). Maximum blade area is usually attained under 20–50% illumination, but in certain species maximal cell size and area require full light (386).
   b. Stomata smaller and closer together (334, 608).
   c. Smaller vein islets (208).
   d. More hairs per unit area, * provided the leaves are pubescent (208, 282).
7. Leaves more deeply lobed.
8. Walls of chlorenchyma cells in *Pinaceae* less folded.
9. Thicker cuticle and cell walls (113, 310, 334, 419). Because full sunlight tends to produce small blades of hard texture, tobacco grown for wrappers, as well as tea bushes, may be given shade to economic advantage.
10. Chloroplasts fewer and smaller.
11. Better-developed palisade, which frequently occurs on both sides of the blade (113, 208, 310, 386, 419).
12. More weakly developed sponge mesophyll.
13. Smaller intercellular spaces (310, 419).
14. Greater ratio of internal/external leaf surface (largely a result of 6a above).
15. Lateral walls of epidermal cells less wavy.
16. Leaf blades not flat, less compound (863), and oriented at other than right angles to the path of incident radiation (113, 282) (Fig. 36).
17. Lower ratio of total leaf area to vascular tissue of the supporting stem (608).
18. Roots longer, more numerous and more branched, with a higher root/shoot ratio (693) (Fig. 56).
19. Greater fresh weight and dry weight of both roots and shoots. A few plants attain maximum dry weight under less than full sunlight, and many show but a narrow margin of benefit with additional light above 50%. The effect of a unit of light energy becomes more pronounced the nearer the approach to the limit of shade tolerance. Since sun plants usually gain weight more rapidly than

---

* Leaf blades with a dense covering of hairs on the upper surface, as in *Verbascum thapsus*, resemble shade leaves anatomically even if grown in full sunlight.

shade plants, yet gain height more slowly, the criterion used to evaluate "growth" in plants is critical!

20. Larger and more numerous nodules on legume roots; better development of ectotrophic mycorhizae.

Physiologic Features

21. Usually a lower chlorophyll content, with carotenoids consequently more apparent and the leaves greenish yellow (282, 693).
22. Higher photosynthetic rate per unit surface in bright light, but a lower rate in weak light (853).
23. High respiration rate (89) and consequently high compensation point.
24. Lower percentage of water on a dry-weight basis (310, 693, 693).
25. More rapid transpiration (310, 608) (related to items 14 and 17 above).
26. Higher optimum fertility level.
27. Higher salt content, sugar content, and therefore more negative osmotic potential (419). The protoplasts of shade leaves exert so little distending force against their cell walls that they wilt when their water content drops only 1–5%, whereas sun leaves can endure a loss of 20–30% without wilting.
28. Decrease in acidity of cell sap (474, 718).
29. High carbohydrate/N ratio (791).
30. Low K, Ca and P content (509).
31. Greater vigor of flowering and fruiting (209, 282, 693, 791).
32. Earlier appearance of flowers (282, 693), but later maturation of leaves (19).
33. More calories per gram dry weight of seeds (479).
34. Greater resistance to:
    a. Temperature injury (related to items 6 and 27 above).
    b. Drouth (related to items 6 and 9 above).
    c. Parasites (related to item 9 above).

The morphologic differences described above not only characterize the same species grown under shade and under bright sunlight but likewise tend to distinguish heliophytes from sciophytes when they are brought together in the same habitat. The two phenomena may be distinguished by the terms *helioplastic* and *heliomorphic,* respectively. However, it is to be noted that the parallelism between heliophytes and helioplastic modifications does not apply to physiologic characteristics to the same extent as to morphologic attributes. For example, heliophytes are capable of more efficient use of high light intensities than are sciophytes, whereas this relationship is reversed in helioplastic individuals, i.e., in series of individuals of a sciophyte grown under different light intensities, those in the denser shade make the more efficient use of light (542).

Because light sources usually emit heat as well as light rays, and because when light is absorbed it is mostly converted into heat, the extent to which differences in structure and function associated with insolation are really results of heating and drying influences can never be determined with certainty. Certainly exposure to dryness and to bright light tend to have closely similar influences on plant structure and function.

In addition to the usual differences in form between plants grown in sun and in shade, light intensity frequently affects the erectness of plants, but the nature of the influence is entirely unpredictable. In many species individuals grown in full sunlight are prostrate and those receiving shade grow erect (443). Reduced light stimulates bush lima beans (*Phaseolus lunatus*) to become climbing vines, whereas other species that are climbing vines under bright lights lose their power to twine when shaded (791).

A green plant from which light is completely excluded responds somewhat differently from the reactions to suboptimal light discussed above. The stems become extremely long, leaves remain in an immature and unexpanded condition, and the plant loses its green chlorophyll pigments. A plant exhibiting such symptoms is said to be etiolated. Because etiolation improves the flavor and results in crisp tender tissues, plants such as celery (*Apium graveolens*) and endive (*Cichorium endivia*) are usually covered to exclude light after reaching the proper stage of growth.

Germination

The seeds of most plants become sensitive to light when wetted. In certain instances germination (165, 232) is benefited; in others, it is retarded. *Verbascum thapsus, Lactuca sativa,* and *Paulownia tomentosa* will not germinate without light stimulation, and *Daucus carota, Rumex crispus,* and *Picea abies* germinate better with exposure to light. In contrast, *Vanilla fragans,* many *Liliaceae,* and *Primula spectabilis* require darkness, with *Bromus tectorum, Ulmus americana,* and many *Cucurbitaceae* germinating better in darkness.

The amount of light needed for stimulation is considerable for bluegrass (*Poa*), but for tobacco (*Nicotiana tabacum*) even 0.01-second exposure allows some germination (408). Seeds requiring light obviously must not be completely covered with soil when planted, but it has been found that, if seeds are soaked, given adequate light treatment, then dried again, light stimulation is retained and germination will take place even if the seeds are completely covered with soil (408, 754).

Although germination responses to light are controlled by a pigment sensitive to red light, environmental influences are difficult to interpret because so many other factors also influence the relationship (232). For example, the light requirement of many kinds of seeds gradually becomes less important during dry storage (761). Also, a number of experiments have shown that the effect of light on seed germination can be reversed one way or another by temperature manipulation or by supplying oxygen, nitrates, or weak acids. These complex interrelationships have caused much confusion in the literature on this subject.

Reproduction

Earlier the fact was mentioned that insufficient light represses flowering and sometimes holds vascular plants indefinitely in the vegetative condition. Similarly the mycelia of many fungi grow well in darkness, but light is essential for the production of functional sporophores, and mosses growing in dimly lighted caves usually remain vegetative.

Because low light intensity favors vegetative development at the expense of flowering and fruiting, crops grown for vegetative parts are favored by climates with a high percentage of cloudiness, whereas fruits, grains, and seeds are favored by bright sun. The economic return on field and greenhouse crops is considerably influenced by the direction and degree of deviation from the seasonal norm of sunshine. Positive deviations in oceanic climates or during normally cloudy seasons are favorable, whereas negative deviations in the dry season of continental climates are beneficial. In regions where one or more seasons is cloudy, the lighting of a glasshouse during the cloudy season is a critical matter. The most benefit from sunlight is obtained when the glasshouse is oriented in an east-west direction, with the glass on the south side at such an angle that the sun's rays strike it at right angles when the sun is low in the sky.

Influence of Radiation on Nongreen Plants

For a long time it has been known that an exposure of bacteria to direct sunlight kills the cells. The lethal effect is due chiefly to the ultraviolet rays between 254 and 280 millimicrons, although violet and blue light have some effect. To a certain extent this knowledge has been put to practical use in sterilizing the air of public buildings with special ultraviolet lamps directed upward so as not to cause "sunburn" or "snow-blindness" of the occupants. In addition, ultraviolet irradiation has been used to sterilize drinking water and swimming pools, but the application here has been limited by the fact that these rays have very little penetrating power so that to be effective they must be played directly upon substances spread in thin films.

The same wavelengths that inactivate bacteria also have an inhibitory effect on fungi (717). Very short exposures sometimes have stimulating effects, but long exposures are commonly lethal. When disease-producing fungi are more sensitive to ultraviolet than their hosts, irradiation can be used in controlling them. Thus certain powdery mildews of plants (342), as well as fungal infections in human skin, can be checked by irradiation.

Ultraviolet and light of short wavelengths also affect the growth form and pigmentation of fungi, and light has various effects on their reproduction. *Agaricus campestris* will fruit in either darkness or light, but certain other agarics will fruit only in the light. In some fungi light is necessary only to initiate the primordia of reproductive organs, but in *Pilobolus microsporus* the sporophores may be initiated in darkness but require light for development. In that fungus also spore discharge occurs only in the light.

Photoperiodism

The length of the photoperiod * is of considerable importance to most plants, and their varied responses to this aspect of the light factor are designated by the term *photoperiodism* (74). Twelve to fourteen hours of daylight is a critical duration for most plants. Daylengths in excess of this range have about the same effect, as do days of varying length but shorter than this range. We can therefore speak of "long days" or "short days." On the equator day and night are of equal length throughout the year but at even slightly higher latitudes plants have developed sensitivity to the small annual variation (568). The stimulus is received by a pigment (phytochrome) in the buds or leaves then transmitted to other parts of the plant. Sometimes only a single cycle of appropriate length starts a process in motion.

Photoperiodism as a timing mechanism. During the year as the length of the photoperiod changes different processes in the plant may be

* Actually it is the length of the uninterrupted dark period in each 24-hour day that is critical, but since the ecologist is concerned principally with natural conditions where the light and dark periods are reciprocally related, continued usage of the older terminology, based on a misinterpretation, is not too objectionable.

activated or stopped as the daylength, hence season, becomes appropriate. Such responses to daylength have developed as a means of regulating activities according to calendar dates. In spring the plant must not respond to the first warm days by resuming growth, or late frost is likely to be damaging. This dependence on daylength to trigger the breaking of dormancy is a means of countering the vagaries of weather by delaying activity until severe frost is no longer likely. *

In addition to initiating and breaking dormancy (204), the initiation of flower primordea (867) and of germination (64) may be timed by photoperiodism.

Photoperiodism and plant distribution.   By the process of natural selection plants tend to become adjusted so that at least some of their normal processes are timed by the annual cycle of photoperiods where they grow. This adaptation is shown very well by the differences that exist in the photoperiod requirements of different populations in species that extend over wide ranges of latitude (579, 530). Plant migration in a north-south direction must be conditioned by the rate at which marginal populations can adapt themselves to slightly different daylengths. Until they make genetic changes, species with flowering timed by the photoperiod are restricted to latitudes in which they retain the advantage of efficient dissemination. Even among those species which possess highly satisfactory methods of vegetative reproduction, daylength may prove unsuited to the proper coordination between seasonal changes and the accumulation of adequate food reserves or may bring on other types of maladjustments.

Some practical applications.   The importance of photoperiodism in relation to the artificial extension of the ranges of economic plants has two aspects. On the one hand photoperiodism may serve as a check upon the degree of north-south displacement possible. For example, beets change from the useful biennial habit to the useless annual habit at high latitudes. The fact has been well established that great longitudinal extension of crop plants is a failure, even when temperature is compensated for by altitude, unless new genetic stocks are developed.

On the other hand, abnormal photoperiodism may be desirable. The ornamental value of *Sedum telephium* is considerably enhanced by

---

* An alternative timing mechanism is the requirement of chilling mentioned in the preceding chapter. Seeds or established plants that go into dormancy during a dry summer could be damaged or killed by winter freezing if they resumed activity in response to the first autumn rains. However, if a number of weeks or months of chilling are required to break dormancy, germination is postponed until spring and this hazard is avoided.

growing it below latitudes of 50° where daylength is too short for it to flower. In like manner, subtropical varieties of maize are grown in high latitudes when fodder is the desired crop. Finally, the classical example of making good use of abnormal behavior due to photoperiodism concerns Maryland Mammoth tobacco. This variety does not flower in the latitude of Maryland, and because of sterility and giantism it produces fine crops of leaves at such latitudes, but other plants must be grown in the latitude of northern Florida where the quality of the foliage is very low but where the plants flower vigorously and produce seed.

Plant breeders who desire to grow two or more generations of plants per year sometimes find it necessary to control light in order to bring about flowering and seed setting in the glasshouse in winter. In the same manner interfertile varieties which have different flowering seasons can be made to bloom simultaneously, thus facilitating desirable crosses. Under certain combinations of species and market prices, floriculturists find it profitable to alter flowering seasons by altering daylength (447).

Wherever it is economically feasible or experimentally desirable to supplement normal daylength to obtain certain desired results, good advantage can be taken of the fact that only very weak supplemental light, often less than 11 L, is adequate to produce the desired stimulation. Even bright moonlight (0.2 L) is strong enough to effect photoperiodic response in certain plants. On clear days the effective photoperiod is approximately an hour longer than the period between sunrise and sunset (465), whereas for plants growing in dense shade the photoperiod is shorter than for those fully exposed to the sky. A difference of only 10 min day^{-1} may determine plant behavior (867).

When plants are grown under distinctly short or distinctly long daylengths, the degree to which this factor controls structure and function is so absolute that it overshadows most minor genetic variation. For this reason plant breeders, in attempting to make selections with respect to adaptation to daylength, should grow their materials under near-critical daylengths, for this allows maximum expression of minor genetic differences.

Light Influences as Modified by Temperature and Other Factors

Temperature and light influences are inextricably related in their influences on plants (Fig. 58). Suitable intensities of one compensate in part for deficiencies in the other. For example, photoperiodism can be altered somewhat by the intensity or quality of light, and it can be reversed by the manipulation of temperature (58). Thus, vernalization allows winter wheat to flower during long photoperiods, whereas without temperature stimulation these varieties are distinctly short-day plants (631). Again, if moistened grains of sorghum and millet are kept in darkness for 5 to 10 days at temperatures between 27 and 29°C, the need for short photoperiods of the plants they produce is removed.

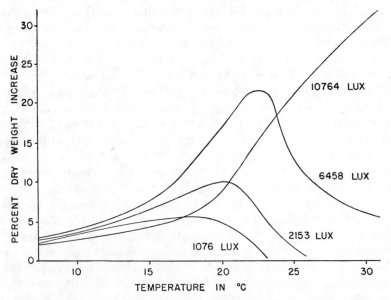

FIGURE 58. Interrelations between temperature and light intensity, with respect to growth in young tomato plants. [After Bolas (70a).]

Possibly the efficiency of light energy increases with rising temperature. This is one explanation for the fact that in general the colder the climate the more intolerant a species is of shade (448). Nevertheless there is a strong likelihood that the controlling factor here is really the heating rather than the lighting effects of insolation. It is significant in this connection that at least certain plants exhibit the same structural characteristics when grown in shade at favorable temperatures as when grown in sun at lower temperatures (603).

When bacteria are exposed to ultraviolet radiation the lethal action of these wavelengths is increased either by a rise in temperature or a drop in pH. This fact and others mentioned above emphasize the point that, in experimental work with light, complete temperature records must be kept, for without them different experiments cannot be critically compared.

At the seedling level in forests the CO_2 content of the air may remain 3–5% above the normal 300 ppm even at midday in consequence of vigorous soil respiration, whereas in nearby deforested areas the lush growth of herbs may use CO_2 so rapidly at this time that its concentration drops 10% below normal. Thus in some microsites light deficiency may be compensated by better supplies of CO_2.

Light Measurement and Control

Instruments for light measurement are designated as *photometers, illuminometers,* or simply *light meters.* Those that measure other wavelengths of radiant energy in addition to light are most appropriately called *radiometers.*

Radiometers

The characteristic appearance of black objects results from their complete absorption of all visible wavelengths, whereas white and to a certain extent silvery surfaces reflect essentially all light. Owing to this difference in properties, black objects assume higher temperatures than white when both are subjected to equal illumination, and the difference in their temperatures is proportional to the intensity of illumination. A number of instruments utilize this principle. Since these are especially sensitive to infrared as well as to light, they must be employed in conjunction with a filter which cuts out infrared if the data are to be used as light measurements. A 1-cm layer of 0.1 N $CuCl_2$ is fairly satisfactory for this purpose (85). Instruments of this type usually have the disadvantage of lacking sensitivity under low light intensities, but the enclosure of the sensitive elements in an evacuated glass chamber overcomes the difficulty. Because photosynthesis is most strongly influenced by the longer wavelengths of light, and radiometers are more sensitive to the longer wavelengths, these instruments are best suited to light measurement where this function is being studied.

A simple radiometer may be made by sealing two thermometers in an evacuated glass tube, one with the bulb covered with flat black paint, and the other with flat white or left naked. The instrument is kept in a dark box until a measurement is to be made; then it is removed and held in a horizontal position over the area where radiation is to be measured. The degree of difference registered by the two thermometers at the end of 5 minutes may serve as a measure of radiant energy, or the instrument may be calibrated against another type of instrument which expresses light in terms of luxes (7).

A more complicated instrument working on the same principle is the thermocouple, one junction of which can be attached to a black metal disc and the other to a reflective disc, the difference in heat absorption as indicated by the galvanometer being proportional to the incident radiation (85). If a series of such thermocouples are connected so as to multiply this effect the instrument is called a *thermopile*. These instruments are rather expensive, but they yield instantaneous readings as well as continuous records and are considered by some the best type of instrument for general ecologic work (694). There are many instruments of similar nature, such as the *pyrheliometer,* the *pyranometer,* and the *bolometer,* but they have been more widely used by meterologists than by biologists.

Another type of radiometer is activated by the distortion of a blackened bimetallic strip which is exposed to radiation. These instruments make continuous records for periods of a week on paper charts.

The sensing elements of most of the above instruments have flat surfaces which are usually exposed in a position normal to the earth's surface. However, a plant even on the equator gets considerable illumination from the side in both morning and afternoon. Therefore some have pointed out the desirability of spherical or at least hemispherical sensing units for plant ecology (260). Two such instruments have been used.

A pair of spherical atmometers, one of which is black, can be mounted near each other and their evaporation rates compared (468, 576). These have been called *radioatmometers.* The members must be standardized to each other in the dark. Difference in water loss between black and white atmometers is closely related to pyrheliometer measurements (304).

The Bellani radiation integrator consists of a blackened copper sphere containing alcohol or water which is distilled into a burette. At the end of the day the meniscus of the accumulated liquid is read and the instrument reset by briefly inverting it (790).

Radiometers are the only light-measuring instruments that integrate all wavelengths of radiant energy. Each of the types to be mentioned below is sensitive chiefly to some particular segment of the solar spectrum.

The Spectrophotometer

The spectrophotometer is the ideal instrument for measuring light since it evaluates separately the energy in different segments of the spectrum. Portable instruments are available that give readings directly and integrate direct with diffuse light. These instruments are quite expensive and since light quality is of minor importance in most ecologic field work, relatively inexpensive instruments giving instantaneous readings or integrating light over a time interval are most commonly used. However most of these have sensitivity curves which differ materially from each other and which change under different intensities of light and with temperature. Light readings in lux are not closely comparable among such instruments.

The Secchi Disc

The Secchi disc method is an ocular method based on the sensitivity of the eye; it is used to measure the penetration of light through water. A white disc about 20 cm in diameter is lowered horizontally during midday when the sky is clear until it just disappears from sight. This depth is recorded. Then the disc is raised until it reappears and the depth is again recorded. The mean of these readings is then taken as a measure of light penetration. Readings vary from about 59 m in clear ocean water to 1.5 cm in muddy rivers. The amount of error to be expected can be judged from the fact that the 2% illumination level in freshwater lakes varies from about 1 m above to 1 m below the Secchi disc reading (580).

Photochemical Methods

When it is desirable to compare summations of light energy received over a long period, solutions that slowly decompose in the presence of light can be exposed continuously, then analyzed from time to time. The decomposition of oxalic acid in the presence of uranium acetate, and hydriotic acid in dilute sulfuric acid, have been used for this purpose (332). A concentrated solution of anthracene in benzene slowly polymerizes when exposed to light, so that the amount of polymer accumulating during a few hours or days serves as an integrated expression of the light energy received (643). An important disadvantage of these photochemical techniques is that they are more sensitive to ultraviolet than to light.

The degree of darkening of panchromatic film with standard exposure can also be used as a measure of light intensity, and through the use of filters the sensitivity can be controlled.

Photoelectric Cells

Of all the types available for general ecologic work, no photometer yields absolute values of light intensities as quickly and conveniently as those employing photoelectric cells as sensitive units. These are portable, read directly, and when encased in a suitable water-tight cell can be used under water. The technique is not without serious limitations, however. Both temperature and light intensity affect the sensitivities of the cells, and it is not yet possible to get cells that are accurately standardized. Also, most types of cells are sensitive chiefly to the blue, violet, and ultraviolet rays. However, in measuring light with respect to its effect on structure and functions other than photosynthesis, this method is fairly satisfactory, for the shorter wavelengths to which the instruments are particularly sensitive are the chief ones involved. Also, suitable filters can be employed to control wavelength sensitivity. Some have cemented half or two thirds of a plastic table-tennis ball over the flat sensing surface to serve as an integrating sphere.

Bio-assays

If water containing a known population density of the unicellular alga *Chlorella* is exposed in a flask for a week, then the population density again determined, the increase may be interpreted as proportionate to the supply of light energy that was available during that period (356). This must be used only as a comparative technique, for different species of algae could respond very differently.

Chlorophyll may be extracted and exposed, using the change in its optical density as a measure of light energy received (613).

Field Technique

From the above discussions it may be surmised that absolute measurements of light intensity using different instruments yield widely discordant results. This problem can be considerably reduced where it is feasible to express light as a percentage of full insolation, for on this basis the differences among light meters are not so great.

Some investigators, in order to obtain a satisfactory value for average light intensity under vegetation, have directed the sensitive element of the light meter upward and placed it successively at equal intervals along a tape, taking an average of the readings made in the open at the beginning and end of the series as a basis of comparison. Obviously the day selected to obtain such measurements should be cloudless or, even better, uniformly overcast with the sun completely obscured during the entire series of measurements, and all measurements should be obtained in as short a period in midday as possible. The readings obtained under the plant canopy are best expressed in the form of a frequency curve, with an adequate sample requiring 20–1200 readings.

Photoelectric cells are damaged ("solarized") by intense light; therefore it is always advisable to measure light as received through opalescent material or ground glass, or as reflected from the surface of heavy white paper. The cells should also be protected from jarring as much as possible.

Light Control

In studying the effects of one factor it is always imperative that others be kept as nearly constant as possible, but the light factor is so complicated that this presents a very difficult problem in ecologic research. One investigator can seldom compare his experimental results with those obtained by another with the assurance that light conditions were essentially identical, and small differences in light conditions affect most plant functions.

Ordinary window glass transmits only 82–90% of daylight and screens out wavelengths shorter than about 330 millimicrons, so that special glass must be used when it is desirable to grow plants under glass but exposed to complete radiation (857). The ability of this glass to continue to transmit ultraviolet should be checked at intervals.

The quantity of light may be increased during dull days by supplementary light from fluorescent lamps mounted under reflectors. However, the bulky reflector of a bank of fluorescent lamps in a glasshouse is detrimental while the electricity is turned off, since the light most useful to a plant is that coming from directly above.

In growth chambers fluorescent lamps have been widely used because they give off relatively little heat as compared with tungsten-filament lamps. However, the brightest artificial light obtainable with fluorescent lamps is of an intensity that is too low for the normal growth of obligative heliophytes. Still other types of lamps are available which provide illumination of approximately full sunlight intensities. With these it is necessary to mount them above a 2cm layer of running cold water to remove objectionable heat (166).

Outdoors different degrees of shading, starting with full sunlight, can be obtained by means of different numbers of layers of cheesecloth, muslin or black sateen. Also, slat frames can be constructed in which different spacings of the slats allow different amounts of illumination to penetrate, or bronze or aluminum screening of different-sized mesh may be used (260). If slats are used they should be narrow and well elevated, otherwise plant parts immediately beneath will be exposed alternately to periods of shade and superfluous light. The use of slats or screens has much to recommend it, since all wavelengths are affected equally. Slats have been widely used not only in experimentation, but also in providing suitable environment for the commercial growing of the sciophytic drug plant *Hydrastis canadensis.* A cruder method of commercial light control is practiced in coffee and cacao plantations where these low trees are grown under open stands of tall trees.

The development of seedlings with different supplies of light energy can be studied by setting opaque tubes of varying heights over them, so that varying amounts of skylight reach the bottoms of the tubes (291).

Some success has been achieved in constructing small chambers to set over plants in the field to determine their photosynthetic rates in terms of CO_2 uptake (660). Here it is very difficult but not impossible to maintain air temperature, relative humidity and air movement within the chamber, equal to these conditions outside.

Wherever it is practicable, methods that make use of natural sunlight are most desirable in ecologic experimentation aimed at understanding the light relations of plants in the field. Although electric light has the advantage of being constant and reproducible, it is very different from sunlight, and therefore data obtained from plants grown under artificial light must be applied to natural conditions only with reservations (228).

5

The Atmospheric Factor

Metabolic
Aspects

Wind

From an ecologic standpoint the term atmosphere must include not only the thick gaseous envelope surrounding our planet but also the small and highly important masses of gas which penetrate or originate in the soil and in plant tissues.

An atmosphere is essential for life as we know it. In the first place, the atmospheric blanket surrounding the earth prevents such wide diurnal fluctuations in temperature as occur on other planets, fluctuations which would quickly extinguish all known forms of life. Second, there must be a continuous exchange of gases between an atmosphere and the living protoplasm of nearly all organisms.

The manifold effects of atmosphere on plants are in part direct, such as supplying CO_2 for photosynthesis and O_2 for respiration, and in part indirect, such as influencing the distribution of heat and light, and promoting transpiration, pollination, and dissemination.

Metabolic Aspects

Composition of the Atmosphere

When the earth was young the composition of its atmosphere is believed to have been totally different from its present mixture of gases. As the hot planet cooled the early gases were replaced by H_2O, CO_2 and N_2(439). Most of the water vapor then condensed to form the oceans, and most of the CO_2 became fixed in rock minerals. With the rise of photosynthetic organisms the CO_2 content declined still further as it became tied up in the reduced C compounds of their bodies, and at the same time the O_2 byproduct of phytosynthesis became a new and important constituent of the atmosphere. At

245

present the principal gaseous constituents of the free atmosphere exist in fairly constant proportions by volume as follows:

N_2, approximately 79%
O_2, approximately 21%
CO_2, approximately 0.03%

Other constituents, the proportions of which vary greatly depending on time and place, are:

Water vapors.
Salt crystals from the oceans.
Dust from arid lands, glacial streams, volcanoes, etc.
Microorganisms (largely carried on dust particles).
Pollen grains and spores.
Disseminules.
Terpenes and other volatile products of natural vegetation.
Pollutants:
  Smoke, fly ash and dust.
  Industrial gases.
  Herbicides and insecticides.

Green plants increase the proportion of O_2 at the expense of CO_2 in the atmosphere. Although they continually produce CO_2 in respiration, whereas the release of O_2 in photosynthesis is limited to daylight hours and favorable seasons, the photosynthetic process is so much more vigorous than respiration that the net effect is to increase the percentage of O_2 in the atmosphere.

Animals and most nongreen plants always take in O_2 and liberate CO_2, so that their metabolism tends to offset the effect of green plants on the atmosphere. It is this complementary metabolism of animals and green plants that makes it possible to establish *balanced aquaria* which require almost no care. Indeed, if the balance is carefully adjusted, one of these biotic communities can be maintained in a stoppered bottle.

The CO_2 content of the air is also affected by the decomposition of carbonate rocks, by combustion, and by volcanic activity, all of which liberate this gas, and by the decomposition of feldspars, which absorbs it. Because the oceans have a great capacity for dissolved CO_2, carbonates, and bicarbonates, and at the same time present a great surface to the atmosphere, the constancy of the CO_2 content of the air seems in a large measure due to the influence of the oceans.

Judging from the lack of O_2 on other planets that have an atmosphere, and from the fact that geologic processes tend to use up O_2 but not to release it, we may assume that the photosynthetic process has been chiefly responsible for the accumulation and maintenance of O_2 in our atmosphere. Also, it seems reasonable to conclude that, since the percentage of O_2 appears to be remaining constant, an equilibrium has been established between the rate of photosynthesis and the rate of O_2 consumption by respiration plus geologic processes.

Exchange of Gases between Plant Shoots and the External Atmosphere

Gases enter vascular land plants by diffusing through the stomata and lenticels, then going into solution in the water contained in the hydrated walls of the parenchyma cells, and finally diffusing through the walls and plasma membranes into the protoplasts. Waste gases of metabolism leave the plant by the same route in reverse. Because this system of gas exchange involves a wet cell surface that is in contact with the atmosphere, the loss of large quantities of water by evaporation is inevitable.

Not only does this mechanism prove hazardous at times from the standpoint of excessive transpiration, but gases contained in air that has been polluted by industries diffuse into leaf tissues and sometimes have toxic effects.

Atmospheric Pollution

Several types of pollution are important in developed countries, and each has distinctive influences on plant life (726).

The earliest type to elicit concern was a result of using soft coal containing much S to heat buildings in large cities in winter. Here, and also in the vicinity of mines where ores were roasted in open fires, the S was oxidized to SO_2, which gas diffuses through stomata and hydrolyses to form the very caustic H_2SO_4. In dilute amounts sulphates are beneficial and the S is assimilated, but as the atmospheric concentration rises above 0.3 ppm for 8 hours the chlorenchyma of plants is injured, with the leaves often assuming a water-soaked appearance followed by a bleaching of the intervenous areas. Immediately around a smelter plant life has been exterminated (Fig. 59), with concentric zones of decreasing damage extending as far as 25 km in a leeward direction (752). Sulphur dioxide also pollutes ponds and lakes with effects up to 10 km from the source where the water pH may drop to 3.3.

During a nocturnal temperature inversion relatively cold and dense air lies next the ground, with warm, less dense air above. This is a stable condition and if pollutants are emitted they spread out in the layer of relatively warm air lying between cold air above and below. In cities where fog was associated with coal smoke and the mixture became trapped by inversion for a period of days, the mixture of smoke and fog came to be called *smog.* Since then the term has been extended to other types of visible pollution where pollutants are trapped in an inversion.

FIGURE 59. Area near Copper Hills, Georgia, where smelter fumes long ago killed all the plants of the original forest. Because high temperature and rainfall favor rapid erosion the year around, plants have not yet been able to revegetate the area since modern refining methods have come into use.

Aside from the caustic effect, to which species differ widely in their sensitivities, smoke is also deleterious from the standpoint of soot particles that reduce light intensity while they are in suspension, then accumulate as a gummy film on leaves to screen out light even more completely, and increase the heat load when the sun shines. Plants with deciduous leaves suffer the least from such surficial coverings, evergreen species being dwarfed by enfeebled photosynthesis and early death of leaves, if not killed. Roadside dust has a similar effect on the evergreen desert shrub *Larrea* (48), and evergreen leaves of coffee bushes in Costa Rica deteriorated alarmingly during the 1960s as a result of fine ash settling out of the air after a series of volcanic eruptions.

Factories where aluminum, steel, phosphate and ceramics are processed emit HF which enters stomata, moves with the transpiration stream to leaf tips and margins, where it accumulates to toxic levels. This gas is invisible and not accompanied by soot, but effects are often strikingly manifest by a denuded area in the immediate vicinity, with damage to crops and forests extending radially for several kilometers.

Plants are useful indicators to recognize and evaluate the above two gaseous pollutants. *Gladiolus, Tulipa,* and *Zea* are sensitive to HF but not to SO_2, whereas *Cichorium, Geranium,* and *Medicago sativa* are sensitive to SO_2 but not to HF (868). Furthermore, HF characteristically causes necrosis of just leaf tips as damage becomes apparent.

Injury from SO_2 or HF is likely to be confused with winter drouth injury (329). However, the latter affects only such woody plants as project above the snow cover, it cannot damage the foliage of deciduous plants, and is noticeably more severe on the windward sides of stands or isolated individuals. Also, winter drouth injury is sustained within a few days, and if the plant recovers, a slow return to normality is shown by the thickness of xylem layers, whereas poisonous gases only weaken the plant, gradually reducing annual wood formation until death.

In addition to the use of vascular plants differentially sensitive to HF and SO_2 as indicators, epiphytic cryptogams have proven very useful in preparing maps to show different intensities of pollution and thus aid in planning urban development (324, 712). These plants are especially sensitive since they are evergreen and absorb over all their exposed surfaces. Severe pollution eliminates nearly all mosses, lichens and epiphytic algae except *Pleurococcus vulgare.*

Another type of atmospheric pollution is the smog in areas where automobile traffic is heavy. Fog is not necessarily involved here, but the atmosphere is hazy. The exhaust from gasoline engines contains hydrocarbons (especially CH_4), oxides of N and some SO_2. Ultraviolet radiation converts these into secondary compounds among which O_3 and peroxyacetyl nitrate are especially toxic to plants. The leaf surfaces may develop brown flecks or bleached spots, or become bronzed or silvery, leaves drop immaturely, growth is retarded, etc. The California State Department of Agriculture reported $25.7 million crop losses in 1970 from pollution of this type, most of it in the Los Angeles region, but with damage recognized in 22 counties (25). Orange trees are partly defoliated, the sale value of leafy vegetables and cut flowers is reduced, and ornamentals rendered unsightly. Ozone damages forest trees up to 35 km eastward from Los Angeles, making them more sensitive to fungi and insect attack (710). In fact, 40,300 ha of surrounding forest has been affected sufficiently that in reforestation only a few alien species tolerant of such pollution are recommended. Similar problems have been reported from the broad urbanized strip extending from Philadelphia northward into Canada (856), and the smog plume from Mexico City extends 200 km downwind (173). Aside from effecting a small reduction in the toxicity of exhaust fumes in California, palliative measures have included efforts to breed smog tolerant races of economic plants, spraying plants or cloths suspended above them with antioxidants, and "planting" imitation shrubbery made of chemically resistant plastic in the landscaping of highways.

A large percentage of the herbicides spread from aircraft does not fall on the target area, but instead drifts laterally or rises aloft to be carried up to 12 km before causing unintentional damage to plants (752).

In the vicinity of paper mills H_2S emissions commonly reduce the pH of rainwater to 3.5, which is directly or indirectly injurious to plants.

Many other types of pollution have significance for plant life, and much has been written on the unfavorable effects on man, both direct and indirect. Rainfall, of course, cleans the air of these substances, but this only transfers the toxins to soils, streams, lakes and estuaries. By absorbing the toxins vegetation has an appreciable effect in purifying the air, but this leads to a replacement of crop plants and ornamentals by species whose main value is their tolerance of pollution (801).

One more aspect of pollution with entirely different possible consequences for plants must be mentioned. In the last century the geometric increase in man's combustion of fossil fuels (coal, oil, gas) appears to have increased the CO_2 content of the earth's atmosphere by at least 10%, despite the buffering effect of the oceans, which act as a CO_2 sink. In a direct manner this should have increased the photosynthetic production of O_2 in the air, but the progressive destruction of plant cover by man has probably offset this influence. On the other hand CO_2 in the air transmits the relatively short wavelengths of incoming solar radiation, but absorbs the relatively long wavelengths of reradiated energy, thus tending to hold heat in the biosphere. Mean temperatures have actually risen approximately 1°C, and we have enough fossil fuel to raise the temperature 10°C more. Whether increasing smoke and dust will reduce incoming radiation sufficiently to offset the warming influence of the increasing CO_2 is debatable. Whether the continued increase in CO_2 at the expense of atmospheric O_2 will eventually cause world-wide impact on present forms of life is likewise unknown.

The CO_2 Cycle in Relation to Photosynthesis

The normal concentration of CO_2 in the free atmosphere is suboptimal for photosynthesis, for glasshouse experiments show that the rate of accumulation of carbon compounds can be materially increased by raising the CO_2 concentration to at least 20 times what is normal. However, high concentrations that are beneficial at first may prove toxic if sustained. Therefore, the atmospheric conditions under which an experiment is performed have considerable bearing on its significance in autecology.

The CO_2 cycle involves absorption by green plants from the air or water, fixation in photosynthesis, then eventually the oxidation of the organic compounds which returns the CO_2 to the air or water by one of three principal routes: (a) respiration by the plant itself, (b) respiration by animals which have digested the plants producing the photosynthate, (c) respiration by animals which have digested the plant products, or (d) respiration by microorganisms which humify and mineralize organic matter. The CO_2 content of the air at any one time thus represents a dynamic equilibrium among a number of factors in time and space which affect the speed of various stages of this cycle.

At times when the photosynthetic process is very active, i.e., during the day and in warm moist weather, the concentration of CO_2 declines considerably below average, whereas at other times the reverse is true (724). Obviously, determinations of the CO_2 content of the air with the purpose of showing the extent to which this factor is limiting photosynthesis should be made in late morning when photosynthesis is proceeding most rapidly. Also, sampling should take into account the vertical decrease in CO_2 content from the soil upward which results from absorption by the shoots, and from dissipation by wind currents of the supply of this gas that continually emanates from the soil. Great quantities of CO_2 diffuse from the soil at those times and places where the organisms have plenty of organic substrate and temperature and aeration are favorable for their activity. Measurements (547) have shown as much as 20 kg ha^{-1} hr^{-1} being liberated from soil.

In the lower layers of the atmosphere in a forest the CO_2 content may be nearly 6 times greater than the average concentration. Over bare but moist and fertile soils in which microbial activity has been stimulated by cultivation, the concentration may be 10 times normal. Soil organisms may liberate more CO_2 per day than green vegetation uses up, thereby almost offsetting the usual tendency for a diurnal drop in CO_2 content and proving the importance of this source of carbon for low-growing plants. Herein lies an often unappreciated benefit derived from manuring soil, although excess stimulation of microbial activity may reverse the net benefit by creating supraoptimal concentrations of CO_2 in the soil itself. It has been pointed out that the high concentration of CO_2 near the soil in forests offers considerable compensation for the suboptimal light intensities there and may account for the greater efficiency of photosynthesis with respect to the supply of energy.

Factors Affecting Soil Aeration

In their respiration all soil organisms, including roots, absorb O_2 from and release CO_2 into the pore space of their medium. This results in a relatively high (up to 13%) concentration of CO_2 and a low concentration of O_2 in the soil, so that there is a steep diffusion gradient between the gases contained in the soil and those in the free atmosphere above. Gas moves through soil very slowly, however, so that CO_2 remains superabundant and O_2 deficient in most soils. Since the amount of gases and liquids in a soil are complementary to each other, the addition of a given volume of water displaces an equal volume of soil gases, and subsequently fresh air is drawn into the soil as the moisture drains away or is used up in evaporation and transpiration. Therefore the addition of water to a soil partly alleviates aeration conditions, provided the water drains out or is used up within reasonable time. The oxygen-saturation of rain water that infiltrates is likewise important. Changes in temperature, barometric pressure, and wind velocity also contribute to the renewal of soil air, but diffusion is considered the most important single factor. The CO_2 content of soil air tends to remain relatively constant, although the O_2 content fluctuates widely. Loams at field capacity contain about 19% O_2 in the principal horizons of root activity, but rain brings about considerable reduction, even though it is temporary (588). The degree to which the O_2 content of the soil air is depressed below that of the free air above depends on four major factors:

Rate of respiration of soil organisms and roots. Where soil organisms have an abundance of suitable substrate, and other factors are favorable, their activities alone keep the soil-oxygen concentration very low, and the CO_2 concentration quite high. In fact, variations in the rate of evolution of CO_2 from a soil are commonly used as a measure of variations in the rate of decomposition (547, 805). However, during the season of their active growth, the roots of plants may produce more CO_2 than the microorganisms.

Amount of pore space. Fine-textured soils have more pore space than coarse soils, common values being 60% and 40% respectively, but the former tend to have a higher water content along with smaller pores, and both factors interfere with diffusion, thus bringing about relatively poor aeration. The swelling of colloids when fine-textured soils are wetted contributes further to their inferior aeration.

Porosity resulting from good aggregation or crumb structure, root channels, and animal burrows all increase aeration. Also, proper plowing greatly increases porosity temporarily in the plow sole; indeed one of the chief reasons for plowing is to provide a porous structure which will favor germination and seedling root growth. However, this benefit of plowing is gained at no little cost, for the same favorable aeration increases microbial activity to the extent that it becomes impossible to maintain a high organic content in cultivated fields. The equilibrium of organic content attained under cultivation is therefore lower than in virgin soils, the chief exception being that crop residues may raise the organic content of the upper horizons of forest soils that were low in the virgin state. Fallowing is most destructive to organic matter. Here good aeration is accompanied by higher temperature due to lack of shade, and by higher moisture due to lack of transpiration loss.

Another aspect of plowing that is important from the standpoint of structure and aeration is the timing of the operation in relation to soil moisture content. If plowing is done when the soil is at or a little below field capacity the structure is improved, but one plowing when the soil is too wet may destroy structure so completely that a decade of careful management may be needed to restore it.

Size of pores. The importance of pore size in aeration was illustrated strikingly by an experiment that showed a coarse sand with a total porosity of 55.5% to be 1000 times more permeable to air than a fine sand with a pore volume of 37.9%. It can readily be seen that the permeability of the coarse sand is far out of proportion to the increase in total pore space and can be attributed only to some difference in the character, in this case the size, of the pores.

Pores so large that they will not hold water against the pull of gravity (except in the angles) remain filled with air almost all the time. These are the *noncapillary pores.* They are the most important from the standpoint of aeration because their aggregate volume represents the *minimal air capacity* of the soil, except for brief periods after a rain or an irrigation, and during such times these are the pores which allow gravitational water to percolate. Smaller pores, which remain occupied by water after a wetted soil has had time to drain, are called *capillary pores.* They are very important because they determine the maximum water a soil can retain, i.e., the field capacity.

High porosity is of no advantage if most of the pores are so small that they hold capillary water, for under such conditions the soil is essentially waterlogged at field capacity. Noncapillary porosity of 12% or less creates this unfavorable state. The roots of most upland plants will not penetrate horizons in such condition (Fig. 60); therefore the growth water they contain must be discounted (684).

An ideal soil should have high total porosity, which is about evenly divided between capillary and noncapillary pores. * Sands generally have too little capillary porosity; clays tend to have too little noncapillary porosity.

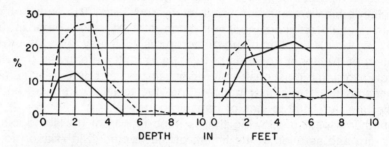

FIGURE 60. Distribution of rootlets 2 mm or less in diameter (broken line), as related to the vertical distribution of noncapillary pore space (solid line), in two soils. (*Left*) A soil unfavorable to trees, with roots confined to shallow layer because of poor aeration below. (*Right*) A favorable soil with aeration permitting a more uniform distribution of roots. [After Schuster and Stephenson (684).]

Drainage. Regardless of the numbers and sizes of pores, if drainage is obstructed, the soil is poorly aerated because the pores remain filled with water that excludes air. The presence in a soil profile of blue-gray horizons, sometimes mottled with rusty brown, is a fairly reliable indicator of poor aeration caused by the rising of the water table to the top of that horizon for a considerable part of the year. The blue-gray color is due to reduced ferrous iron, and red or yellow color is limited to pockets or streaks of coarser material where ferric iron forms, these colors having rather reliable indicator significance for aeration in most soils.

As an example of the great importance of drainage in moist climates, on the Piedmont Plateau of southeastern North America the rate of height growth of pines bears a useful degree of correlation with the depths to the least-permeable soil horizon (148).

---

* Pore-space analysis is most conveniently made on small samples of undisturbed soil that are wetted and then subjected to standard moisture tension in the laboratory (451). Since the continuity of pores is almost as important as their total volume, infiltration rates may be of more significance than pore-space analysis (588).

Effects of Suboptimal Aeration on Vascular Plants (283, 285, 666)

Species differ widely with respect to their requirements for aeration. For certain plants there has been demonstrated an optimum degree of aeration above and below which plant functions are impaired (475). The outline below indicates many of the types of responses to reduced aeration which have been reported in the literature; it also serves to show the considerable extent to which responses differ among species.

Morphologic Effects
1. Cell walls in roots remain abnormally thin (106, 475).
2. Root-hair formation is usually suppressed (190, 225).
3. Roots may be more numerous (106, 819).
4. Root branching remains less complex (190, 475).
5. Dry weight of roots may be reduced or increased.
6. Submersed organs may thicken as a result of formation of lacunar tissue (705) or may become more slender thereby presenting much greater absorptive surface (819).
7. Groups of stem cells may elongate radially to form conspicuous knobby swellings (512). This behavior represents physiologic disturbance, and is not adaptational as item 6.
8. Roots are shorter, and root systems occupy less space (106, 190, 475).
9. Root system becomes shallow, and sometimes root branches extend upward into the atmosphere.
10. Shoots occupy less space (106, 201).
11. Leaf area and number of chloroplasts may be reduced.
12. Xylem vessels are more numerous, but smaller.

Physiologic Effects
13. Roots change from aerobic to anaerobic respiration, at least in part, with a consequent accumulation of toxic by-products (164).
14. Permeability of plasma membranes decreases.
15. pH of plant sap declines (475).
16. Rate of absorption of water and nutrients (especially K, Ca and Mg) is reduced (475).
17. Rate of transpiration is reduced (owing at least in part to item 16 above).
18. Rate of respiration of leaves is increased (131).
19. Foliage becomes discolored.
20. Carbohydrate content may increase (106) or decrease (475).
21. Reproductive processes are delayed or repressed.

22. Reproductive structures that have been initiated may absciss prematurely (8).
23. Shoot growth terminates earlier.
24. Susceptibility to root diseases increases.

Physiologic Aspects of Soil Aeration

Toxicity of carbon dioxide. Carbon dioxide becomes toxic when its concentration in the soil atmosphere rises to about 10%, and 30–50% becomes lethal. Only under exceptional conditions do the detrimental effects of excess CO_2 exceed those due to O_2 deficiency (127, 456).

Production of other toxins. Excess CO_2 and deficient O_2 favor the formation of H_2S and bicarbonates of ferrous iron and manganese, and these may increase to toxic concentrations. Normally roots secrete only H_2CO_3, but under conditions of deficient O_2 they may secrete formic, acetic, oxalic, and other toxic acids. Similar toxic products of metabolism are formed by microorganisms of decay forced to live under anaerobic conditions. Aeration inadequate for normal processes of decay is the primary cause for the accumulation of peat and muck, but the accumulation of toxins soon becomes a secondary factor of at least equal importance.

Unfavorable changes in the chemical states of nutrients (588, 797). The compounds of Ca, Mg, K, Mn, and Fe which exist under anaerobic conditions are not very available to plants, and nutrient deficiency results. Also S exists as sulfides rather than sulfates, and N as ammonium or N_2 rather than nitrates. When the O_2 in a soil drops below the critical level certain bacteria begin to reduce nitrates to nitrites in order to obtain O_2, thereby further depleting the N fertility. * CO_2 is partly replaced by CH_4.

Deficiency of oxygen for root respiration and germination. The O_2 content of free air is always far enough above the minimal requirements of the shoots of terrestrial plants to be of no significance as an ecologic factor, but for seeds, roots, and soil-inhabiting organisms O_2 beneath the soil surface is usually suboptimal for metabolism (155). Some important microbes, especially *Clostridium,* have become so adjusted

---

* The ratio of reduced to oxidized substances in a soil affects its electric potential, hence by inserting platinum electrodes into an undisturbed soil its *redox potential* can be determined (514). The measurements provide a graduated scale extending both up and down from a point at which reduction and oxidation are balanced. Other techniques for evaluating soil aeration involve measurement of (a) O_2 diffusion rate, or (b) rate at which O_2 is supplied to a porous absorber (378).

to life in media lacking O_2 that they remain inactive in environments where there are appreciable quantities of this gas, but most soil organisms are aerobes and must have at least a small amount of O_2 to respire. Likewise among higher plants certain hydrophytes have become so well adapted to waterlogged conditions that they grow better there than where the substratum is well drained (54, 819). Other plants native to wet habitats (453), as well as most mesophytes and xerophytes, are very adversely affected by a substratum that furnishes but little O_2. It has often been demonstrated that ordinary potted plants grow under conditions of suboptimal aeration even though the pots may be porous and the roots are most abundant at the surface of the soil next to the pot, for by forcing air up through the bottom of the pot growth is usually markedly stimulated. This problem of O_2 deficiency in the soil increases with rising temperature, for rising temperature increases the rate of respiration so that O_2 is used up more rapidly (121).

Roots usually do not begin to show definite injury until the O_2 content of the soil atmosphere drops as low as 10%. Ordinarily the O_2 content of drained loams lies somewhere between this lower critical value of 10% and the 21% characteristic of free air, with the highest values nearest the soil surface. In general the roots of terrestrial plants are concentrated in the upper, better-aerated levels of the soil. In any position they are so nicely adjusted to the horizon in which they develop that disastrous effects frequently follow any disturbance of the aeration conditions, either by a rise in the water table or by the deposition of a smothering blanket of new soil upon the old surface (Fig. 61). When roots are suddenly deprived of soil O_2, absorption and transpiration decline strongly, leaves wilt and become discolored, and if conditions do not soon improve the plant dies (54). These manifestations seem to result not only from a reduction in the roots' ability to absorb, but also from their inability to produce hormones necessary for the normal functioning of the shoots (827). Often, however, a decrease in degree of aeration is survived if the change is brought about so gradually that new roots have time to form and the distribution of absorbing rootlets becomes altered in accordance with the new conditions (387) (Fig. 62). When landscape engineers need to raise the ground level around a desirable tree, very coarse materials are used, at least for the lower part of the added layer.

Normal functions of most roots begin to be impaired as the O_2 content of soil drops below about 10%, and at 2% they can barely remain alive. Because the O_2 content of the soil drops abruptly to about 1% just above the water table, the roots of most land plants are restricted to soil horizons above this level (Fig. 63). Many can produce shallow root systems and thus survive where the water table is high (54); but, where the root systems are thus altered to any great extent, the amount and vigor of shoot growth, fruit production, and longevity may be reduced (351). On the other hand there are many plants such as *Typha, Scirpus, Salix,* and *Taxodium* which are unaffected by poor aeration and regularly send their roots well below the water table. Plants that can neither tolerate poor aeration nor develop a shallow root system are unable to grow where the water table is high.

The O_2 content of the waters of ponds and lakes is much higher than that of ground water but is still well below that required for normal growth of most mesophytes. Along the margins of ponds, streams, and seas, various species show by their distribution their differing tolerances of high water tables, or of submergence during spring floods and high

FIGURE 61. Lee slope of a dune that is encroaching on a forest of *Quercus velutina* and killing the trees. Southern shore of Lake Michigan.

FIGURE 62. *Tilia americana* showing adventitious roots that were formed during the period when the lower stem was covered by dune sand. An advancing dune usually kills trees that lack this ability to produce adventitious roots at successively higher levels.

tides (394). Apparently the maximum depth of water coverage in submergence is less important than the duration of the interval when oxygen is unavailable to the roots. Some experimenters have found that a longer period of submergence is tolerated during the dormant than during the growing season, but in many cases water tables are high enough to be injurious only during the dormant season. The endurance of submergence at different seasons becomes a matter of considerable economic importance in cranberry culture where temporary flooding serves as a means of controlling insect pests (804).

It has often been noted that in rainy climates above-average rainfall is associated with reduced plant vigor and crop yield, but, since poor soil aeration is accompanied by reduced insolation, lower temperatures, and increased fungal parasitism, the relative importance of the aeration factor is not easily assessed. However, some influences of excessive soil moisture can easily be demonstrated by observing the effects of applications of a superabundance of water to lawns or crops. Under such treatment the roots of plants develop in the upper layers of the soil where aeration is best, thereby making them so vulnerable to drouth that the frequency of watering must be maintained. It is an important

FIGURE 63. Root system of *Populus trichocarpa,* which grew on a low river terrace until undercutting by the stream removed the supporting soil. Numerous secondary roots had penetrated vertically downward from the laterals, then ended abruptly in a series of short branches at a level which corresponded closely with the normal level of water in the river.

FIGURE 64. Forest interrupted by the alluvial soils of an aggraded valley bottom. The coarse residual soils of the slopes favor trees because only a little water is required to raise the moisture content above the permanent wilting percentage, moisture penetrates rapidly below the reach of surface evaporation and shallow-rooted plants, and good aeration allows seedling roots to penetrate deeply and early. The heavy soils of the valley floor are occupied by xerophytic grasses which can endure protracted soil drouth, and which on account of their relatively shallow roots, are favored by the high water-retaining capacity of the heavy soil. Estes Park, Colorado.

objective in irrigating to wet the soil only to the maximum depth to which the roots of the particular plants grown can be expected to penetrate. If irrigation is then suspended until growth water is nearly exhausted, roots obtain sufficient O_2 to extend to their maximum depth. Thus, the effective water storage capacity of the soil is at maximum, and the roots have access to nutrients in the largest possible volume of soil.

Attention was drawn earlier to the fact that, although roots do not elongate in the absence of growth water, root systems are most extensive where the moisture content approaches this critical minimum from time to time and the soil is never waterlogged. This is explained by the superior aeration of a soil when relatively little of the pore space is permanently occupied by water. Well-aerated soils in dry regions are a boon to deep-rooted plants, for they allow roots to penetrate to greater depths where the moisture supply is permanent (Fig. 64).

One of the conditions necessary for germination in most seeds is an abundance of O_2. When this gas is deficient respiration proceeds very slowly and dormancy is prolonged, a fact that explains the longevity of seeds buried by plowing and earthworm activity, as well as the superior keeping quality of thoroughly dry seeds stored in sealed containers. Buried seeds may remain viable without germinating for decades but germinate promptly when brought back to the surface, this apparently holding true for lotus seeds buried in a Manchurian bog for 1000 years. Although darkness may be an additional controlling factor with certain seeds, it is generally believed that a lack of sufficient O_2 prevents germination while the seeds lie buried.

Plant Adaptations to Suboptimal Aeration

Adaptations to suboptimal aeration may be morphologic, such as shallow root systems, special aerating tissues, and special aerating organs, or physiologic, such as low O_2 requirements and a special ability to respire anaerobically.

It has been pointed out that a special aerating tissue is not possessed by thallophytes, probably because they originated in water and have been thoroughly adjusted to the oxygen conditions there from the beginning (705), or because nearly all cells contain chlorophyll. Vascular hydrophytes, however, became so adjusted to the high oxygen content of the air that when they again take to the water they must make special provision for aeration. The meristematic tissue of land plants is at first lacking in intercellular air spaces, and it is significant that the growth of such tissue progresses slowly until the air-space system develops, at which time growth increases rapidly. Most vascular plants possess a continuous intercellular air-space system extending through their mature parenchymatous tissues and communicating with the external atmosphere by means of stomata and lenticels. In hydrophytic vascular plants these internal air passageways are elaborated as described earlier (Fig. 31), and in certain mesophytes similar aerating tissues can be developed by individuals growing in wet soil or in water (44, 106). The efficiency of O_2 conduction from shoots is well illustrated by paddy rice

(*Oryza sativa*) in which the roots may contain 18% O_2 while the surrounding mud contains none. Submersed organs may not contain air channels if they are thin enough that nearly all cells are in direct contact with the water, or if the organs are located near the water surface where O_2 deficiency is least (125).

In a number of emergent hydrophytes, such as black mangrove (*Avicennia nitida*), special root branches grow erect until they project above the poorly aerated rooting medium (Fig. 65). These structures, called *pneumatophores,* usually have a well-developed intercellular system of air spaces continuous with the stomata so that they are of unquestioned value in gas exchange (577, 678).

Certain trees that grow on soil subjected to long periods of inundation produce excrescences that rise vertically from their lateral roots at points where the roots occasionally rise to the soil surface (295, 431, 609).

FIGURE 65. A young plant of *Rhizophora mangle,* a mangrove shrub, growing between tide levels in a muddy lagoon in Florida. The prop roots of this plant have lenticels that permit entry of O_2 into porous parenchyma extending downward. The pneumatophores of another mangrove shrub, *Avicennia nitida* (shoot not included in the photo), are shown projecting above the mud sufficiently to extend a short distance out of water at high tide.

They are typically cone-shaped but laterally flattened, and in North America they are commonly called *knees*. The possibility that knees are useful in permitting gas exchange between submerged roots and the free air is doubtful since (a) the only anatomical peculiarities of the xylem of which they are mainly composed is the prevalence of thin walls and parenchymatous elements, (b) the structures are frequently absent where the water is deepest and aerating organs would be of most service, (c) the surfaces are covered with bark and frequently with mats of cryptogams and other epiphytic organisms as well, and these coverings would seriously interfere with gas exchange, and (d) experiments have shown that there is little gas exchange between knees and other roots (427).

Bald cypress (*Taxodium distichum*) is a good example of a tree possessing these knees, and it is also peculiar for the enormous enlargement of the trunks at the water line (Fig. 66). It has been noted that at the water line a woody organ is kept wet by the constant lapping of waves yet has access to an abundance of O_2 not usually associated with such a wet condition. This special combination of environmental factors is closely associated with excessive cambial growth and seems to be the explanation for the production of both buttress patterns and knees (431). Knees are produced only where the soil is periodically exposed to the air, and their height growth depends on wetness and aeration to the extent that, although they may attain a height of 3 m, they never project above the highest level of wave action. Well-established bald cypress trees can endure partial submergence for periods of many years, but since seedlings are killed by submergence it is evident that stands owe their origin to periods of low water levels during which establishment is possible (192).

Many plants that grow best in muddy soils where the O_2 concentration is extremely low have developed especially low O_2 requirements for germination (548). For example, the O_2 requirement for the satisfactory germination of rice is only one-fifth as great as that of wheat (378). In part this adaptation enables the embryo to respire anaerobically for at least a while. This type of adaptation is also necessary for certain upland legumes, the seed coats of which are relatively impermeable to O_2. Only after germination has progressed far enough to rupture the seed coat can normal aerobic respiration of the embryo begin.

The ability to respire anaerobically for a short time without injury is possessed to a limited degree by the mature tissues of most plants, but it is especially well developed in certain hydrophytes which grow in still water or wet soil. Anaerobic respiration begins when the oxygen content of the intercellular spaces drops to about 3%, and it assumes

FIGURE 66. *Taxodium* growing along the Wakulla River, Florida, showing the swollen bases of the stems and the spire-shaped "knees."

increasingly greater importance, as indicated by alcohol production, as the O_2 content drops still lower (434). Possibly anaerobic respiration accounts for the ability of many anatomically unspecialized roots, such as those of willows, to extend into waterlogged soil.

It is clear that many factors change as soil aeration decreases, and this makes it extremely difficult to assess their relative importance. The problem is compounded by the dynamic status of aeration, by the varying abilities of plants to conduct O_2 from their shoots to their roots, and by their varying abilities to endure low O_2 concentrations about their roots (532).

Aeration and Current Effects in Aquatic Habitats

The photosynthetic activity of green hydrophytes in still water may bring about a supersaturation of water with dissolved O_2 that often exceeds 500% in their immediate vicinity. But in many aquatic habitats the amount of green vegetation is so small, or the conditions for photosynthesis are so unfavorable, that O_2 deficiency is very important. The surface of a lake contains less than 1% dissolved O_2 by volume, in contrast with the 21% of the air above, and downward diffusion is very slow. This low concentration, coupled with the influence of so many forms of life competing for it promotes O_2 deficiency whenever there is a lack of convection currents to distribute it throughout the mass of water.

The O_2 contents of lakes and deep ponds are closely related to their thermal stratification, as described earlier (825). In summer the dense, cold water of the hypolimnion is immobile, with the result that the organisms that oxidize organic sediments keep its O_2 content almost zero. This is called *stagnation.* Wind friction causes a circulation of water in the epilimnion, the surface moving in harmony with the wind direction, and a reversed current flowing back through the lower epilimnion or thermocline. Thus in summer only the epilimnion tends to be saturated with oxygen, and consequently it is the principal region of biologic activity, although a few bacteria and other organisms live in the hypolimnion.

In spring and again in autumn there are transitional periods, when the direction of vertical gradient in temperature is reversing, during which there is no temperature stratification. At these times, the periods of *overturn,* convection currents set up by wind affect all water levels to the bottom of the lake, a condition which lasts the year round in very shallow bodies of water. All the decomposition products that have accumulated in the hypolimnion are dissipated during the overturns, and the entire body of water takes on the maximum and uniform load of dissolved O_2. Turbidity also increases as a result of current action on bottom sediments.

The hypolimnion again stagnates in winter, and, if an ice covering prevents wind action on the water, hypolimnion conditions extend up to the surface.

Most organisms of decay require O_2 to be active, and when abundant substrate is available they use up O_2 at a rapid rate. Thus the release of sewage into streams reduces oxygen and depresses the populations of all organisms except those that decompose the material, but before the sewage is carried far the effect of turbulence in mixing oxygen from the surface has permitted nearly complete mineralization of the organic compounds, and the resultant high fertility of the waters allows plant and animal life to flourish in great abundance. In the absence of pollution, organisms of the natural biota are fewer but are more evenly distributed along the stream course (95). Controlled biologic disposal of sewage is accomplished in a very short time by artificial aeration in receiving tanks, so that streams, even at the point where they receive the resultant effluent, are spared the effects of extremely low O_2 content.

Running water, if it is not too deep, can approach the maximum O_2 content of surface films. Many algae and some vasculares such as the *Podostemonaceae* are confined to riffles that alternate with pools in streams, but this is not a reflection of a better O_2 supply there. It has been demonstrated that until water speed exceeds 15 cm sec^{-1} there is a layer of quiet water (the "boundary layer") less than 1 mm thick over the surface of an aquatic, and the effect of currents seems to be that of thinning this layer and so facilitating the diffusion of nutrients toward the cells as well as wastes away from them (838).

Although CO_2 is quite soluble in water, if there are no currents to replace that which is vigorously taken up adjacent to an aquatic when it is photosynthesizing rapidly, the supply can be exhausted. The cells then use CO_2 ions from any soluble bicarbonates present, which results in a precipitation of $CaCO_3$ or $MgCO_3$ on their surfaces, or free in the water. The absorption of CO_2 during daylight hours temporarily raises the pH of water.

Wind

As the atmosphere rotates with the earth, circulation systems are set up within it as a result of the uneven heating of land and water areas, and the large differences in temperature, hence atmospheric density, between the poles and the equator.

Wind is an ecologic factor of considerable importance, especially on flat plains, along seacoasts, and at high altitudes in mountains. It affects plants directly by increasing transpiration, by delivering heat to relatively cool leaves or dissipating the heat load of insolated leaves, by causing various kinds of mechanical damage, and by scattering pollen and disseminules. Less direct effects are numerous, including the transportation of hot and cold masses of air, moving clouds and fog that change water relations and alter lighting conditions, modifying temperature on shores to lee of bodies of water, mixing air and thus preventing temperature inversions, etc. Some of these aspects of wind influence have been discussed previously; others remain to be considered.

The rate of movement of air is generally expressed as average velocity over an interval of time, such as an hour or month, but, since the flow of air is not nearly as constant as the flow of water in a stream, mean velocities for long intervals of time are very misleading. Wind of gale proportions may blow for a few minutes yet not be suspected from the magnitude of the hourly total.

The velocity of wind is affected by the configuration of topography and vegetation masses, by position with respect to seashores, by major paths of wind movement, and by regions of calms. Also the rate of air movement increases regularly with increasing height above the ground, a fact that must always be taken into account in collecting and interpreting records.

Where the significance of wind lies in its effect in increasing transpiration, evaporation measurements are more useful than separate measurements of wind and saturation deficit because of the difficulty of evaluating separate environmental factors. Also, in interpreting wind measurements, it should be remembered that wind velocity at a particular level in the open is always lower than the velocities on all but the leeward side of the canopy of an isolated plant.

Windbreaks (641)

Even a cover of low herbaceous plants strongly reduces the velocity of wind along the ground, and forest cover, especially where the canopies of individual trees are staggered at different heights, reduces the velocity by as much as 80%. The effect of a forest in ameliorating wind and other aspects of microclimate may extend 100 m beyond its edge, with breezes diluting forest influences a like distance inward. By means of this action in slowing down wind currents near the earth's surface, plant cover prevents wind erosion of the soil and causes winds carrying particles to drop their loads. The accumulation of dune sand and loess is thus promoted by plants.

Special plantations of trees or shrubs are often made with a view to providing protection against wind for fields, orchards, buildings, or livestock. A *windbreak* or *shelterbelt* is a densely planted strip of tall vegetation, usually between 15 and 60 m wide, oriented at right angles to the direction of prevailing winds in order to reduce their velocity near the ground.

The effect of a natural grove of trees or of a windbreak, especially one flanked by shrubs, may extend to the leeward for a distance 100 times the height of the trees. However, at this distance the influence is never more than barely detectable, and significant benefits do not extend more than about one-fifth of this distance. By deflecting currents upward there is also produced an effect in a windward direction which may be detected up to about 7 times the height of the trees. If windbreaks are spaced at about 25 heights they provide significant though variable protection throughout a large area.

The effectivity of a windbreak depends on the denseness of the vegetation. If too open, insignificant resistance is offered to the wind. If too dense, undesirable turbulence results. Optimal denseness filters air equal to that of lath or boards offering 50% obstruction to movement (275).

When the plantations consist of deciduous trees the effectivity is considerably reduced during the leafless season, so that theoretically it is best to use evergreens where suitable species are available. Wind velocity itself also has something to do with the effectivity of a windbreak, for the percentage reduction in velocity varies under different wind conditions.

Five major classes of benefits can be expected from windbreaks: (a) a reduction of evaporation and transpiration, and consequently a more efficient use of soil moisture and less severe winter drouth injury; (b) a reduction of damage by breakage, lodging, and abrasion; (c) a reduction in the movement of soil by wind; (d) a more even distribution of snow over undulating topography; and (e) a saving of as much as 40% in the heating of dwellings.

Other influences of windbreaks are detrimental, so that the plantings must be made judiciously in order to accentuate the beneficial aspects and minimize the following detrimental influences.

Windbreak plants exhaust the soil of moisture and nutrients, and they cast shade in their immediate vicinity, thereby considerably reducing the total crop area. With some trees root competition may be evident for a distance greater than twice the height of the tree, but usually roots do not cause much damage beyond the spread of the crown. Root-competition losses are minimized by: (a) using species that tend to root deeply, (b) promoting deep rooting by breaking up subsoil hardpans with dynamite before planting the trees, (c) cultivating deeply near the trees to prevent the development of long shallow laterals, (d) growing highly competitive crops near the trees during the period of their establishment, (e) avoiding the use of species that spread by suckering, (f) planting the windbreaks as narrow as is feasible and spaced as far apart as is consistent with the desired results, and (g) using part of the space adjacent to the trees as a lane.

The economic loss due to shading can be offset to a certain extent by using trees that produce narrow crowns, and by growing adjacent to the windbreak strips of special crops that do not demand much direct light, especially hay crops where flowering is not essential. If rows of trees are planted along every stream, outcropping, and roadside, a relatively small amount of the arable land will have to be relegated to windbreaks. Also, maximum effectivity demands narrow, long, continuous rows of trees with shrubs.

Very dense windbreaks may promote snow drifting to the extent that moisture derived from snow is reduced in the fields and increased in the vicinity of the windbreak where only the trees can benefit much from it. In some instances drifting is so extreme that the water table is actually raised for a time after the spring thaw. This influence can be counteracted to some extent by leaving the windbreak as open as possible near the ground level.

Insect pests of crops and weeds may be more troublesome in consequence of their needing forest microclimate to complete their life cycles, or maintain a seed source. On the other hand desirable predators of insect pests may need the shelter of a windbreak.

Finally, windbreaks interfere with breezes which would otherwise keep the air stirred up and at a more uniform temperature; thus they tend to increase temperature extremes in their vicinity. Although this increases the frost hazard, the high midday temperatures hasten the ripening of grain. Soil temperatures are also affected by windbreaks, but it is not so easy to classify these effects as beneficial or detrimental.

The net effect of well-planned windbreaks is definitely to increase total crop production, although this statement applies chiefly to arid and semiarid regions and seacoasts, and for seasons that are not extremely dry. The balance is thrown still further in a favorable direction when the species used for windbreak planting have some value as a source of food and cover for game animals, or as fence posts or fuel.

Wind Influences on Plants (480)

CO$_2$ replenishment.The boundary layer of still air that covers the surfaces of a shoot to a depth of about 1–2 mm impedes the diffusion of O$_2$ from the stomata and CO$_2$ toward the stomata. This gas exchange is favored by vigorous air movement which thins the boundary layer.

Cooling. The boundary layer over an insolated epidermis accumulates heat by conduction, and if a breeze thins this layer the leaf's heat load is dissipated more effectively. The larger the leaf the greater the wind velocity needed to thin the boundary layer to a given thickness.

A layer of hairs thickens the boundary layer considerably, adding to the transpiration resistance of the plant, but reducing the rate of CO$_2$ uptake. Such hairs probably do not interfere much with the cooling of a leaf since they may act as fins aiding conductive cooling.

In cold environments the cooling effect of wind may be quite detrimental by limiting physiologic activity during the few periods of warm weather. Leaves in such environments can be cooled as much as 7°C if well exposed to wind, and if temperatures are already suboptimal this could be a major factor in the dwarfing of tundra plants (850).

Desiccation. In still air evaporation is simply a process of diffusion, but when air is in motion the process becomes strongly affected by convection. Wind causes evaporation even when the saturation deficit is zero.

With plane surfaces, the rate of evaporation increases with the square root of wind velocity. Thus, while breezes have a strong influence, the influence is not proportionately greater at high velocities. Herein lies the explanation for the practice of whirling a psychrometer. Evaporation from the wet bulb is affected very little by wind velocities in excess of 3 km per hour, and the instrument can be conveniently rotated by hand at about 6 km per hour. Therefore, if rotation is rapid, wet-bulb depression is effected almost entirely by relative humidity.

Wind tends to increase transpiration by thinning the boundary layer, which under still conditions approaches saturation and so impedes transpiration. Wind also bends thin leaves, causing alternate expansion and contraction of the intercellular space, which forces saturated air out and draws drier air in. The efficiency of the cuticle becomes a very important factor in determining a plant's resistance to wind desiccation, for at velocities exceeding approximately 3 m sec^{-1} the tissues lose turgidity sufficiently that the guard cells usually close, and thereafter almost all transpiration is cuticular. These influences are offset to varying degrees by the cooling effects of air movement, so that wind has a small net effect on transpiration (Fig. 18).

Previous mention has been made of the physiologic consequence of wind in increasing the evaporative power of the air during cold weather when both the water-supplying power of the soil and the movement of water through the plant are slow. As westerly winds off the Pacific Ocean cross mountains in the vicinity of the Canada-U.S.A. border they lose moisture progressively as they are forced up over successive ranges, so that by the time they have moved well inland their absolute humidity has been much reduced. On descending the last few ridges the warmth generated simply by descending and being compressed (the *adiabatic* temperature change, approximately 1°C 100m^{-1}) lowers their relative humidity to such an extent that they have remarkable drying power. These are *chinook* or *foehn* winds that fatally desiccate the needles of coniferous trees along the sides of mountains where the warm dry air flows over the surface of cold air lying on the valley floor, adding to the distinctiveness of the thermal belt (511).

Even when the weather is warm and the soil moderately moist, the protracted influence of dry winds can kill leaves and twigs and can injure fruits or cause them to be abscissed (645). The taller the plant the more subject it is to desiccation, as well as to other forms of wind injury. Some plants, normally tall and erect, become low and spreading as a result of wind action. Others are normally low of form and thereby escape

FIGURE 67. *Arenaria sajanensis,* a typical cushion plant of high, windswept ridges in the central Rockies.

much wind damage. The most efficient life form on windy habitats, and one which is especially prevalent there, is that of the *cushion plants* in which streamlined contours are presented by the uniform tips of numerous crowded branches (Fig. 67). The action of wind on many erect plants molds their shoot systems into this cushion form (Fig. 68). When branch tips are killed, the lateral buds produce vigorous growth so that each dead terminal is replaced by several shoots of lateral origin. The resultant complex system of branching gives the shoots of one plant or of a small group of plants a dense streamlined contour. Wind injury to the terminals occurs frequently enough to perpetuate these contours. Each time the twigs that have elongated the most vigorously are killed back the farthest (Frontispiece).

Dwarfing. Plants developing under the influence of drying winds never attain a degree of hydration, and consequently of turgidity, that enables them to expand their maturing cells to normal sizes. As a result all organs are dwarfed, without necessarily being deformed, because constituent cells become fixed at subnormal sizes. Obviously dwarfing attributable to wind desiccation can result only from those winds that blow during the period when cells are expanding and maturing.

At the approach to a seacoast, or arctic or alpine timberline, or the edge of a forest that is adjacent to an extensive prairie, the stature of the trees is gradually reduced, and the farthest outposts are usually confined to those parts of the topography that afford the most protection from wind. Here the dwarfing is generally thought due to an unfavorable internal water balance, possibly aggravated by cooling, and along sea-coasts by salt spray, and the confinement of individuals to protected areas is usually due to death of seedlings that germinate on exposed places. Trees may be so dwarfed that specimens a century old are no larger than a small shrub (Fig. 68).

Deformation (wind-training). When developing shoots are subjected to strong wind pressure from a constant direction, the form and position of the shoot may become permanently altered (Fig. 69). Deformation is not necessarily accompanied by dwarfing, for moist winds can mold the form of a shoot without appreciably reducing its size. Trees with inclined trunks are commonly observed on ridges, in prairie groves, and

FIGURE 68. *Picea mariana* growing at upper timberline near the summit of Mt. Katahdin, Maine. These trees are at least a century old, and, though still sterile, are about as large and as old as the habitat will permit. Dwarfing and deformation by winter desiccation are illustrated. Meter stick is marked off in decimeters.

FIGURE 69. *Pinus flexilis* at upper timberline on Long's Peak, Colorado. Wind pressure has caused the tree to develop a prostrate position from the beginning. Blasting ice particles have eroded the bark from a large area at the base of the trunk.

along coasts where exposure to unidirectional winds is common and severe.

Trees vary greatly in their response to the force of strong wind. *Pinus flexilis, Picea mariana,* certain live oaks (*Quercus*), etc. often grow flattened against the ground while other trees in the same habitat remain erect, although they show other types of wind influence. Tree branches often develop only in a leeward direction. Sometimes this is due entirely to pressure effect as is shown by the fact that branches emerging from the windward side of the trunk remain alive but are all strongly and permanently bent around until they are pointing in a leeward direction (Fig. 70) (449). In other instances asymmetrical crowns result from the death of all buds that form or twigs that emerge on the windward surface, as a result of desiccation, erosion, or breakage (Fig. 71).

Constant winds that cause a woody plant to sway chiefly in one direction evoke an adaptational response in many plants consisting of a flattening of the trunk, roots, and branches in a plane parallel to the direction of flection (267, 388). Thus the organs assume a plank shape and in *Cupressus macrocarpa* along the California coast some trunk sections may be over six times as long as broad, the growth rings showing that the cambium has been active chiefly on the leeward side of the stem (Fig. 72). Here, as well as in other types of asymmetrical xylem deposition, the thickest layers are associated with that side of the bole with the most foliage and hence the greatest photosynthetic capacity. Because wind deformation lessens the value of timber, reforestation or afforestation near windy coasts often necessitates the establishment of a special shelterbelt of trees to prevent deformation which would reduce the marketability of the main plantation.

Anatomical modifications. When a coniferous tree becomes inclined from wind deformation or landslide, a dense, reddish type of xylem called *compression wood* forms on the compressed side and tends to offset

FIGURE 70. Protracted unidirectional wind may keep branches bent until they become permanently oriented in a downwind direction, with the branch bases then becoming imbedded in secondary growth. Dalles, Oregon.

FIGURE 71. The tip of this *Abies lasiocarpa* remains symmetrical for the first few years, but ice-particle abrasion eventually kills all branches except those directly leeward of the bole.

further bending in this direction (484). The corresponding anatomical response of angiosperm stems is called *tension wood.* It develops on the upper side of the inclined stem, where the abnormally thickened xylem layers have gelatinized tracheids which resist gravity by exerting tension (370).

In herbaceous plants wind-sway may stimulate the formation of more collenchyma than usual (787).

Lodging. Lodging is the name given a form of wind injury sustained especially by grasses, such as wheat, maize, and sugar cane. Wind frequently flattens these plants against the ground, but, if the stems are not too mature, the prostrated plants become partially erect once

FIGURE 72. *Cupressus macrocarpa* on the coast of California, showing the effects of landward winds. Xylem layers tend to be extremely thin except on the lee sides of the stems so that the trunk and branches become strongly flattened in a direction parallel to the wind.

more by means of differential growth at the lower nodes which remain meristematic. Such injury is never important except where some factor promotes rapid growth and soft tissues, and it is important only in crop plants. The crop suffers mainly because of mechanical injuries sustained by the stem tissues.

Lodging as well as breakage, to be discussed below, is due to unusually violent winds, especially the tornadoes of temperate regions and the typhoons or hurricanes of tropical regions.

Breakage. The susceptibility of plants to breakage by wind (Fig. 73) depends somewhat on the type of anatomical structure and on whether or not the tissues are frozen. Trees such as *Populus* and *Acer saccharinum* have soft, brittle wood and are especially susceptible to breakage. *Tilia* has equally soft wood but is considerably less susceptible because of the strong sheath of phloem fibers surrounding the wood. The brittleness of any twig is increased greatly when it is frozen, and buffeting at this time may cause severe pruning.

The severity of breakage is also magnified by glaze storms (449) or

FIGURE 73. Breakage and uprooting by the 1938 hurricane, on Mt. Washington, New Hampshire. During the hurricane the wind velocity exceeded 100 km per hour on the summit of this mountain.

wet snows, both of which leave the branches bent to the limit of flection by a heavy burden of solid precipitation. Also, trunks weakened by disease, or fire scars at the base, break easily.

Trees may be uprooted even if the stems successfully resist breakage by high winds. Single trees or groups of trees which are uprooted in a forest are often referred to as wind throws or windfalls. This type of damage, which involves breakage of at least the smaller roots, is favored by anatomical structure that does not resist much tensile force and by shallow rooting. Sometimes artificial drainage channels are dug when a forest is partially cut, with a view to promoting deep rooting and thereby reducing wind throw among the residual trees.

The conditions under which a tree develops has a very important bearing on wind throw. Trees grown in dense stands, and thus not exposed to full wind velocity as they grow, lack the stimulus to become sturdy which open-grown specimens have. When one or two trees are blown over in a wind storm the ones left standing around the edge of the opening are very vulnerable to the next strong wind, and in this way an area of wind throw may increase in size for some time. Trees that are blown only until they lean permanently soon lose much of their value as lumber because compression or tension wood forms, and the bole becomes elliptic in cross section.

When a forest is logged a few trees are often left to supply seed for the next generation. Foresters have tried leaving these seed trees in groups for mutual protection against wind throw. However, this procedure is usually not very effective. Wind throw among such seed trees is almost as great where the groups are subject to mild winds as where

FIGURE 74. Banana leaves, entire as they unfold, tear easily between the parallel veins. Although this reduces their photosynthetic capacity, the heat load of the leaf is much more easily dissipated.

winds are strong, showing the universal weakness of trees developing under forest conditions (822).

The root system of a wind-thrown tree usually lifts a disc of soil which subsequently falls in a heap at one side of the basinlike depression from which it was removed. In this way the surface of the mineral soil over a forest is kept from being flat, and small mounds of mineral soil free of organic layers are made available for the establishment of those species which cannot get started on organic layers. The unevenness of the mineral surface results in very uneven accumulations of organic matter, for such material fills in the depressions before accumulating over the small knolls.

The leaves of large-leaved monocots such as banana (*Musa sapientum*) are quite vulnerable to shredding by wind. Numerous tears between parallel veins (Fig. 74) significantly lower the photosynthetic capacity, and this is reflected in reduced crop yield (480).

Abrasion.   When wind carries particles of ice or soil it is a very potent abrasive force. Bark and buds may be eroded away from the windward sides of woody stems when the plants are situated in exposed places. This abrasive action is strongest a few centimeters above the ground

FIGURE 75. An area where wind erosion of sand has destroyed by undermining all the original vegetation except one bunch of grass which has held intact the soil encompassed by its root system. Central Minnesota.

or snow, and frequently it results in an evident cutting zone on woody plants (Frontispiece, Fig. 69). Crops grown on sandy soils in windy climates are often damaged in this way.

Effects of erosion and deposition. An undisturbed plant cover is very effective in preventing the movement of soil by wind, but when even at one point the cover is thinned or destroyed, wind may scour out the soil so as to expose the roots of the adjacent living plants, bring about their death, and thereby increase the area of devastation (Fig. 75).

The eroded material likewise becomes a hazard to the existence of plants in other places where it is deposited, for few species can tolerate the sharp reduction in aeration about their roots that follows the deposition of new soil upon the old. Those that can survive this change develop adventitious roots at successively higher levels on the stem as deposition takes place.

Salt spray (81, 581). Along seacoasts the spray that dashes into the air in the breakers is carried ashore by wind, and in the immediate vicinity of the ocean wind-borne salt spray may have considerable injurious effect on certain sensitive species of plants (Fig. 76). By means of special salt traps it has been shown that the quantity of air-borne salt diminishes with increasing distance from shore, and by spraying plants it has been demonstrated that species differ in their tolerance of salt, the most tolerant generally growing nearest the sea (481, 569). Plants sensitive to salt spray can grow near the ocean only if they are sufficiently short-lived to complete their life cycles between storms when most of the spray is deposited. Salt damage from severe storms has been observed as far as 27 km inland (552, 592).

The chief damage seems to follow those storms that are not accompanied by rain, so that as the salt spray dries it leaves a film of salt on the plant surfaces. The exact manner in which this film causes injury is not known, but the withdrawal of guttated water back into leaves may well bring into them hypertonic solutions formed while the extruded droplets were in contact with the incrustation on the cuticle (171).

Two chief methods of reducing salt-spray injury to economic plants are available. By planting a windbreak of salt-tolerant species along the ocean, much of the spray is combed out of the wind (704). Second, the toxic effects of salt on the soil can be reduced through the application of potassium salts which have an antagonistic effect on the sodium ions.

Snow cover. In areas of uneven topography, snow is swept from the windward slopes of the prominences and deposited on lee slopes and

FIGURE 76. Shrubs streamlined by winds blowing from the right off the ocean at Roosevelt Beach, Oregon. Probably wind desiccation and salt spray are both important here.

in hollows. Winter after winter the same areas remain thinly covered, while others near by accumulate excessive drifts and are covered for longer periods. The boundaries between vegetation types on habitats where accumulation is scanty and those where it is excessive are often very sharp. Woody plants may be excluded from habitats lacking a protective snow cover, they may be regularly killed back to the snow surface where accumulation is moderate, or they may again be excluded from areas where deep drifts accumulate and persist far into the normal growing season (837). This wind transfer of snow frequently produces very dry soils along the top of a sharp ridge oriented to get maximum exposure. The direct measurement of precipitation here is not only difficult, but snow transfer after falling makes the measurements rather meaningless from the standpoint of plant ecology.

On level topography the irregular drifting of snow has great effect on the thermal characters of environment. When the same series of habitats are compared over a series of years the temperature relations are occasionally reversed simply by the vagaries of snow drifting.

Wind Pollination (Anemophily)

The most primitive pollen-producing plants are believed to have depended on wind to transfer their pollen from anther to stigma. In cool and cold climates the vast majority of trees, shrubs, and herbs are still wind pollinated. When atmospheric turbulence carries pollen to considerable altitude it may be transported several hundred miles.

Although air currents are omnipresent, and a few pollen grains are regularly carried for distances up to 1500 km (654), this method of pollination has certain inherent disadvantages. Because of indiscriminate scattering by wind, the chances of any one pollen grain's alighting on the proper kind of stigmatic surface at the opportune time are very remote. Therefore great quantities of pollen must be cast on the wind if the chances of pollination are to be increased to a point where an adequate seed crop is assured. In consequence wind pollination is very wasteful, as is shown by the sulfurlike films of pollen that form on the soil and water surface when conifers are pollinating, and by the abundance of pollen in the air which so frequently causes hayfever in man. Nevertheless anemophily has proved quite successful, for many of the plants that dominate the earth's vegetation are wind pollinated, especially the *Coniferae,* the *Glumiflorae,* and the *Amentiferae.*

Anemophilous plants have developed certain morphologic adaptations that facilitate wind pollination. The flowers have underdeveloped perianths, but long stamens which are extruded and which sift their pollen into the passing breezes, and stigmas that are well exposed and often feathery, with the result that they literally comb pollen out of the wind. The flowers are typically unisexual, and they are located high on the shoot where they are never shielded from free air movements by the foliage. The pollen has a smooth exine that is not adhesive so that the pollen readily falls from the anther sacs as they open. In certain gymnosperms the buoyancy of the pollen grains is increased by a pair of attached air bladders which greatly increase the frictional surface without adding much weight.

Tests have shown that the dispersal distance bears little or no relation to size of grain, rate of fall, distance of fall, or average wind velocity.

Wind Dissemination (Anemochory)

Wind is the most efficient agent of dissemination, and most terrestrial plants depend on it to scatter their disseminules. Among these plants (called *anemochores*) there are recognized six common types of adaptation which tend to facilitate wind dispersal.

Minute disseminules (854). The seeds of the *Orchidaceae, Ericaceae, Orobanchaceae,* and a few other higher plants are very small, some weighing no more than 0.002 mg. Because of their small sizes, such seeds, together with spores, bacterial cells, etc., are readily picked up and carried by wind or animals. The facts that viable spores are abundant in the upper atmosphere well over 1 km high, and may be found in mid-Atlantic air, may account for the widespread distribution of most fungi and algae and many mosses. The rapid spread of the fungi causing chestnut blight and blister rust of white pines after they were introduced into North America furnishes adequate proof of the efficiency of this type of disseminule.

Comose disseminules. The friction surface of many disseminules is greatly increased by the production of hairs which do not materially increase the weight. The disseminules of the *Salicaceae, Asclepiadaceae,* many *Compositae, Clematis, Epilobium,* etc., all possess sufficient fuzziness so that they are readily picked up and carried by wind. A velocity of 1.2 km hr^{-1} will keep *Taraxacum* achenes afloat, so that this type of disseminule can be blown hundreds of kilometers over land or sea (713). Experiments indicate that comose disseminules may be more buoyant than the most minute glabrous types.

Winged disseminules. Many seeds (*Bignoniaceae, Betulaceae, Pinaceae*) and one-seeded fruits (*Ulmus, Acer, Fraxinus*) produced by trees possess a wing that greatly retards the speed of descent of these structures when they are cast. This allows the wind to carry the disseminule a considerable distance to the leeward of the tree before it strikes the ground. The taller the tree and the slower the descent of the disseminule, the more effective a given wind velocity (707). Studies with seed traps of the lateral distribution of seeds of *Pseudotsuga menziesii* in Oregon have shown that an abundance of seed can be expected to fall as far as 300 m from the trees (384), but updrafts and strong winds can carry seeds of this type for a distance of several kilometers.

Saccate disseminules. In *Physalis, Ostrya,* and many *Chenopodiaceae* the seed is enclosed in an inflated papery structure which can be rolled over the ground or snow surface by wind.

Tumbling disseminules. The entire aerial shoot of *Salsola pestifer, Sisymbrium altissimum,* as well as the isodiametric and finely divided inflorescence of *Panicum capillare* and the globular fruiting bodies of certain puffballs (*Bovista, Disciceda*), break off from the remainder of the plant when mature, and as the structures are rolled along before the wind seeds or spores are knocked loose from time to time. This type of disseminule is common only in grassland or desert vegetation.

The catapult mechanism. Incompletely dehiscent dried fruits, which remain in a vertical position at the stem tips when they mature, often retain seeds that are shaken out only a few at a time when buffeted by wind. Examples are provided by *Delphinium, Iris, Oenothera,* and *Papaver.*

In these plants the disseminule itself is in no way modified, but the mechanism functions as a catapult which develops small but effective centrifugal force. Obviously this mechanism is not very efficient for covering great distances rapidly.

The direction of dissemination when wind is the agent is strongly influenced by the degree to which the winds blow from a constant direction. If winds are unidirectional most of the disseminules will be carried in that direction, and the only chance for plants to migrate in the opposite direction is provided by occasional eddies. It is significant that in forests anemochory is most prevalent among the trees, and relatively uncommon among the undergrowth plants that live in a microclimate with little wind movement.

Wind Measurement

Wind is usually measured (544) with a Robinson *anemometer* consisting of a series of three or four horizontally rotating arms terminated by cups. The arms are fixed to a vertical shaft which operates a train of gears motivating a dial. If dial readings are used, only average velocities can be computed for the interval between readings, but electrical recording apparatus can be attached which records the time required for each mile of wind to pass. This provides a much more useful index of velocities, for transient wind conditions can have effects that operate over a long subsequent period. In microclimatology a hot-wire anemometer or thermistor is often useful (53, 590).

A simple and inexpensive means of comparing windiness at different places is to make a number of unhemmed flags of the same size and cut from the same piece of cloth, placing one of these at each of a series of locations to be compared. After a suitable period of exposure, visual estimates of degrees of tattering are considered proportionate to wind influence (667).

6

The Biotic Factor

The average green plant is commonly referred to, or at least thought of, as an "independent" organism, in contrast to other plants and animals that cannot synthesize their own food. In reality, however, not even green plants are independent, for they are considerably influenced by other organisms, although in ways that may not be obvious. For example, the blueberries and huckleberries (*Vaccinium* spp.) depend on nectarivorous insects for pollination, upon fructivorous birds and mammals for dissemination, and, like many other green plants, they depend in part on fungi associated with their roots to absorb nutrients.

Even the CO_2 used by a green plant in photosynthesis has been released in respiration by other organisms, and the oxygen used in respiration has been accumulated largely by previous generations of green plants. The amounts of heat, light, moisture, and nutrients available to one plant are all conditioned by the proximity of other plants, and most of the soil nitrogen, it will be recalled, is organic. Furthermore, at least some injury from disease-producing organisms and herbivores is sustained by almost all plants. Our "independent" plant, on analysis, therefore turns out to be no more than a figment of the imagination, and it becomes clear that the environment of any organism is always in part biologic as well as physical.

Physical factors are usually the true governing forces of environment, one organism ordinarily affecting others by its ability to modify the physical environment; the numerous means of doing so have been pointed out in many places previously. However, some biotic influences are truly direct, as illustrated by grazing, animal pollination, and animal dissemination.

Classification of
Symbiotic Phenomena

The terminology concerning the relationships of one organism to another has long been in a state of confusion. Three terms are in common use where plants are involved; parasitism, commensalism, and symbiosis. All biologists seem agreed that parasitism is a term that should be applied to a situation in which one organism, the parasite, benefits nutritionally at the expense of another, the host, but there is considerable disagreement on the application of the other two terms.

The classification developed by McDougal (508) involves no special terminology. It makes provision for practically all types of relationships, from the most remote to the most intimate. The ecologically important fact is emphasized that no organism lives in an environment where it is free from numerous influences of others. And finally, the term symbiosis is used as De Barry intended when he proposed the term in 1879, i.e., to embrace all kinds of interrelationships between organisms (Gr. *syn* = together, *bios* = life). This classification with some examples involving plants, is as follows:

I. *Disjunctive symbiosis* (i.e., associated organisms not in constant contact).
   A. *Social* (i.e., no direct nutritional relationships).
      Includes the effects of one plant on another with respect to shade, air movement, soil moisture, etc.; also birds nesting on trees.
   B. *Nutritive.*
      1. *Antagonistic.*
         Herbivorous animals and their food plants; carnivorous plants and their prey. The term predation is commonly used when the victim is killed, as when insects destroy a seed crop (392).
      2. *Reciprocal* (Reciprocal here does not imply that the two symbionts are mutually helpful, but rather a condition of reciprocal parasitism in which the advantages of the relationship exceed the disadvantages for both organisms).
         Animals effecting pollination or dissemination incidental to their activities in obtaining food from plants; fungi cultured and disseminated by insects that use them for food (777, 788); the agricultural pursuits of man; ants inhabiting hollow thorns or other organs of vascular plants which they defend against the attacks of other insects (777).
II. *Conjunctive symbiosis* (i.e., dissimilar organisms living in contact with each other).

A. *Social.*
Lianas and epiphytes using other plants for support; algae inhabiting the hollow interiors of *Azolla* leaves, the roots of cycads, the ventral cavities of the thallus of *Anthoceros,* the stem cortex of *Gunnera,* the surface layers of the hair of sloths; the cryptogamic flora of certain Papuan weevils (288), etc.

B. *Nutritive.*
1. *Antagonistic.*
Parasitism of plants by bacteria, fungi, Protozoa, nematodes, insects, mistletoe; galls formed by bacteria, fungi or insects (237, 525); parasitism of animals by bacteria and fungi.
2. *Reciprocal.*
Lichens; mycorhizae; nitrogen-fixing bacteria and blue-green algae in roots, stems, and leaves; algae inhabiting *Protozoa,* coelenterates, molluscs, flatworms, the egg membranes of amphibians, etc. (107, 562.)

Herbivorous Animals

Grazing and Browsing *

Many ecologic problems are involved in the management of natural or artificial types of plant cover so as to provide good forage for livestock and game animals. First there is the matter of food preferences. For each kind of animal, the plants available to it may be classified in a sequence from those that are very palatable to those that are strictly avoided. Palatable species tend to be the most severely injured, and unpalatable ones may escape injury and even be benefited by the release from competition. Among the palatable species differences in nutritive value must be taken into account in those planting and management practices that benefit certain species at the expense of others (137). Owing to differences in seasonal development different types of plant cover may have different periods during which they are grazed to best advantage.

Grazing may injure a plant either because the frequency or degree of removal of its photosynthetic organs curtails its assimilation or because of its susceptibility to trampling. Less direct but even more important consequences of intense grazing result from erosion and deposition

---

* *Grazing* refers to the use of unharvested herbs as forage by animals; *browsing* refers to a similar use of shrubs or trees, although there is not much significance to this distinction.

that result when plant cover is so thinned that the soil is no longer protected from the erosive influences of wind and water. Grazing damage to certain species is magnified by drouth cycles, the plants being unable to cope successfully with such a combination of adverse conditions. On the other hand, plants the shoots of which are kept clipped to small size by continual grazing may be better able to withstand drouth. They do not keep the moisture reserves depleted to the normal extent, and furthermore, during the drouth the smaller plants make correspondingly smaller demands on the supply of moisture stored in the soil.

The effects of grazing on different species of plants depend to a large extent on life form. Annuals that are palatable quickly disappear from an. area if grazed so much that they cannot set many seeds. Among herbaceous perennials, palatable grasses and sedges withstand grazing much better than forbs, for on account of their basal meristems the leaves are not destroyed when only distal segments are removed; in fact the plants are stimulated by mild grazing. Also, trampling is least apt to injure the vegetative buds of these plants, for they do not extend above the ground surface and are protected by the dense tufts of foliage. Shrubs are less easily injured by browsing than herbs are by grazing because browsing, generally confined to new growth, removes but a small portion of the shoot, and because the greater longevity of shrubs increases the chances of survival of sufficient seedlings to replace old-age mortality. Low shrubs may be kept reduced to dwarf "hedged" mats, and tall shrubs and trees, if they can attain a height above the reach of animals, finally become essentially free from direct injury.

The least damage is sustained by perennials if the new foliage is not removed until the plant has had sufficient time to restore part of the underground food reserves which were nearly exhausted in building up the new photosynthetic organs (Fig. 77). If grazed heavily and early the plant is weakened and killed because the roots are starved and consequently become inefficient in performing their normal functions. By grazing different parts of an area at different seasons, altering this seasonal sequence, and allowing each unit to be completely relieved from grazing pressure every few years, vegetation can be maintained at a high level of productivity. The details of such rotations need to be adjusted for the type of vegetation, its condition at the start, the type of herbivore, etc., but the principle has wide application where man can control the activity of any class of herbivore, or even the use of campgrounds.

The aspect of vegetation may be markedly changed as a result of these differences in reaction of individual species to the grazing factor. Grazing in shrubby vegetation often increases the numbers and sizes of the shrubs by removing the competitive grasses that would otherwise use up a large share of the water and nutrients available. Grazing in purely herbaceous vegetation generally results in a sparser plant cover consisting of fewer species, these being unpalatable or having such

FIGURE 77. (*Upper*) Late winter picture showing completeness of utilization of *Agropyron spicatum* in the pasture to the left of the fence. (*Lower*) Later picture of the same area showing that the close utilization of cured forage during the dormant season had no significantly detrimental effect on subsequent growth. The bunches in the ungrazed area to the right are denser and lighter in color owing to the presence of old bleached shoots. *Sisymbrium altissimum,* an annual forb which has been favored by cattle grazing, is conspicuous in the grazed area. Southeastern Washington.

a short growing season that livestock does not have time to cause serious damage. The kind of animal that produces most of the grazing influence is also an important factor. Sheep normally prefer forbs, horses and cattle prefer grasses, and goats and deer prefer broad-leaved woody plants.

Although rodents are much smaller than the principal livestock animals, their numbers are often so great that they exert an important degree of grazing pressure. Even mice can keep vegetation in a depauperate condition. Each family among the burrowing rodents affects but a limited area about its burrow, but if communal, large areas may be nearly denuded. The jackrabbit, in contrast, depends on speed rather than a burrow for safety, and therefore affects wide areas rather uniformly. In Utah it was found that 5.8 jackrabbits consume or waste as much forage as one sheep (167).

The grazing and browsing of certain wild rodents of positive economic importance are not wholly detrimental. In heavily populated areas the value of cottontail rabbits as game has stimulated considerable effort in creating and preserving vegetative conditions favorable to this animal. Also the beaver, valued for both fur and erosion control, deserves to have those species of trees the bark of which forms the bulk of its diet planted and conserved.

Many plants have developed adaptations that reduce their susceptibility to grazing. Some, like *Chrysothamnus*, *Gutierrezia*, and *Vernonia*, have objectionable taste, and other have developed stout thorns. Certain *Acacia* spp. have developed a highly specialized relationship with stinging ants whereby in return for providing the ant with food and housing accommodations at low energy cost, the *Acacia* is very effectively protected from all other classes of herbivores as well as from competition from nearby plants (391). Other characters that tend to repel insect attack more directly are pubescent ovaries, alkaloids and high silica content.

Animal Destruction of Seeds and Seedlings

The high energy value of the particular types of foods stored in seeds makes them very satisfactory as a source of food for animals. Insects, birds, squirrels, mice, and other rodents annually consume tremendous quantities of seed (716, 748, 799). As with grazing animals, the factor of food preference enters strongly here. Small seeds may be neglected when larger ones are available (57). The amount of seed produced by a particular type of perennial varies considerably from year to year, and when the seed crop is small, seed-eating animals may destroy it entirely. Since a perennial can well afford years of barrenness, such irregularity

FIGURE 78. (*Upper*) Virgin sugar maple-beech forest in central Indiana, showing an abundance of reproduction and a forest floor carpeted with leaf litter. (*Lower*) Similar forest type in the same region that has been subjected to heavy grazing and browsing by cattle. Forest reproduction has been eliminated, so there will be no young trees to replace the old ones when they die. The ground is covered by a *Poa pratensis* sod which provides good pasturage. A scattering of the unpalatable *Vernonia altissima* indicates that the sod has been weakened by overgrazing.

may be a significant adaptation keeping these semevore populations low, for the animals must have food every year without fail to survive. However, to be effective the barrenness must affect the entire plant population simultaneously, and at the same time the insect must have strong dietary restrictions.

Reforestation has often been tried by scattering seed over the ground, but with rare exceptions this method has proved unsuccessful chiefly on account of the biotic factor. Caged forest mice have been observed to consume 2000 seeds of *Picea glauca* in a day. A degree of success can be achieved when the seed is broadcast on areas within a few months after they have been denuded by fire. The barren condition created by the burning drives most rodents and birds out of the area, and seedlings may have a chance to become established before vegetative cover attractive to these animals is reestablished. Under certain circumstances the extensive use of rodent poisons may be warranted in good seed years to insure the survival of adequate crops of seedlings.

Not only are seeds consumed by animals, but during their first summer seedlings remain so succulent that they are subject to attack by rodents, wireworms, etc. Also, grazing animals may cause much damage to seedlings by trampling and browsing them (602). Where snow cover limits the availability of winter forage, and the populations of game animals or livestock are heavy, all woody plants within reach may be browsed until killed. This is most detrimental to young trees and shrubs that protrude but slightly above the snow, and it is obvious that continued browsing of these plants eventually destroys woody vegetation for lack of replacement of old trees and shrubs as they die (Fig. 78).

Insects destroy many seeds, especially by feeding on the developing ovules, and plants have developed several characters which serve to minimize such damage, including toxic compounds, repellants and indurated seed coats or fruits (392).

Aquatic Plants as Animal Food

The economic importance of ducks, muskrats, fish, and other aquatic animals has stimulated considerable interest in the hydrophytes that provide them with food and shelter. Techniques for making inventories, for planting desirable species, and for destroying worthless kinds have been developed (499).

Management of the margins of bodies of water impounded for irrigation presents an especially difficult problem. Because depth of water is so important a factor in the autecology of rooted plants, most of them cannot grow where there is an appreciable annual "draw-down" of water. Instead of supporting a dense growth of food and cover plants as would ponds of equal size, the margins of such reservoirs are characterized by barren aprons which are somewhat of an erosional problem and consequently contribute to turbidity and sedimentation (Fig. 79). Birds and mammals are largely excluded by these conditions, though fish do not fare so badly because their food chains are based upon plankton organisms that are not disastrously affected by the annual draw-down.

FIGURE 79. Eroded apron about a mountain lake dammed up as an irrigation reservoir. Southern Idaho. It is possible to reduce erosion and make such areas both productive and attractive by transplanting suitable nursery-grown stock.

Exclosure Technique

A simple but effective method of studying the effects of herbivorous animals on plants is by *exclosures.* A fence or other type of barrier is constructed to exclude a particular class of animals from a small experimental plot which can thereafter be compared with adjacent but unprotected areas. The size and method of construction of an exclosure depends on the type of animal to be excluded. Small conical tents of hardware cloth are effective to protect seedlings and other small experimental plants against mice or other rodents (800) and birds. Shallow boxlike cages of chicken wire have been used to exclude birds. Suitable types of fencing material have been widely employed in studying the influences of large rodents, sheep, cattle, deer, etc. (730).

Exclosures that are maintained for a considerable length of time permit an evaluation of the net effect of animals on the surrounding vegetation and provide a basis for making beneficial adjustments in range use (158). These areas also show which changes in vegetation are the result of climatic cycles and weather rather than of grazing pressure. When an exclosure is established after an area has been badly overgrazed recovery may be slow in spite of release from grazing pressure because the current phase of a climatic cycle may be unfavorable to seedling establishment or because certain species have been so completely eliminated that no adequate seed source remains.

Temporary exclosures which are moved from place to place permit the measurement of plant production during each season as well as an evaluation of the quantity of forage consumed by animals. On such plots forage is commonly measured by clipping and weighing the ovendry material. When movable exclosures are used, it is important to select different parts of the grazed area for experimental and control plots each year in order to rule out effects carried over from preceding years. Some attempts have been made to simulate grazing by clipping, but the natural forage preferences of animals, together with their trampling and manuring effects, greatly reduce the significance of such studies. However, clipping grasses and the new growth of shrubs allows a study of the extent to which the plants can endure regular loss of photosynthetic organs and still remain in a fair state of vigor.

It should be obvious that the fence or screen used in constructing an exclosure will promote the drifting of snow and deposition of dust, will provide some shade, and will interfere with normal air movement (175). Since the object of exclosure studies is the control of but a single factor of plant environment, namely herbivore pressure, the utmost attention should be directed to the problem of minimizing the effects of the barrier on physical factors. This is accomplished by using exclosures of the largest size and of the lowest and most open structure that will possibly give the desired type of protection. The degree of success in this respect can be estimated by comparing evaporation rates and other factor intensities within and without exclosures (160, 844).

Enough toxic zinc can wash off rustproof metal netting to account for most of the vegetation changes induced by a small cage used as an exclosure (316). Therefore new netting should be allowed to weather a long time before use.

Large exclosures tend to divert the movements of excluded livestock, so that areas selected outside for comparison must be situated where they will be free of paths used in passing around the exclosure.

A final difficulty that must be guarded against is that the establishment of an exclosure, especially of the small fenced type, may encourage the development of an abnormal rodent population on the area.

Carnivorous Plants

Pitcher plants are any of a number of species belonging to the genera *Sarracenia, Darlingtonia, Nepenthes,* etc., the leaves of which are somewhat pitcher-shaped and partly filled with liquid (Fig. 80). This liquid is partly or entirely an excretion from the leaf surface and, in some species at least, has been shown to be essentially an aqueous solution of proteolytic enzymes (550). Insects and other small animals wandering into the leaf commonly drown in the liquid because their efforts to escape are thwarted by downward-pointing hairs or an extremely smooth vertical surface. Enzymes dissolve the softer parts of the bodies, thus providing soluble amino acids that can be absorbed by the leaf. Wholly unrelated to this nutritional phenomenon is another relationship that pitcher plants commonly maintain with other kinds of animals. Certain protozoa as well as the larvae of a few dipterous insects are able to resist the enzymatic action and live within the pitcher fluid (617).

Sundews (*Drosera* spp.) comprise a genus of small herbs which are widely distributed in boggy habitats. The leaf blades are somewhat orbicular or spatulate, reddish, and covered with prominent gland-tipped hairs, each of which bears a glistening droplet of a sticky fluid. Small insects alighting on the leaves become stuck to a few of these hairs, and this stimulus causes other hairs to bend over so that most of the glands touch the body. The same stimulus causes the glands to secrete proteolytic enzymes that digest certain parts of the insect body, then when the digestible parts are absorbed the hairs resume their former position and are ready for another victim. It is interesting to note that materials containing no protein fail to stimulate the movement of the hairs or the secretion of enzymes!

The aquatic bladderworts (*Utricularia* spp.) are delicate herbs that bear bladderlike traps 5 mm or less in diameter. These traps have trigger hairs attached to a valvelike door which normally keeps the trap tightly closed. The sides of the trap are compressed under tension, but when

FIGURE 80. Pitcher plants. (*Left*) *Sarracenia flava* growing in Georgia. The opening to the pitcher is directed upward and is sheltered
from rainfall by a bright-yellow, umbrellalike flap supported at one
side by a maroon-colored stalk. (*Right*) *Darlingtonia californica* growing in Oregon. The opening is on the bottom surface of the curved,
bright-yellow upper part of the leaf.

a small form of animal life touches one of the trigger hairs the valve
opens, the bladder suddenly expands, and the animal is sucked into
the trap. The door closes at once, and in about 20 minutes the trap
is set ready for another victim. A single plant bearing a number of traps
that work at this speed can capture a great many small animals each
day.

The leaf blade of the Venus flytrap (*Dionaea muscipula*) superficially
resembles a steel trap, and the red upper surface bears six sensitive
hairs, any two of which, if contacted, cause the leaf to close suddenly.
Subsequently digestion and absorption take place; then the leaf blade
resumes its original open position.

None of the above carnivorous vascular plants seems in the least
dependent upon its animal prey for nitrogenous compounds; therefore

it must be concluded that the carnivorous habit is only an incidental feature of their nutrition. However, many fungi that parasitize man and other animals are obligate carnivores. Certain aquatic fungi (e.g., *Dactylella bembicoides*) possess looped hyphae that constrict and hold any wormlike animal that crawls through the mycelium; in others (e.g., *Zoophagus insidians*) hyphae serve as bait and suddenly expand when engulfed by a feeding animalcule, thus effecting its capture. *Arthrobotrys entomophaga* captures insects by means of adhesive secretions from the ends of specialized hyphae (205).

Pollination by Animals

The cross-pollination * of most showy or odorous flowers is secured by flying insects, chiefly bees, butterflies, moths, and flies. These insects are by far the most important group of animal pollinators, but other kinds of animals occasionally perform this function. Hummingbirds and the sunbirds are the ecologic equivalents of insects in this regard. Bats (780), rats, and snails also appear to be the pollinators of certain flowering plants.

Many species of plants, including such important crops as fruits, legumes, onions, cucurbits, and crucifers, absolutely depend upon insects for pollination. Fruit growers and glasshouse men commonly rent hives of honeybees during the proper seasons to insure pollination. In fact, the monetary value of all the honey produced commercially is far exceeded by the value of the honeybee in pollination. Honeybees are especially efficient on account of their habit of visiting only one kind of flower on each collecting trip, thus carrying nearly pure burdens of one pollen from one flower directly to another of the same species.

Insects visit flowers either for nectar or for pollen, both of which are a source of food. It is believed that long ago some insects developed the habit of feeding on the abundant pollen of primitive flowering plants, which in the beginning were presumably anemophilous. Certain of those species of plants developed protective measures against insect marauders, such as unpalatable pollen or a very short period of pollination, and thus became obligate anemophiles. Others found the insect visitations beneficial in that the accidental distribution of pollen by the insect as it flew directly from flower to flower proved far more economical

---

* The broad evolutionary significance of cross-pollination will be brought out by the discussions in Chapter 9. However, at this point it may be stated that self-pollination in wild plants is not particularly detrimental, although cross-pollination is of distinct advantage in permitting a maximum degree of adjustment to existing environmental conditions.

than the wind method. These plants became *entomophilous* by developing adaptations favoring insect pollination: showiness due to the color, size, or aggregation of the flowers into clusters; fragrance due to volatile oils; perfect flowers with stamens and pistils that are scarcely if at all exserted, but instead are strategically located with respect to the insect's path. Their pollen is always sticky and often rough or spiny so that it adheres to an insect's body but is not readily picked up by wind. This pollen is not produced abundantly unless, as in poppy, it remains as the principal inducement to visitation. Although there are many flowers that are regularly cross-pollinated by insects without possessing devices more specialized than the ordinary features just enumerated, others exhibit an infinite variety of highly specialized features that tend to insure this service. Some of these special adaptations are as follows (415).

Berberis and *Kalmia* have stamens that are reflexed until an insect scrambling over the flower causes them to spring inward and dust pollen on its body. In alfalfa and other legumes the stamens and stigma are held in the keel under tension and are freed ("tripped"), striking the under part of the body, when a bee depresses the keel to get at the nectar. When pollen dusting is as indiscriminate as in these flowers, partial or complete self-sterility or the maturation of stamens and pistils at different times (*dichogamy*) is necessary if self-pollination is to be prevented.

Catalpa, Tecoma, and *Utricularia* have lobed stigmata with the stigmatic surfaces on the inner faces. A few seconds after an insect contacts these stigmata they close together, thereby guarding against self-pollination and possibly aiding the germination of the first pollen received (564).

The perianths of *Aquilegia, Delphinium,* and *Viola* have deep spurs into which nectar is secreted. These are always located so that insects normally must crawl into the flower and brush against the stigmata, then the anthers, to get the liquid. Though the nectar contained in the long spurs is normally available only to those insects possessing long proboscises, certain others, including insects and hummingbirds, have learned to steal the nectar without entering the flower by puncturing the lower end of the spur from without!

In *Cypripedium* insect visitors enter the inflated lip through the opening in the top but leave by means of openings in the back of the corolla; in pursuing this route they contact first the stigma and then the anther.

In *Salvia* a remarkable lever device, against which the insect's head strikes, brings the anther down on the insect's back as it enters the corolla tube.

The stamens and pistils of *Mitchella, Decodon, Pontederia, Primula,* and *Linum* are of two or more different lengths, a condition referred to as *heterostyly*. Pollination is effected chiefly between those organs of the same length, because that part of the insect's body that receives pollen from a long stamen comes into contact only with the long type of pistil, which occurs only on short-stamen flowers, and vice versa.

The relations between insects and flowers described thus far are the result of the insects' visitations for food, but the attractiveness of flowers to insects may result from other types of relationships. In the genus *Yucca* the sole pollinator, the female *Tegeticula* moth, carries a ball of pollen from one flower to another, thrusts this down the tubular stigma, then uses her ovipositor to place eggs among the ovules whose fertility she has just insured. A few seeds always escape destruction by the larvae, so that in the long view the *Yucca* is benefited by this relationship. Carrion insects, likewise searching for a suitable medium upon which to lay their eggs, are attracted to the ill-smelling carrion flowers such as *Smilax herbacea, Symplocarpus, Lysichitum,* and *Stapelia.* Here the insect is tricked into performing its valuable function, for in its fruitless wandering over the flower or inflorescence it accidentally transfers the pollen. Nor does the insect benefit in the following circumstance. The lip of certain orchids (*Ophrys* spp., *Cryptostylis* spp.) is constructed, colored, and scented so that it resembles the female of certain hymenopterous insects, thereby inducing males to alight and attempt copulation, which act, when repeated on different flowers, brings about cross-pollination (430, 855).

Specialization connected with insect pollination has tended toward parallel and complementary evolution between insects and flowers, leading to the complete dependence of one kind of plant on one kind of insect. When such an interdependence involves an economically important plant, the insect deserves special consideration. For example, numerous attempts to grow marketable quality Smyrna figs in California were unsuccessful until finally the pollinating wasp *Blastophaga grossorum* became established. Again, the production of red clover seed in Australia became possible only after the bumblebee was introduced, for the flowers are usually too deep for honeybees to reach the nectar. *Vanilla* must be pollinated by hand if grown out of the range of its special pollinating insect.

Extreme specialization is as dangerous in connection with pollination as with other functions in the biologic world, for not only must the ranges of the two symbionts coincide but also the extinction of one heralds the doom of the other. *Yucca* and *Tegeticula* appear to have reached the final stage where each is dependent on the other, but such a degree of interdependence is not common, perhaps for the simple reason that the situation is too precarious to endure for long.

The interrelations between flowers and insects is a fascinating topic on which there is much literature but little consistency in experimental results (145).

Dissemination by Animals (Zoochory)

Edible Disseminules

Disseminules ingested but not digested (endozoochors). The seeds of berries and other small fleshy fruits are encased in succulent tissues which are used by birds as food, yet in many of these fruits the seeds (*Ribes, Rubus*), achenes (*Fragaria*), or stones (*Elaeagnus, Prunus*) have thick coverings and so pass through the digestive tracts uninjured to be carried some distance before being excreted (500). This is, of course, the fate of only those hard seeds or stones that are swallowed whole, for such birds as grosbeaks that crack the hard coverings to extract the soft endosperm and embryo destroy rather than disseminate seeds. The passage of seeds through a digestive tract facilitates germination in certain species.

Cattle readily eat the sweet pods of *Prosopis*, but many of the hard seeds escape injury during mastication and pass unharmed through the digestive tract. Cattle are known to disseminate *Berberis vulgaris* (405) and herbaceous weeds (661), and jackrabbits disseminate *Opuntia* (757).

The spores of stinkhorns pass uninjured through the digestive tracts of the carrion insects that feed on the hymenium of these fungi.

Disseminules with only accessory tissues ingested. Rodents commonly carry the fleshy fruits of *Diospyros* and *Crescentia* to a place of relative safety from predators before consuming the flesh and discarding the seeds. The seeds of *Trillium, Cyclamen, Pedicularis*, etc., have fleshy appendages eaten by ants that carry them about before gratifying their hunger (571).

Disseminules cached but not recovered. Rodents and birds are suspected of hiding many seeds which they subsequently fail to recover, and which may be situated in a position favorable for germination. Apparently this is the only manner by which plants with heavy disseminules such as *Juglans, Carya*, and *Quercus* can extend their distribution up-slope (293). The value of rodents in this respect is easily overestimated, for their ability to detect the presence of buried seeds under experimental conditions is remarkable; probably very few seeds are not recovered except in years of exceptionally abundant supplies.

Inedible Disseminules (Epizoochores)

Specialized for clinging. Burrlike fruits or inflorescences, such as those of *Arctium, Bidens, Cenchrus, Desmodium,* and *Xanthium,* are especially well adapted for clinging to the fur of animals which carry them some distance before chewing them from their fur or shedding them with it. Ecologically related but certainly less pleasant types of disseminules are found in species of *Opuntia* and in *Stipa,* in which backwardly barbed spines cause the stem segments and florets, respectively, to become anchored in the flesh of animals that come in contact with them. Adhesiveness in *Adenocaulon bicolor* is accomplished by glandular hairs covered with a gummy excretion. In fungi belonging to *Phallales* and mosses in the genus *Splachnum,* the spores are transported by insects which are attracted to the plants by foul odors.

Unspecialized. Mud clinging to the feet of migratory water birds has been found to contain a wealth of tiny disseminules of aquatic plants—a fact that explains the appearance of such plants within a very short time after any suitable body of water accumulates. Spores of many kinds, including those of fungi causing important tree diseases, have been isolated from the plumage of migratory arboreal birds (327). Many plant pathogens are carried from host to host by insects. Such important diseases as the Dutch elm disease and cucumber wilt are transmitted by these organisms.

Man's activities have allowed many plants to extend their ranges in the last few centuries. The facts that rare weeds first make their appearance along railroads and that many plants are common only on ballast dumps indicate important means of dissemination. Weed seed unintentionally included with grain, and diseases in imported nursery stock, constitute other important methods by which plants are spread. Examples of the effectiveness of these two methods are provided by the history of Russian thistle (*Salsola pestifer*) and of white pine blister rust (*Cronartium ribicola*) in North America.

Lianas

Vascular plants that are rooted in the ground and maintain their stems in a more or less erect position by making use of other objects for support are called vines or *lianas.* This habit of growth has the advantage of enabling the plant to get better light with a maximum economy of supporting tissues. Those lianas that climb over other plants are, in a sense, parasitizing the other plants' supporting tissues, but there is no direct nutritional relationship between true lianas and the trees upon which they grow. Lianas, however, do show all degrees of intergradation with climbing parasites (338).

Anatomically liana stems usually exhibit two peculiarities: the woody cylinder occurs as strands separated by vertical partitions of parenchymatous tissues, and the xylem vessels are long and wide. The first of these characters may be looked upon as an adaptation favoring suppleness, which may be of advantage in accommodating flection resulting

FIGURE 81. Sectioned stem of *Amelanchier alnifolia* showing the effects of constriction by the woody liana *Lonicera ciliosa*. Interference with the downward translocation of foods in the tree has allowed a greater amount of wood to form on the upper side of the constriction.

from unrigid supports. The superior conductance of the wood may be necessary because the stem remains relatively slender yet supports considerable foliage at its summit.

The abundance of lianas varies directly with the humidity and warmth of the climate, so that this ecologic class is most conspicuous in the moist tropics. They are of minor economic importance in forestry in three ways. In the tropics the trees may be so thoroughly woven together by myriads of woody vines that an individual will not fall when cut, thus necessitating the felling of trees in groups. Also, by the constriction which develops as both tree and vine grow, a woody twining liana inhibits downward translocation with the result that the rate of increment in the supporting tree is increased above and decreased below the point of stricture (Fig. 81). Although the tree may be killed in some instances, in others new conductive tissues may form that have a spiral orientation in harmony with the spiral course of the liana stem, or possibly natural root grafting may prevent starvation until the constricting stem is engulfed (493). Woody vines frequently spread their canopies over those of the trees they climb and thus interfere with their normal growth (235).

Lianas may be classified according to the following scheme which emphasizes increasing degrees of morphologic specialization:

1. *Leaners.* Those plants that have no special devices for holding onto a support, e.g., *Plumbago capensis.*
2. *Thorn lianas.* Lianas possessing thorns or prickles which are hardly specialized for the function of climbing but which prove of definite passive value in this regard, e.g., rambler roses, *Bougainvillea, Euphorbia splendens, Galium.*
3. *Twiners.* Lianas in which the entire stem twines about the support. These are mostly herbaceous plants such as *Phaseolus* and *Ipomoea,* perhaps because woody plants are at a distinct disadvantage when twined about trees with secondary growth. As mentioned above, woody twiners such as *Celastrus scandens* and *Lonicera ciliosa* become deeply imbedded in the tissues of the supporting trees so that the continued growth of both tree and vine is impaired (Fig. 81). With but few exceptions the direction of twining in this class of lianas is either constantly clockwise or constantly counterclockwise among the individuals of a taxon.
4. *Tendril lianas.* Lianas possessing special organs, the tendrils, which are modified to facilitate climbing either by twining about the support (many *Cucurbitaceae, Vitis*) or by adhering to its surface (*Hedera helix, Vanilla planifolia*). Tendrils are generally weak organs until they become attached, but subsequently their mechanical tissues develop to such a striking degree that an individual tendril

becomes capable of supporting great weight. Before they have made contact with a support they are very sensitive and may respond to a stimulus of touch lasting only a few seconds. Tendrils may be parts of leaves, such as the modified terminal leaflet of a compound leaf (*Pisum, Vicia*), or the modified tip of an acuminate blade (*Gloriosa*), or the petiole which takes a turn about the support (*Clematis*). In *Vitis* the tendril is of stem origin. Adventitious roots emerge from the stems of *Vanilla* and *Parthenocissus* to serve as tendrils.

Lianas may also be classified on an ecologic basis into heliophytic and sciophytic categories. The former spread their foliage over the canopy of a supporting shrub or tree, often achieving this position by growing upward with the host. When a tall tree has been used as the support, the liana stem hangs free from its trunk as a "monkey rope." Sciophytic lianas (*Monstera, Lygodium, Vanilla,* etc.) climb the stems of shrubs or trees but complete their life cycles without ever reaching the sunny surface of the host's canopy.

Epiphytes

Epiphytes are plants growing perched on other plants, which differ from parasites in not deriving water or food from the supporting plant and from lianas in not having soil connections. Of all ecologic classes of plants these are the most directly dependent on precipitation for their water supply, and unless rains or heavy dew fall at frequent intervals they must be able to endure drouth. Their nutrient supply is derived in part from rainwater, which always contains some dissolved substances (333, 383), in part from accumulated wind-borne particles, and in part from the decaying bark surface of supporting plants.

Epiphytes include cryptogams, herbs, shrubs and trees. They may grow on trees, shrubs, lianas or submerged plants. Rhizosphere organisms are technically epiphytes.

On woody plants epiphytes may perch on the trunk, limbs or on the upper surfaces of evergreen leaves. When in the last position they are called epiphylls (177, 614, 626). They are especially abundant in the forks of trees and on horizontal limbs, on which habitats a considerable depth of soil can collect (Fig. 82). They are least abundant on vertical and smooth surfaces.

As indicated above, the greatest single vicissitude of the habitat of aerial epiphytes is drouth, and these plants are most abundant where drouth is never protracted. In cold or dry climates epiphytes are few

FIGURE 82. Luxuriant rain-forest vegetation with an abundance of epiphytes and lianas. Tamazunchale, Mexico.

and consist chiefly of algae, lichens, liverworts, and mosses. In warm wet climates these groups are augmented by a wide variety of vascular plants, especially ferns and species belonging to the *Bromeliaceae* and *Orchidaceae*. The rich epiphytic flora of a tropical rain forest shows a remarkable gradation from sciophytic hygrophytes, confined to the lower trunks of the trees, to xerophytes (including cacti!) that demand bright light but can endure the occasional desiccation of the treetop habitat. Frequently epiphytes show marked preferences for particular supporting species (59). The moss *Tortula pagorum* is abundant in southeastern North America only on the bark of trees located in cities, apparently requiring some air-borne constituent of smoke (18). Some epiphytes grow on rocks or insulated wires as well as on plants (Fig. 83).

FIGURE 83. *Tillandsia usneoides* (the pendant Spanish moss) and *T. recurvata* growing on insulated wires in Sebring, Florida.

True epiphytes may harm their supporting tree only through shading effects (150), or by adding weight to limbs that favor breakage. However, there are all gradations between epiphytes and parasites (618), and also between epiphytes and lianas. Some, as the fern *Nephrolepis,* are at first rooted in the soil, then by extending their rhizomes up a tree trunk and later losing connection with the ground they eventually become epiphytes. Other plants, such as the hemlock *Tsuga heterophylla* (Fig. 84) and the strangling figs (Fig. 85), germinate on trees but live as epiphytes only until their roots grow down over the surfaces of the trunks and establish connections with the soil.

True epiphytes have soil connections at no time during their life cycles, and they frequently show marked adaptations favoring their mode of living. Xeromorphy, such as thickened cuticles, sunken stomata, and succulence, is common. Some epiphytes spread their roots over the plant surface in such a manner as to be well situated to absorb water from the film deposited by even a small shower. In others, the *nest epiphytes* (e.g., *Platycereum*), the roots accumulate large masses of

FIGURE 84. A young tree of *Tsuga heterophylla* that germinated approximately 5 m (see meter stake) above the ground on the trunk of an older tree of *Pseudotsuga menziesii.* This phenomenon is limited to regions such as the Olympic peninsula, Washington, where the bark surface is continually moist.

debris that hold water like a soil. The roots of many tropical epiphytes in the *Orchidaceae* and *Araceae* extend outward into the air, appearing as thick, unbranched, whitish organs. Covering the surface of these roots is a special layer of empty, whitish cells that can take up water rapidly

FIGURE 85. (*Left*) Strangling fig becoming established as an epiphyte in the canopy of a palm. (*Right*) An old strangling fig, the roots of which have fused to form a hollow "trunk" about the lower part of the palm. The fig is beginning to shade out the palm and will eventually bring about its death. Mexico.

from even a brief shower; afterward the living core of the root absorbs the water from this storage layer. The special absorptive tissue is called *velamen.*

Among the epiphytic species of *Tillandsia* (*Bromeliaceae*) the roots serve chiefly as anchorage organs, the leaves and stems taking over the function of absorption. In *T. usneoides,* "Spanish moss," the finely divided shoot system is covered with peltate scales that collect capillary water and allow it to be absorbed by the small uncutinized spots on the epidermis which they shelter during times of drouth (Fig. 83). The leaves of other species of *Tillandsia* may be shaped like gutters (Fig. 86) and so collect water in their axils. These are *tank* or *cistern* epiphytes. Numerous aquatic invertebrates, including mosquito larvae, live in these tiny cisterns, so that they have been described as a vast interrupted marsh extending across the tropical forest, rendering the problem of mosquito control difficult (432). Among the most specialized of epiphytes is *Dischidia,* the leaves of which form pitchers that accumulate water and leaf litter, and into which special absorptive roots extend!

FIGURE 86.   A tank epiphyte (*Tillandsia* sp.) growing on the trunk of a palm. Florida.

Parasitic Vascular Plants

Dodder (*Cuscuta* spp.) is a widespread, threadlike twining herb growing on other vascular plants. Its seeds germinate to produce slender green stems that grow over the soil surface until they reach a suitable host about which they twine. Adventitious roots emerge from the dodder stem to penetrate the stem of the host, where they make contact with both xylem and phloem. These specialized roots, as well as other kinds of absorptive organs of parasites, are called *haustoria*. Once established on its host the dodder loses connection with the ground but continues to grow by extracting water and foods from the host plant, and eventually it produces flowers and seeds. Dodder is leafless and has a yellowish-brown color, but the stems contain chlorophyll so that the plant is completely dependent upon its support only for water. It is therefore most properly called a *partial parasite*. Dodder becomes a troublesome weed at times, especially in fields of *Linum* or *Trifolium pratense.*

The genus *Cuscuta* is usually placed in the *Convolvulaceae,* so that it appears that dodder represents a further step in specialization beyond that of the green but lianoid members of that family. Remarkably similar to dodder in form and nutrition is *Cassytha*, which, however, belongs to the Lauraceae.

Broomrapes (*Orobanche, Conopholis, Epifagus,* and others of the *Orobanchaceae*) are herbs completely parasitic upon the roots of seed plants. The roots of their seedlings are connected to the roots of suitable host plants, and in some cases the seeds will not germinate unless they are in contact with the root of a suitable host. The aerial parts are no more than brownish inflorescences lacking chlorophyll. This family is closely related to the *Scrophulariaceae,* in which many common genera are facultative (e.g., *Castilleja, Pedicularis, Melampyrum,* and *Gerardia*) or obligative (*Striga*) root parasites.

Striga, an Asiatic genus of annuals that are partial parasites on the roots of other herbs, including crop plants, has recently been introduced into the U.S.A. Since its seeds are stimulated to germinate by contact with the roots of species they cannot parasitize, crop rotation helps in controlling the plant (184).

Rafflesia is a genus of Malaysian plants parasitic upon the roots of *Vitis.* This parasite has become so extremely degenerate that it resembles a fungus, for the vegetative parts are mycelioid and wholly contained in the roots of its host. One species, *R. arnoldii,* is famous for bearing ill-smelling flowers about a meter in diameter, these being the only organs to appear above the ground.

Mistletoes (*Loranthaceae*) are dwarf, shrublike plants growing rooted in the branches of trees. Their haustoria are connected with the vascular tissues of the host, but since the plants possess an abundance of chlorophyll they are only partial parasites. They are none the less distinctly injurious to their hosts in many cases.

Viscum album is the European mistletoe, *Phoradendron* spp. are the large leafy mistletoes so common for Christmas decorations in North America, and *Arceuthobium* spp. are the dwarf mistletoes common on coniferous trees in the northern hemisphere (270). These partial parasites, as well as the broomrapes and dodders, generally are able to parasitize more than one species of host plant, but their appearance may vary with the type of host.

The adaptations between an obligate parasite and its host tend to become delicately balanced in that the parasite obtains as much benefit as possible without interfering with the life cycle of the host. When a parasite finds a new host that does not have a degree of resistance commensurate with the virulence of the parasite, the effect is usually disastrous.

Lichens

A lichen is a dual organism formed by the close association of one or more fungi, usually an ascomycete or rarely a basidiomycete, with one or more unicellular green or blue-green algae. The alga is usually a species unknown in the free-living state, and the lichen fungi are found only in lichens. The algal cells are enmeshed within the fungal mycelium, and each combination of the two types of organisms produces a structure sufficiently unique, morphologically and physiologically, to warrant a botanical name of its own. Reproduction is accomplished chiefly by *soredia,* which are bits of fungal tissue enclosing a few algal cells; thus the peculiar relationship is perpetuated. Ascospores are also produced, but they usually perish unless they come in contact with acceptable species of free-living algae so that a lichen can be formed anew.

The fungus obviously gets all its food from the alga. Some believe that the algae may benefit from certain excretions of the fungi, and in at least certain lichens their blue-green algal symbiont can fix atmospheric N (71). Truly reciprocal benefit is demonstrated simply by the fact that the combination has made possible the exploitation of a wide variety of habitats otherwise without vegetation.

Lichens exhibit all degrees of variation from complete independence to parasitism, in which cases the fungus not only uses the alga but even sends haustoria into the living tissues of the vascular plant that supports it.

Symbiotic N Fixation

The fact has long been known that the roots of most legumes produce gall-like nodules inhabited by bacteria (*Rhizobium* spp.) which fix atmospheric N into organic compounds that eventually benefit the legume. The legume in turn furnishes the bacterium with foods, nutrients, and water (573). Because the *Leguminosae* is the second largest family of flowering plants and contains so many species useful as food and forage, the nodule phenomenon has been studied most intensively in these plants, but nodules of similar form and equally efficient in N-fixation (although here involving *Actinomyces* rather than *Rhizobium*) occur also on the roots of *Alnus, Alopecurus, Casuarina, Ceanothus,* Cycadaceae, *Elaeagnus, Hippophae, Myrica, Podocarpus, Shepherdia,* etc. (49). Also in the leaves of about 370 species of nonleguminous plants there are colonies of bacteria some of which are definitely known to fix N (373). In most of these instances neither the bacterium nor the seed plant can live without the other symbiont (373).

Legume nodule bacteria gain entrance into the root through root hairs and are purely parasitic in early stages of nodule formation. Furthermore, they revert to this relationship under anaerobic conditions. In well-developed nodules the bacteria normally make use of N to synthesize amino acids. Just how these acids are obtained by the host is unknown. Nevertheless the facts have been well established that (*a*) legumes that are inoculated will grow well in N-deficient soils, (*b*) the N content of the soil is greatly increased as a result of its occupancy by legumes, and (*c*) in the presence of available N nodule formation is weak.

The various species and strains of *Rhizobium* have limited ranges of leguminous symbionts. At present 12 such strains have been recognized. So much benefit is derived from the bacteria that the seed of the leguminous field crops is commonly inoculated by immersion in an aqueous suspension of the proper kind of cultured *Rhizobium* to insure maximum symbiotic development. Also most crop rotation systems include some legume in the cycle for the sake of maintaining N fertility. Legume bacteria have similar importance in forestry as well, for it has been demonstrated that planting the leguminous *Robinia pseudoacacia* with other trees is an economically feasible method of increasing timber yield (128). *Alnus* has similar utility.

Blue-green algae are associated with the roots or other organs of many higher plants, and in some of these relationships there is evidence that the algae can fix N (71).

Mycotrophy

The nutrition of most vascular plants and bryophytes is directly tied up with the nutrition of higher fungi, a phenomenon called *mycotrophy* (Gr. *mykes* = fungus, *trophia* = nourishment). The fungal mycelia combine with certain underground parts, particularly roots, to form a special compound structure called a *mycorhiza* (Gr. *mykes* + *rhiza* = root), a term that for convenience is commonly extended to include organs other than roots. Such dual structures are homologous with lichens and with but few exceptions are found in almost all taxonomic groups above the thallophytes. In fact, vascular plants lacking mycorhizae are not common. The fungal component is either a basidiomycete such as *Amanita, Boletus, Lactarius, Russula,* etc., a phycomycete, or the ascomycete *Cenococcum.*

Mycorhizae are of two chief types, although these are really extremes between which there are all degrees of intergradation: (*a*) *ectotrophic,* in which the mycelium forms a dense mantle over the surface of the root with many hyphae extended outward into the soil and many others extending inward, forcing their way between the cells of the epidermis and cortex (e.g., *Amentiferae, Pinaceae*); and (*b*) *endotrophic,* the major category, in which there is no surface mantle but some of the hyphae inhabit the protoplasts of parenchymatous tissues and others extend outward into the soil (e.g., *Compositae, Ericaceae, Orchidaceae* and all annuals).

There is great variation in the anatomy of mycorhizae on different plants and in the nature of the interrelationship between the two organisms. Apparently every gradation exists from outright parasitism to conditions of controlled parasitism in which the higher plant turns the fungal invasion to good advantage. The three groups of plants that have been studied most intensively from this standpoint are the *Pinaceae, Orchidaceae,* and *Ericaceae,* and these will be discussed briefly.

Pine Mycorhizae (62)

Pine rootlets, as well as those of many other trees, are divided morphologically into the long roots (the main axes) and the short roots, the simple laterals emerging from the former. In fertile soils the rootlets are not mycorhizal, the root-hair zones of the dominant long roots absorbing almost all the water and nutrients used by the tree (323). On the other hand, in soils deficient in N, P, K, or Ca, which is a common condition, the absorptive powers of the long roots seem to be inadequate, and the short roots are invaded by fungi that take over the absorptive functions. As a result of the fungal invasion these short roots are stimulated to branch repeatedly so that they become coralloid in appearance, the degree of branching being inversely proportional to the degree of fertility (Fig. 87). These short roots thus become mycorhizae which are definitely more efficient than an uninfected root because of the tremendous aggregate surface of the long hyphae extending into the soil. Where the fungus is lacking, as in soils of artificially maintained high levels of fertility, absorption by the short roots is limited to the small area of their unsuberized tips.

Apparently the pine takes water, nutrients, and possibly some elaborated compounds from the hyphae, and the hyphae obtain foods and growth substances from the roots (539). In pines, at least, the relationship appears to be typically obligative, but the same dependence has not been demonstrated for most dicots. Unless attached to a root, the fungi live only as saprophytes in the soil and cannot fruit.

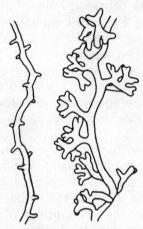

FIGURE 87. Rootlets of *Pinus strobus*. (*Left*) Non-mycorhizal rootlet. (*Right*) Mycorhizal rootlet. Hyphae and root hairs have been omitted. [After Hatch (323).]

The importance of mycorhizae in the silviculture of the Pinaceae is considerable. Attempts to introduce pines into regions where the soil was not highly fertile and the proper symbiont was lacking have ended in conspicuous failures. It appears also that under long-continued deforestation the fungi necessary for tree establishment may die out to the extent that reforestation is hindered (545). Possibly soil inoculation should have as much place in the silviculture of conifers as it has in agriculture where legumes are important crops (239, 323).

Orchid Mycorhizae

The mycorhizae of orchids (Fig. 88) have been favorable materials for experimentation because of the commercial practice of germinating orchids in agar media. The seeds are extremely minute, containing only the rudiment of an embryo with a small amount of fatty reserves. Unlike pines, whose seeds germinate without mycorhizae, orchid seeds will not germinate normally except in the presence of the mycelium of a fungus such as *Rhizoctonia.* However, it has been found that the plants can be grown nonsymbiotically from seed if sugar is supplied and the pH regulated (203, 414, 433). Some believe that the fungus normally digests higher carbohydrates obtained from litter into soluble sugars and likewise performs the function of regulating pH to 5.0 or lower, creating conditions thought to be necessary for the absorption of sugars or possibly iron. Others believe that the true benefit to the orchid lies in vitamins that are supplied by the fungus.

In orchids lacking chlorophyll (e.g., *Corallorhiza*) the vascular plant is parasitic upon the fungus. There is no evidence that vascular plants are ever saprophytes; when they lack chlorophyll they depend on fungi for foods.

FIGURE 88. *Calypso bulbosa,* a highly mycotrophic orchid with a root system (shown in its entirety) that alone would be inadequate for a shoot of this size.

Ericad Mycorhizae

Members of the *Ericaceae* are strongly mycorhizal, the roots, at least, nearly always containing endotrophic hyphae. An especially interesting feature of this group is that the fungus (often *Phoma*) frequently permeates all plant organs, including the seed coats, thereby insuring inoculation of the young seedlings (3, 40). Within the root cells hyphae appear to be eventually digested.

The coralloid root system of *Monotropa hypopitys,* a nongreen erica-ceous herb, is completely sheathed by a mycelial mantle, so that the vascular plant does not make contact with the soil. Parts of the same mycelium form mycorhizae with the roots of *Pinaceae,* and it has been shown that simple sugars are transferred from the trees to the *Monotropa* by way of the interconnecting mycelium. Thus *M. hypopitys* must be considered an epiparasite on the trees (63).

Allelopathy

Certain species of plants are known to release toxins which injure other species in their vicinity, and sometimes even their own seedlings. Most such substances known at present fall into five chemical classes: pheno-lic compounds, aldehydes, coumarins, glucosides and terpenes. Some are released by the decay of litter, others are excreted from living cells or tissues, possibly as a mechanism for keeping these metabolic by-products from accumulating to harmful concentrations in the cells that produce them. Possibly some are of value to the plant producing them by repelling potential pathogens or grazing animals while still retained in the cells, or upon release they may damage potential competitors or even regulate their own populations. HCN secreted by *Linum* roots inhibits a fungus causing root rot (758).

Practically all plants can be shown to contain one or more toxic compounds, but for the most part these either deteriorate as the cells containing them die, or if released they are soon inactivated in well aerated soils by microbial decay or colloidal adsorption.

Where harmful to other plants these toxins tend to have selective effects in that not all species are damaged by any one of them. Their mode of action is unclear, but growth may be retarded, turgidity reduced, germination prevented, and possibly some may simply interfere with nutrient uptake by the injured plant. Rain dripping from *Juglans nigra* leaves has a strong influence on the kinds of plants that can grow beneath the trees (98). Similar interactions among bacteria, fungi and algae, by means of water-soluble excretions, are referred to as antibiosis, and certain of the toxins (antibiotics) have proven very useful in the treatment of infectious diseases.

The leaves of several species of *Salvia* give off volatile terpenes in sufficient amounts that the growth of other plants is prevented or im-paired for distances up to 10 m away (559). These terpenes may enter other plants by dissolving in the cutin of their leaves. Physiologic experi-ments have shown that potentially harmful volatile materials are given

many species, but in amounts too small to be
nt where there is free air movement.

uspected that living roots may also yield toxins that
, but this has been extremely difficult to prove. There
that roots secrete nontoxic materials that are taken
by roots (251, 271), but the problem is to show that
ound are not products of root decay. As the roots
ompose microbes attacking amygdalin produce ben-
s so toxic and persistent that a new orchard cannot
an old one is removed (416). Whether a product of
y, the subterranean organs of *Helianthus rigidus* pro-
tive toxin (170), and perhaps this may account for at
least some of the common fairy-ring phenomena in other vascular plants
and fungi (Fig. 89).

Many toxic compounds can be extracted from plant tissues in the
laboratory, but the significance of such materials in natural environments
always stands in need of critical field tests. Water extracts of plant tissue
can easily have osmotic potentials sufficient to account for the suppression of germination on test plants, even if none of the solutes are toxic.
Maize and barley are relatively insensitive to prepared extracts regardless
of their concentration, so the plant used for laboratory tests can influence
conclusions.

FIGURE 89.   Toxic by-products accumulating beneath vegetatively-
reproducing plants often results in the centrifugal spread of a clone
as the center dies. *Muhlenbergia torreyi* in southern Colorado.

7

The Fire Factor

Except in very wet, very cold, or very dry regions, fire has always been an important factor in terrestrial environment. Charred fossils show that lightning-started fires have periodically ravaged land vegetation since its earliest appearance on earth. To this eventually was added the effect of primitive man who, for a variety of reasons, periodically set fire to vegetation. With the spread of civilization man-caused fires have become far more numerous than those caused by lightning, but the abundance of roads and plowed fields in well-populated regions, together with efficiently organized fire prevention and control agencies operating on forest lands, prevent individual fires from spreading far as a rule.

Fire injures plants directly by subjecting the tissues to lethal temperatures, but many other ecologic aspects of fire do not involve the temperature factor.

Kinds of Fires

Wherever the soil is overlaid with thick accumulations of organic matter, the latter may catch fire and smolder for long periods. This may happen even though the material is moist, for the front of active oxidation moves forward so slowly that the heat from it dries out the adjacent unburned debris at an equal rate, thus perpetuating a zone of combustible fuel. Fires of this type that are flameless and subterranean are called *ground fires.* They kill almost all plants rooted in the burning material, although old woody plants may survive if their large roots have thick bark and descend well below the organic material.

Fire often sweeps over the ground surface rapidly, the flames consuming litter, living herbs, and shrubs, and scorching the bases of any trees it may encounter. These are called *surface fires,* and if there is no thick horizon of litter and duff that contains most plant roots, subterranean organs and buried seeds may escape serious injury.

In dense woody vegetation fire may travel from the canopy of one plant to another. Such fires are called *crown fires*. Usually everything from the ground upward is consumed, or at least killed (Fig. 90), but sometimes the ground is moist enough so that many subterranean organs and buried seeds escape destruction.

When burning conditions change, one of the above types of fire is frequently converted into another. Ground fires especially are likely to smolder for many days before a wind whips up the glowing material into active flame and a surface or crown fire results.

FIGURE 90.   Photograph taken of a forest interior only a few days after a devastating crown fire had swept the area. All large trees were charred and killed; small trees, shrubs, and herbs were entirely consumed. The mineral surface is covered by about 2 cm of fluffy ash.

Adaptations
Related to Fire

In regions where there is a heavy vegetation cover coupled with one or more dry seasons, environmental selection may have included the fire factor for so long that plants have developed special adaptational features that favor persistence under repeated burning. Whether the herbaceous perennial life form of grasslands has developed in response to the fire hazard or not, it has proven invaluable in a vegetation type that is very susceptible to burning.

Germination

Certain taxa of woody plants in *Acacia, Arctostaphylos, Ceanothus,* and *Rhus* produce large quantities of hard-coated seed that tend to lie dormant in the soil until the vegetation in which they occur is burned. Fire stimulates their seedlings to appear in large numbers (632). In the laboratory treatment for a few minutes in boiling water is useful in obtaining maximum germination of these plants.

Rapid Growth and Development

Certain woody plants that are killed by fire but have fire-resistant seeds or fruits have developed an ability to fruit only a few years after germination, so that the life cycle is completed before enough debris accumulates for another devastating fire. *Pinus muricata,* for example, may start to produce cones when it is only a meter tall.

A unique degree of immunity from the effects of surface fires is possessed by the seedlings of *Pinus palustris.* The terminal bud remains close to the ground for about five years after germination, and during this time it is located in the center of a radiating hemisphere of long needles that do not burn readily. Although fire results in high mortality the first and second years, subsequently light fires do no more than scorch the tips of the needles. After the seedling has developed an efficient root system, the shoot begins to grow rapidly, carrying the all-important apical bud above the zone of potential injury. This stage of development constitutes a second critical period, for the slender stem is now exposed to fire, but in about three years the bark becomes thick enough to enable the young tree to survive fire again.

Fire-Resistant Foliage

Although the high resin or oil content of most conifer needles, *Ulex europaeus* leaves, and *Betula* bark allows the living tissues of these plants to burn readily, other woody plants may have so little of these compounds that the water content of the leaves may stop a surface fire where a dense canopy extends down to the ground (22). Strips of plants with fire-resistant foliage may be planted as firebreaks (461).

Fire-Resistant Bark

Quercus borealis, Thuja spp., *Pinus strobus,* and *P. contorta* var. *latifolia* have thin bark and are usually killed when surface fires sweep the area in which they are growing. In contrast, *Quercus macrocarpa, Larix occidentalis* (Fig. 91), *Pinus palustris,* and *P. ponderosa* have thick bark and often escape injury when associated species are heavily damaged or killed. Elevated canopies and evanescent lower branches tend to make the insulating value of thick bark more effective.

FIGURE 91. *Larix occidentalis,* a tree with remarkably thick and fire-resistant bark that can endure rather hot fires.

Adventitious or Latent Axillary Buds

Pinus rigida and certain taxa of *Eucalyptus* (555) have a remarkable ability to regenerate branches along the trunk from adventitious or latent axillary buds, even after a crown fire has swept through the trees (Fig. 92). Many other woody plants, e.g., *Betula papyrifera, Sequoia sempervirens* (Fig. 93), *Chrysothamnus* spp., and *Vaccinium* spp., produce new shoots after fire destroys the old ones, these arising from adventitious buds at the summit of the root. In *Populus tremuloides* killing the shoot stimulates the growth of sucker shoots from adventitious buds arising on shallowly placed lateral roots at some distance from the old trunk.

FIGURE 92. Sprouts arising from the trunk of *Pinus rigida* a few months after a fire. New Jersey.

FIGURE 93. Circle of redwood (*Sequoia sempervirens*) sprouts around the burned-out stump of the parent tree. California.

Lignotubers

In the chaparral vegetation of Europe, North America, and Australia many shrubs (in *Arctostaphylos, Eucalyptus,* etc.) have evolved *lignotubers* which facilitate shoot replacement after a crown fire (555). A lignotuber is a conspicuous turnip-shaped or tabular swelling (up to 4 m across!) of the axis, mostly or entirely below the ground surface, which bears latent buds and has the capacity to produce new shoots quickly because it contains food reserves. It represents a further degree of specialization beyond plants that can sprout from an unmodified root.

Serotinous Cones

In a number of *Coniferae* ripe cones remain on the trees, retaining viable seeds for many years, and the term *serotinous* (meaning late to open) is used for the cones as well as the taxa. In *Pinus contorta* var. *latifolia* most cones open within a decade, but some retain viable seed for 75 years (142). Normally secondary growth eventually causes the pedicel to break, and with the vascular water supply cut off the carpels dry and open, allowing the seeds to be disseminated. Vegetation fires cause these cones to open promptly because the death of the tree stops the ascent of water. Fire may also favor opening by burning surface coatings of resin off the cones. In addition to *Pinus contorta* mentioned above, some other representative North American conifers with serotinous cones are *P. banksiana, P. clausa, P. leiophylla, Cupressus sargentii,* and *Picea mariana* (452).

Indirect Effects of Fire on Plants

The immediate effect of fires in killing woody plants and reducing litter to ash are readily apparent. The indirect effects which ultimately result from burning vegetation are not so evident, and only recently have they been given the attention they deserve.

Removal of Competition for Surviving Species

Those species that are not killed by a fire are obviously benefited by the subsequent reduction in competition and possibly by other changes in environment that are brought about. Thus fire-tolerant species always increase in abundance at the expense of their fire-sensitive associates. In mixed stands of conifers in the northern Rocky Mountains the proportion of *Larix occidentalis* in the stand increases greatly each time the forest is swept by surface fires, for, among the old trees that survive and furnish seed for the subsequent generation, this species is best represented. After two or three fires have swept an area *Larix* may be the sole dominant, provided the intervals between fires are long enough to allow some of the trees to develop a layer of bark thick enough to withstand scorching.

Fire Injury and Parasitism

Woody plants frequently survive fires that leave large scars on their stems. The upslope sides of trees often suffer heavily in this way on account of the great accumulation of litter in the angle between the bole and the ground. Fallen logs lying next to living trees have the same effect. Where organic debris is more evenly distributed scars tend to be confined to the windward sides of the trees (142).

The effects of a given fire may be superimposed on the effects of one or more previous fires in that unhealed scars are greatly enlarged or the trees are burned through. This is true especially where the earlier scar has become covered with highly inflammable resinous exudations (698).

The larger these scars are, the more time is required for their healing and consequently the greater is the opportunity for infection by parasitic fungi and insects (337, 401, 735). In some regions almost all decay in timber trees can be traced to previous fire injuries. Merchantable trees that are seriously scorched should be salvaged soon after a fire if it is economically feasible.

Alteration of Environmental Factors

Almost every aspect of environment is altered when vegetation is burned. Where the rooting medium is mainly organic the habitat may be damaged to the extent that thousands of years are required for sufficient recovery to allow the return of the original vegetation (172). Under other circumstances the effects of burning are not measurably detrimental even though environment is greatly changed, and in many instances fire enhances subsequent plant growth. Some of the chief effects of burning will be listed below, but it should be pointed out in advance that the effects produced vary with the type of vegetation, the kind of soil, the season of burning, the prevailing weather, and other factors.

One of the most evident effects of fire is the increase in light at the ground surface that results from the destruction of plant shoots. The vegetation that quickly invades burned areas is usually dominated by heliophytes.

The lack of shade also allows the soil to heat up and cool off to a greater extent so that the daily range in temperature is increased. The heating effect is augmented by the blackened color of the surface so that maximal and average values are higher (259). Ordinarily the results are a decidedly earlier development of vegetation in spring, especially the first spring following the fire (656), and an earlier desiccation of the upper soil layers.

Rainfall interception is eliminated when the plant cover is destroyed, and, if runoff losses are not increased proportionately, the soil moisture content may rise for lack of plants to use up the water. However, the burning of surface organic matter, coupled with the effects of beating raindrops and the extermination of burrowing microfauna, usually reduces porosity, thus increasing runoff and often favoring rapid erosion (259).

Fire sweeping across grassland after the shoots have matured and seeds cast has little influence other than to remove litter and expose the soil to sun, and perhaps to erosion, before new shoots rise from undamaged subterranean organs. As grass shoots mature their nutrient content declines, with a result that the small quantity of ash returns a negligible quantity of nutrients to the soil (472).

When the more massive bodies of woody shoots burn N and S are volatilized and thus escape from the habitat, but all other mineral nu-

Table 12
Comparison of Certain Properties of a Forest Soil in Western Washington Before and Immediately After a Severe Fire (385).

Property	Horizon	Before Fire	After Fire
Organic matter, % dry weight of soil	01 + 02	88.5	9.7
	0– 3″	5.7	3.5
	3– 6	3.7	3.1
	6–12	3.4	2.8
	12–30	2.2	2.5
Total N, % dry weight of soil	01 + 02	0.9	0.3
	0– 3″	0.1	0.1
	3– 6	0.1	0.1
	6–12	0.1	....
	12–30	0.1	....
C/N ratio	01 + 02	57	17
	0– 3″	27	18
	3– 6	24	18
	6–12	22	....
	12–30	21	....
Field capacity, % dry weight of soil	01 + 02	190	60
	0– 3″	75	50
	3– 6	43	55
	6–12	50	57
	12–30	61	79
pH	01 + 02	4.9	7.6
	0– 3″	5.0	6.2
	3– 6	4.8	5.5
	6–12	5.0	4.9
	12–30	5.1	5.2
Water-soluble salts as parts per million	01 + 02	1116	1330
	0– 3″	370	585
	3– 6	365	164
	6–12	164	222
	12–30	82	142

trients are converted to simple water-soluble salts that are readily available. If these can be adsorbed by the soil colloids or taken up by a new plant cover before they leach away, soil fertility is raised in consequence of burning (Table 12). The warmer soil speeds up mineralization of humus, only the upper few centimeters of which are oxidized by the fire (Table 12), so that N and S are obtainable from this source, Nitrification is also increased by the warmth. The plants that flourish on burned areas of this type are often species with high N requirements that cannot endure the absence of nitrate characteristic of the undisturbed forest. Since the ubiquitous bryophytes *Funaria hygrometrica* and *Marchantia polymorpha* which are such conspicuous invaders of severely burned forest soils, have high N requirements (350, 796), the fire-depletion of soil N is probably much less of a factor in fire ecology than has been thought. Burning is especially beneficial when it removes thick accumulations of organic debris with such high C/N ratios that tree growth has stagnated. Decaying roots of plants killed by the fire act as a green manure to further improve soil nutrition. If the soils were acid before the fire the release of K and other bases brings pH to a level where nutrient availability is improved (Table 12).

Surface accumulations of organic debris in the forest are often physically detrimental to the establishment of seedlings (244, 476, 485). Apparently the reason is that fluctuations in moisture content are much more violent in this material than in bare mineral soil. After fire destroys the organic layers the new seedbed conditions may permit the germination and survival of species whose disseminules had regularly perished on the area previously.

Although the immediate effect of fire is to remove litter, a single burn may kill more vegetation than the dead material consumed, so that the litter cover soon becomes heavier than it was originally. When the fuel supply and fire hazard are thus increased, a secondary burn following a few years after the first may have much the more devastating effect.

The influence of fire on soil below the surface organic layers is usually slight (385, 742). Ordinarily no more than the upper part of the humus is oxidized. However, where burning results in the replacement of fire-sensitive woody plants by grasses, the ultimate effect of the fire may be to raise the humus content above the initial level within a few years. If herbaceous legumes are favored by fire, the N content of the soil is likewise increased.

Finally, of great importance is the change in animal life brought about by fire. The new conditions favor new kinds of animals having different requirements for food and cover. The significance of the temporary exodus of seed-consuming rodents from burned forest areas has already been discussed. Birds that prefer open vegetation congregate on burned areas and carry in seeds from plants growing on other burned areas that they have visited. Thus a dense scrub of *Rubus, Ribes,* and other berry-producing plants may develop rapidly. The herbaceous and shrubby vegetation that dominates burned forest areas before they are again claimed by trees greatly increases the forage capacity of the land so that game animals and livestock are benefited.

The above outline, though incomplete, is sufficient to show that the relations of fire to environment are very complicated and difficult to appraise.

Stimulating Effects

Wholly aside from the matter of release from competition, fire sweeping an area frequently seems to have a stimulating effect on certain survivors. These effects are attributable to one or more of the environmental changes discussed immediately above.

Epilobium angustifolium, commonly called fireweed, is a long-lived forb that is conspicuously stimulated by burning. Dwarf sterile specimens are commonly encountered in forests that have not been burned for centuries, but the plants are so small and scattered as to escape notice. When fire destroys the trees these dwarfed *Epilobium* plants develop to many times their former size and flower abundantly. This species produces such great numbers of comose seeds that, even in the absence of an external supply of seeds, the few stimulated plants on a burned area can give rise to a dense stand of fireweed within a year or two after the forest is destroyed (554).

Populus tremuloides is a small tree peculiar for the fact that seedlings rarely survive, but once one does a grove is usually formed by suckers that rise from shallow horizontal roots radiating from the initial tree. The *Populus* stands thus formed are relatively permanent, maintaining themselves by continued suckering. Fire usually kills the thin-barked shoots, but the roots are only stimulated to sucker the more vigorously. Apparently the additional solar radiation absorbed by the bare and blackened soil surface is responsible for the fact that on the first season after a light burn suckers appear in unusual numbers and attain a greater height (695).

Although a number of grasses are easily killed by fire, some are not injured, and still others are stimulated to produce abnormally large quantities of seed. The latter include *Aristida stricta, Cynodon dactylon, Paspalum notatum,* and *Andropogon gerardi.* It is significant that nitrate fertilizers have been found to have essentially the same stimulating effect as burning in *Cynodon* and *Paspalum* (109) With *Andropogon gerardi* it is simply the mechanical removal of litter that accounts for most of the stimulation (372).

In very arid regions grasses are usually damaged rather than stimulated.

Practical Value of Vegetation Burning

So many variables are involved in the relations among fire, other environmental factors, and organisms that generalizations are hazardous. Sufficient careful research has been done to show that the effects of burning vary from extremely devastating to more beneficial than detrimental. Although uncontrolled burning is usually detrimental to man's interests, the usefulness of properly controlled fire in certain vegetation types can no longer be questioned (311, 772). Some examples of the scientifically justifiable use of fire follow.

Use of Fire to Favor Economically Important Plants

Where economically valuable species that are not destroyed by fire grow associated with relatively valueless fire-sensitive species, fire can be used to advantage in improving vegetation for the good of mankind. The aim here should be to cause the fewest and lightest fires that will accomplish the desired vegetation changes, so that the detrimental effects of fire on the soil will be minimized.

Originally the vegetation of the sagebrush-grass steppe of western North America consisted of a dense cover of perennial grasses among which were scattered moderate-sized shrubs, chiefly *Artemisia tridentata.* When this vegetation is subjected to heavy grazing the perennial grasses are reduced to a scattering of weakened plants, whereas the relatively unpalatable sagebrush increases in size and numbers to an extent that is highly objectionable from the grazier's standpoint. Fire, in contrast to grazing, kills the shrubs but does not injure the grasses significantly when these are aestivating. On areas where the grasses have not been too severely depleted, it has been demonstrated that the forage production can be increased tremendously by burning to kill the shrub, then not allowing grazing for a few years until the grasses have multiplied to form a complete cover (Fig. 94). Sagebrush eventually returns to such an area, but this is so slow a process that burning need be repeated only at intervals of many years. Obviously, if grazing has already removed all perennial grasses, burning must be followed immediately by artificial reseeding to secure a stand of desirable grasses (67).

It is generally true that fire favors grass at the expense of woody vegetation, and other examples in addition to the above can be cited. Grass fires set by the aborigines were responsible for extending prairie eastward into the winter-deciduous forest in the region centering about eastern Iowa. In New Zealand even rain forest has been converted to grassland by means of burning especially.

Since the seeds of a majority of desirable timber trees must have contact with mineral soil to germinate, in cool and cold climates a practical method of maintaining a continuing supply of commercially valuable forests is to burn the debris off an area after logging. The fire is controlled so as to minimize scorching the soil, and still leave the mineral surface bare to favor prompt establishment of a new tree generation.

Fire is especially an important tool of forest management on the Atlantic Coastal Plain of southeastern North America (259). Here pines, especially *Pinus palustris,* are relatively fire-tolerant and valuable in comparison to the relatively fire-sensitive and worthless oaks. Light surface fires at

FIGURE 94. In the distance is unburned sagebrush-grass vegetation. The area in the foreground was burned three years before the picture was taken. Fire has eliminated the sagebrush temporarily and allowed the fire-tolerant grasses to become dominant. Dubois, Idaho.

intervals of a few years are sufficient to keep oaks entirely eliminated, yet the intervals are long enough to allow the pines to germinate and attain a size sufficient to withstand burning. Temperatures during these fires usually remain below 100°C, and the maximum is sustained for only a few minutes, so that humus is not charred and the net effect of burning is not harmful to the soil in this region (Table 13) (345). Fire has no significant effect on the growth of pines large enough to survive burning, and, because grasses thrive under these conditions (Fig. 95), both timber and forage resources are better under burning at intervals of several years than under complete protection or under annual burning. In the Rocky Mountains fire can be used on certain habitat types to thin over-dense young stands of *Pinus ponderosa* (815).

Table 13

Comparison of Certain Soil Properties in a Stand of *Pinus Palustris* That Has Been Subject to Frequent Ground Fires, with an Adjacent Stand That Had Not Been Burned for 15 to 20 Years (345). Trenton, Florida.

Property	Horizon	Rarely Burned	Frequently Burned
Loss on ignition, %	0– 2″	2.925	2.249
	6– 9	1.346	1.152
	16–18	0.990	0.815
Total N, %	0– 2″	0.040	0.032
	6– 9	0.020	0.018
	16–18	0.012	0.011
pH	0– 2″	5.97	5.29
	6– 9	5.67	5.54
	16–18	5.75	5.58
Replaceable Ca, %	0– 2″	0.047	0.024
	6– 9	0.010	0.011
	16–18	0.008	0.009

Use of Fire to Improve Quality of Forage

Deer, elk, and moose depend heavily on browse provided by shrubs and tree seedlings, especially in winter after the foliage of deciduous woody plants and herbs has withered or become covered with snow. Unless these plants are browsed excessively they tend to grow too tall for the animals to reach the nutritious younger twigs. On sites where game is more valuable than trees, occasional burning of patches of vegetation is recommended so that new sprouts from the roots will renew the supply of available browse (720). Not only are the young twigs more available, but they are more nutritious than twigs from older plants (197, 222). In grasslands also an improvement in nutritive quality is among the benefits usually resulting from occasional burning (648). In fact, where mowing is impractical tropical grasses must be burned frequently to keep their protein content up to the minimal level required by livestock (181).

FIGURE 95. Fire-maintained stand of *Pinus palustris* near Baconton, Georgia. Burning has kept out oaks but has not been so frequent that a few pines have become established recently. The trees are being bled for resin.

Use of Fire to Remove Undesirable Organic Debris

The greater the amount of dry litter, the hotter and more devastating a fire is. Under certain circumstances the cumulative effects of occasional small fires are not so destructive as the holocausts that would almost certainly result from fires coming at greater intervals (217, 259, 497).

Insects and fungi that are facultative parasites often must live in plant debris for a period until conditions become propitious for attacking living hosts. As suggested above, whenever the potential hazard of such organisms outweighs the protective value of the plant and debris cover as a source of humus, sanitation by burning is desirable. Many agricultural situations demand the burning of crop residues that harbor such pests as pea weevil, the European corn borer, cinch bugs, viruses, rusts, smuts, and rot-producing fungi. On the Atlantic Coastal Plain one of the benefits of burning the pine forests is the control of the brown-spot needle disease of *Pinus palustris.*

In other regions the benefits of periodic burning may lie in the destruction of ticks and other important parasites of animals. Fire is quite effective in reducing the numbers of tsetse flies in Africa.

The seeds of many plants germinate and become established better in contact with mineral soil than with superficial layers of organic debris. Thus burning has a place in the silviculture of certain trees by preparing the land for planting or natural seeding.

Logging operations invariably leave the forest littered with unmerchantable logs, tops, and branches, which are collectively called "slash" by foresters. This material becomes a fire hazard as it dries out; it may be detrimental to the establishment of seedlings of the next generation of trees, and it may harbor injurious pests. On the other hand, the debris improves soil structure and fertility, reduces erosion, and under other circumstances may favor seedling establishment. Depending on the relative significance of these factors under different conditions, silviculturists recommend piling and burning of slash, scattering it uniformly and burning, or scattering and letting it decompose gradually in the hope that it will not catch fire (604).

Fires set by primitive man or by lightning were more frequent and more extensive than those resulting from the planned burning of small areas in modern techniques of vegetation management. Furthermore, smoke from this source accounts for only a very small percentage of man-made alterations of the atmosphere, and the materials added are relatively innocuous (561).

Measurement of Temperature and Heat Release

The duration of different temperature levels at various heights above the ground can be recorded by the use of thermocouples installed prior to making a planned burn (381).

Maximal temperatures can be approximated simply by noting changes in a series of fusable compounds having different melting points, that are painted on pieces of mica or aluminum and fixed at various heights (52, 357)

Heat release during the passing of a burning front has been evaluated by the loss in weight of 3 l of water contained in a metal can painted flat black, with a 1 cm hole in the lid (45).

8

The Environmental
Complex

The preceding chapters have considered, as far as possible, the separate influences of soil, water, temperature, light, atmosphere, fire, and other organisms on the plant. It is now desirable to focus attention on the important fact, which has previously been suggested in many places, that factors never operate separately or constantly upon the organism, that environmental relations are at once complicated and dynamic.

When ecology was in its infancy as a science, a disgruntled wag characterized it as "the painful elaboration of the obvious." But as more and more information accumulated, exactly the opposite viewpoint had to be accepted—what seems obvious superficially, involves complexity that is at times staggering. It is the purpose of this chapter to point out the main phenomena responsible for this complexity.

Modes of Action of Environmental Factors

All plant behaviorisms cannot be closely related to environmental stimuli. For example, the cessation of cambial activity in many temperate zone trees cannot be related to concomital changes in seasonal weather conditions. Apparently a cycle of physiologic activity is involved that runs its course, its termination being essentially endogenous. Where plant responses have been related to specific environmental antecedents, the mode of action of environment has proven to be quite diverse.

Direct effects of environment involve responses that are induced and sustained by some condition, ceasing when the condition is removed. The appearance of ephemeral leaves on the desert shrub *Fouquieria spendens* exemplifies this situation, for the small delicate leaves appear whenever growth water becomes available, then are abcissed promptly when the wilting coefficient is attained (470).

Inductive effects are those "triggered" by an environmental condition, and once the process gets under way it does not depend on a continuation of the stimulus. Thus a fraction of a second of exposure to light initiates the germination process in *Nicotiana tabacum,* but once illuminated even so briefly, germination continues even in darkness (408).

Conditioning effects require one condition which by itself doesn't stimulate, but rather makes the plant sensitive to another condition. As an example, those seeds requiring stratification are not stimulated by chilling until after they have been thoroughly moistened.

Carry-over effects involve responses that appear long after the responsible environmental factor has ceased its action. In cool climates, a summer with above-normal heat may initiate above-normal flower buds, resulting in a heavy crop of pine cones the following summer (180). So much energy going into a cone crop may not allow needles to attain full length the following year. In the alga *Scenedesmus quadricauda* variable relationships between population growth and the P content of the medium appears to be largely a result of the organism being able to accumulate this nutrient above current needs when it is plentiful (a phenomenon called "luxury consumption"), then to draw upon the cellular store for several weeks after the external supply of P has been cut off (659). Even the seed reserves of minor elements such as Mo may be sufficient for completing the life cycle of the plant developing from that seed (341). At times carry-over effects are transmitted by seed for several generations (349), a phenomenon bordering on the old hypothesis of inheritance of acquired characters!

Multiplicity of Factors

The complexity of vital processes within the organism is far more generally appreciated than the complexity of environmental media which surround the organism (467). We commonly report specific germination percentages for a given lot of seeds, yet the value obtained has been governed by genetic differences in the taxon from one population to another, by the year of collection, stage of plant development, weather preceding collection, methods of cleaning, storing, and testing, date of testing, etc. (450). In the glasshouse we usually assume far greater uniformity of environment from place to place than can be verified by testing, with the result that critical work there demands a systematic rotation of pots. Furthermore, biologists are continually discovering wholly new aspects of environment which have had strong but unappreciated influence on their observations, as attested by the lateness of the discoveries of photoperiodism and thermoperiodism. Apparently some of the contradictory results of different experiments may have been only a consequence of working under different intensities of smog, which was exerting an unrecognized influence upon plants (831). How many experiments have been performed and close interpretations drawn without knowing that the mere handling of leaves may cause a large increase in their subsequent rate of respiration! (276.)

A plant is at once affected by the amount of heat, light, moisture, and nutrients available to it, by the degree of activity of parasites and mycorhizal fungi, etc. The fact that its life processes continue under the numerous and fluctuating elements of environment bespeaks a considerable breadth of tolerance of environmental variation. Yet this tolerance is not unlimited. When the intensity of any one of these factors begins to tax the plant's ability to cope with it, vigor declines, but because of the multiplicity of factors involved the cause of the disturbance is often unapparent, becoming known only after a series of experiments have been directed toward this end.

This situation is pointed up by the problem facing a botanist who is called upon to ascertain the cause of death of a plant. If a parasite was responsible it may have left definite and recognizable symptoms in the form of necroses, galls, etc., but not all parasites do this. On the other hand death may have resulted from any of a wide variety of nonbiologic factors, and in this group also symptoms vary from the manifest to the obscure.

Because numerous factors are always operating on the organism simultaneously, each function is a multiconditioned process. Perhaps no better illustration of this fact is provided than that furnished by the research connected with the Neubauer technique for determining the availability of P and K in soils. Some time ago it was suggested that if a number of rye seedlings were grown in a definite quantity of soil mixed with a definite quantity of sand, and the seedlings were subjected to chemical analysis after about two weeks, the amounts of P and K in the ash would provide a standard measure of the availability of these elements in the soil used. However, subsequent research has shown that mineral uptake is not only affected by the chemical nature of the soil solution (as hypothecated), but by temperature, by soil moisture, by light, and by the degree of change in soil texture brought about by incorporating sand (15). Because so many factors affect plant functions, the method is valid only when a series of soils are tested simultaneously and all groups of plants are subjected to identical conditions except those of soil nutrition.

Environment is highly complex and integrated, but this should not be a matter for despair, for environments are probably still less complex than organisms!

Heterogeneity of Environment

The concept of habitat implies only a portion of space that is characterized by certain temporal and spatial variations in factor intensities. There may be a close similarity in the combinations of conditions obtaining among the members of a class of habitat (e.g., the dune habitat), but the conditions within each of the habitats, especially of light, moisture, and temperature, vary considerably from place to place.

The problem of measuring those physical conditions that really govern plant behavior is much more difficult than is commonly conceived. For example, when an investigator correlates the records made by a battery of weather instruments with the growth of a tree near-by, cause and effect are really not under study. The results, if positive, show no more more than a relationship existing between certain atmospheric and certain protoplasmic phenomena, all of which are members of a galaxy of changes set in motion by unmeasured master forces. Actually only a thin shell of environment adjacent to the organism is of immediate causal significance, and the conditions in various parts of this shell differ materially from the conditions to which the instruments in the weather station are exposed (31).

Environmental factors are subject to marked vertical stratification, so that the axis of the average terrestrial plant extends through several distinct *microhabitats* or *partial habitats* (865). The questions might be raised: When rising temperatures in spring allow a resumption of growth in a tree, are the edaphic or the aerial aspects of temperature the more important? How are we to evaluate the different levels of temperature stratification in each? The temperature of the meristematic tissue seems to be the datum of maximum value here, for air temperature as recorded in the instrument shelter is no more than a second temperature condition dependent indirectly upon solar radiation. Yet, if the growth of one organ is conditioned by hormones produced in another, we cannot single out the temperature of the apical meristem as the only temperature condition affecting its activity! How can we evaluate soil fertility when a root system extends across a series of horizons that differ chemically and physically? Questions of this sort cannot be solved at present, but ecology has at least advanced far enough so that the problems have been recognized.

Environment can be sampled from the standpoint of the various micro-habitats surrounding the different parts of a large plant, or from the broad standpoint of variation in homologous microhabitats within an area of essentially uniform plant cover. When the latter is the objective a great number of records is necessary in order to obtain both a mean value capable of reproduction and a true record of the extremes. Useful information concerning variability can often be obtained by sampling the environment of fixed intervals along a line. When only a few in-struments are available they can be moved about at regular intervals over the area under study, one being left in place as a standard of comparison for the others (753). Even in the glasshouse environment is far from uniform across the benches.

The importance of microenvironment is emphasized by the modern trend in agronomy and soil conservation, as exemplified by the individual consideration given in land-use planning to each part of a farm. The interplay of variations in microclimate and soil results in significant varia-tions in agricultural potentialities within short distances. At some places windbreaks are beneficial; at other contouring and wide spacing of plants are desirable to increase precipitation effectivity; and at still other places drainage or the planting of shallow-rooting crops is advisable.

Another aspect of this problem of environmental heterogeneity involves the fact that variations in the supply of light or moisture become relatively widest at places where the quantities of either are least. Not only are the variations widest here, but also the plant is more sensitive to them, and therefore the same quantitative aspects of environment must be evaluated by different standards in different habitats. At the dry edge of a species' range small variations in moisture are very important in its local distribution, whereas at the cold edge of its range small variations in temperature become critical. Near its center of distribution the individ-uals of a species commonly appear to be insensitive to even wider absolute values of either moisture or temperature.

Dynamic Nature of Environment

In the natural course of events, identical combinations of environmental conditions are repeated only at rare and irregular intervals (467). The intensity of most factors varies with the hour, day, and season, and the *rates of change,* the *durations of particular intensities,* and the *extreme values* are all ecologically important aspects of the same environmental condition. Examples of the importance of these aspects of environment have been provided by earlier discussions of temperature, tides, daylength, etc. Also it has been pointed out that the common practice of integrating measurements taken over a period of time as mean values may obscure very important time aspects of factor variation.

Seedlings are especially vulnerable to the vagaries of environment, and this is undoubtedly reflected in the frequence with which species have evolved the habit of intermittent germination, with respect to the seed matured at any one time. The importance of extremes for seedling establishment is also shown by populations of perennials that are made up of only a few age classes. These plants, especially on suboptimal habitats, produce a successful crop of seedlings only as a result of a rare concomitance of extremely favorable conditions, in proper sequence, for flower bud initiation, for seed maturation, and for seedling establishment.

Ecologic phenomena are often keyed to short-lived extreme conditions. It was concluded that the zonal sequence of species along Japanese seacoasts is a result of storm conditions, and unrelated to the environmental gradients that characterize the long intervening periods of calm weather (569). Once *Helianthus annuus* has been subjected to wilting, its roots appear to be permanently damaged so that they absorb and transmit water less readily (423).

Variable weather frequently confounds field work. It is unsafe to conclude that the order of severity of environmental conditions as measured in a series of habitats during one season will necessarily reflect the order of severity at other seasons (298).

The above phenomena involve short-time fluctuations in factor intensities which recur at irregular intervals. Another class of environmental variations involves unidirectional changes that take place over a series of years, as for example when an abandoned field is allowed to revert to native forest. During this process a drastic change is brought about as the warm sunny habitat is converted into a cool and shady one. Only after the lapse of several decades when the native forest approaches maturity does further evolution of the habitat slow to an imperceptible rate.

Finally, in addition to the recurrent and the evolutionary types of environmental change discussed above, there is a third which may be called the ontogenetic type. The microenvironment of the tiny seedling is usually quite different from that of the developing organism, and the mature plant comes under the influence of still other parts of space.

Thus habitat change is incurred simply through the increase in size of a plant (865).

Variability of Plant Requirements

Ordinarily the combination of factor intensities most favorable to the welfare of a plant differs at different stages of the life cycle. Examples of this variability of plant requirements are numerous. The seeds of certain bog-inhabiting trees find conditions most favorable for germination on habitats that are distinctly too wet for the best growth of the mature trees (453). In peas, growth has been found to be most closely determined by temperature during the early stages of the life cycle, but later light becomes the most important single factor (88). Wheat is completely drouth resistant until the coleoptile is 3 to 4 mm long, after which it becomes increasingly more sensitive to dryness. A plant may require a special sequence of photoperiods to pass successfully through different stages of its life cycle. Also temperature requirements are usually lower for breaking dormancy than for germination, and germination requirements are lower than those for normal vegetative growth; vegetative growth requirements in turn are often lower than the most suitable temperatures for flowering and fruiting. During its annual cycle of development, the optimal temperature for *Tulipa* bulbs varies from 8 to 23°C. It is obvious that all environmental requirements of the different functions at each stage of development must lie within the seasonal variations which prevail during that stage of development. Generally the period of seedling establishment is the most critical for the average plant. Either earlier as a dormant seed, or later as a well-established individual, it is capable of withstanding conditions to which the seedling would succumb immediately.

Not only do temperature requirements vary throughout the year, but at any one date different organs and even different tissues of the same organ may respond differently to temperature. Examples mentioned earlier are the destruction by frost of the gynoecium while the surrounding floral parts remain uninjured, and the killing of the youngest xylem cells while the enveloping cambium escapes damage. Thus even one plant does not behave as a unit. Consequently in experimental work it is usually essential to draw comparisons only between exactly the same portions of homologous organs of equal age (831).

Genetically determined variation from plant to plant within one experiment becomes a problem necessitating replication in proportion to variability, largely so that average differences will reflect treatment more than genetic differences. A better means of overcoming this difficulty is to use plants that can be propagated vegetatively, so that replication becomes desirable chiefly as an insurance against accident. Genetic uniformity is quite high in named varieties of cultivated herbs that are grown from seed, but one can seldom solve problems in the ecology of one species by using another as the subject for experimentation.

Another aspect of this problem, and one that frequently has a strong effect on conclusions in experimental work, is a phenomenon that has variously been described as fatigue, protoplasmic adjustment, physiologic conditioning, or simply the time factor. It has often been observed that the new pace taken by functions when environment is changed abruptly does not usually persist for long. After an initial change in one direction, a secondary change usually sets in in an opposite direction, although the second change is less than the first. Thus, when a plant is placed under more favorable temperature conditions, its rate of growth may be speeded up to a degree which it cannot maintain for more than a few hours. Most graphs in which the rate of functioning is plotted against time are curvilinear rather than rectilinear. When given sufficient time for adjustment, plants can become adapted (*acclimated*) to a surprisingly wide range of environmental variation, but the time element is very importtant, and adjustment must be brought through gradual change in environment or the plant suffers heavily.

Factor Interaction

It is a commonplace in ecology that one factor compensates for deficiencies in another (663). A high water table, or an abundance of fog, or low temperature, each may compensate for low rainfall. Light influences plant requirements for nutrients such as Zn (586) and K, and there is good evidence that the effect of every nutrient element in the soil depends to a certain extent on the quantity of other elements present at the same time. Dry climate may compensate for the lack of bumblebees to pollinate *Trifolium pratense* by so reducing the size of the flowers that smaller bees can perform this function. Apparently the wide range of environment tolerated by many species is attributable at least in part to the fact that deficiencies in one factor may be compensated for by others as they arise. The same ecologic sum can be derived from different combinations of individual factor intensities.

Plants may become predisposed to one injurious factor by being exposed to another. Frequently a plant succumbs to parasites only as a result of debility incurred by growth under unfavorable climatic or soil conditions. But sometimes the reverse is true, the plant being subject to parasitism only when conditions favor lush growth.

It has often been noted that the separate actions of two factors do not have the same influence where they are combined. For example, in one fertilizer trial the addition of $NaNO_3$ increased yield 10%, and in another trial the addition of K_2SO_4 had the same amount of influence on yield. But when both were added, the increase was not 20%, but 100% (90).

Extrapolation is also dangerous where one might wish to predict a certain ecologic relationship basing the prediction on experimental work performed at a remote location, for climate exerts a strong control over the plant's requirements for nutrients, water, energy, and resistance to parasites, salinity, flooding, and other injurious forces.

Because environmental factors are interrelated and dynamic, and because they often exhibit delayed effects, an alteration of one factor frequently initiates a series of adjustments of far-reaching and often unpredictable consequences. Thus an experiment in fertilizing forest soil resulted in increasing the density of the tree canopy, which in turn had a marked effect on the epiphytic flora.

A particularly difficult problem is posed by groups of factors that are inseparably related. Thus heat, light, and moisture relations vary simultaneously with every change in the intensity of solar radiation, and a number of soil characteristics always vary with change in pH. In these instances it is very difficult to design experiments to show the relative effects of different phases of the factor complex.

A difficulty of experimental ecology not encountered in physical or chemical work is the change that a plant undergoes when it is subjected to different conditions. If in a photosynthesis experiment one plant is grown under bright light and the other in shade, with all others factors equal, the morphology and physiology of the sun plant will soon become so modified that one is no longer following the safe rule of comparing systems differing by a single component. He will be comparing the photosynthetic rate of *thick leaves* under *bright light* with that of *thin leaves* in *shade*. This problem cannot be solved by growing plants under identical conditions and then suddenly transferring them to the experimental conditions, for the responses thus obtained may be so abnormal as to lack true ecologic significance (105).

The above facts show that not only are the factors of environment interrelated, and the functions within the organism interdependent, but the two complexes are practically inseparable. Another phenomenon which obscures the line of division between organism and environment is that for a limited time environmental deficiencies can be compensated for by adjustments within the plant. Thus, the characteristic effects produced by the permanent wilting percentage of the soil are postponed in a succulent plant by a redistribution of water contained in the succulent tissues.

Because so many experimental data of the above nature have accumulated, it is now clear that mathematical expressions of the relations between environment and response are of value only as an aid in the interpretation of field observations. Another reason why the concept of definite tolerance limits has had to be depreciated is that most species are not genetically uniform. Instead, they are composed of races many of which are indistinguishable except for physiologic tolerances. This point will be discussed in detail later. For the two reasons just stated, field ecologists attempting to explain plant behavior are often forced to confine themselves to inductive methods, seeking to establish concomitance between behavior and environment, and realizing that the phenomena may never be capable of expression in any but broad generalities.

An awareness of the inextricable interrelationships among all the physical factors and all the organisms of a particular habitat has led ecologists to recognize the ecosystem as a natural unit in the study of landscapes. This is not taken to imply that meticulous study of one ecologic relationship is undesirable, but that its meaning is enhanced a great deal when the results are viewed against a background of the biophysical complex of which it is a part.

Optima and Limiting Factors

Enough has been said above to show that a fixed optimum intensity of any one factor does not exist. For every change in one factor, a different optimum of all other factors comes into existence. The concept of optimum is further complicated by the fact that optimal requirements differ for different processes in the same organism. Thus it is clear that when the term optimum is used it should always be understood that "apparent optimum" with respect to one type of response is really implied (11). Optimum conditions for longevity (resulting in trees reaching an age of 4100 years) depend on suboptimal conditions for nearly all plant functions! (682).

A plant extends its range in all directions until some detrimental aspect of environment prevents a completion of the life cycle either by vegetative or sexual means. As one aspect of environment approaches the extreme limit of tolerance, the welfare of the plant comes to depend closely on this condition, and the expression *limiting factor* is used. For example, where forest abuts a dry grassland region, trees become confined to habitats that have a water-balance less severe for them, suggesting that moisture deficiency is the primary environmental limitation. In this way the behavior of plants at the edges of their ranges usually provides valuable clues to the most important factors limiting distribution. However, there is a definite danger in attaching too much importance to apparent limiting factors (Fig. 96). For example, at the dry timberline referred to above, rainfall would cease to be inadequate if temperature were lowered, and possibly neither temperature nor precipitation would have to be altered to allow forest extension if herbaceous competition with tree seedlings were eliminated. Therefore, the term limiting factor must usually be defined as the factor of most immediate importance in the existing biophysical complex. The concept is justified in that often one can remove the limitation by a single act. Thus trees will usually grow in a dry region either if irrigated, or if clean cultivated so as to eliminate competition from other plants. Similarly, on sandy outwash in New York the poor growth of trees can be corrected merely by fertilizing

FIGURE 96. Although wind is the most evident factor operating at this natural edge of the forest, the immediate cause of the difference in vegetation appears to be edaphic. Forest is confined to areas where a layer of silt overlays the gravelly substratum. San Juan Island, Washington.

with K (330), and dormancy can often be broken by altering temperature alone.

Liebig's law of minimum, that the size of a crop is determined by the essential nutrient that is present in minimal amount, does not apply to environmental relationships so regularly or so closely as might be expected. Sometimes factor compensation mitigates the effect of a severe environmental condition without lessening its intensity, whereas at other times or places the same condition untempered by compensation would have a much stronger effect. Another possible explanation of exceptions to the law of minimum is as follows. Optimum environment demands the most favorable levels of all factor intensities; therefore an improvement in any single environmental condition, even if it is not the most detrimental one, tends to favor plant growth.

Since Liebig was interested only in nutrient deficiency, whereas excessive amounts of nutrients and other aspects of environment may be equally critical, and whereas plant success is conditioned by the concerted action of all factors, a much broader restatement such as the following is needed. The more nearly a factor approaches the extreme condition tolerated by the individual at a given stage of its life cycle, the greater the relative impact of change in that factor on plant behavior. This implies that wide variations in factor-intensities are of relatively little consequence in the region of the optimum.

Extrapolation from Controlled Environment to the Field

Owing to the multiplicity of interacting factors that influence plant function and structure, the possibility of working under conditions where all factors are under control has had strong appeal. Glasshouses provide a large measure of control, but growth chambers and phytotrons allow complete control, and so have enjoyed great popularity in recent decades. However, from the standpoint of an ecologist attempting to understand plant responses in natural habitats, such control is achieved at a cost that was at first grossly underestimated.

Constant conditions are abnormal for most organisms, and the need for a 24-hour cycle of light and temperature conditions, with night temperatures the lower, is generally taken into account. The most advanced control systems make the change from "day" to "night" conditions gradually, as in nature, but none accommodates the fact that plants may respond to small changes in light and temperature conditions that occur at few-minute intervals during the illuminated period. As examples, tomatoes grow better if day temperatures fluctuate 2.5°C at 2-minute intervals; intermittent cloudiness reduces maize growth disproportion- ately to the energy loss.

In natural habitats leaf temperatures rise well above air temperatures during the day then drop below air temperatures at night, but under fluorescent lamps leaf temperatures remain below air temperatures con- stantly (635). In a sense this is of minor importance since the light intensities in most growth chambers are inadequate for the normal growth of any but sciophytes. Even for these plants the important sunfleck influence is lacking.

The axis of a plant growing in the field cuts across environment that is markedly stratified as to temperature and dryness, from the apical bud to the root tips. Few growth chambers permit even a differentiation between root and shoot temperatures, although the importance of this has been well demonstrated. Furthermore, the direction of temperature differential between root and shoot is reversed day and night under field conditions. It was observed that the root temperature of a *Trifolium* plant in the field remained at −1°C while the temperature of the stolon 7cm above varied from −10 to 22°C. Translocation at least must be strongly affected by these vertical temperature gradients.

A soil deficient in N or P in the field may not prove deficient when tested with growing plants in a glasshouse or growth chamber, since the elevated soil temperature greatly increases the rate of release of these nutrients from humus. Even the size and shape of the pots used in plant culture has been shown to influence P uptake.

The behavior of isolated plants may provide no clue as to their relative success in association with others in the field. The ranking of yield of four strains of *Trifolium* under sward conditions proved the reverse of that for yield when individuals were grown free of other plant influences. Plants growing alone may not grow rapidly in height, but the capacity to do so under crowded conditions could go far in explaining success under field conditions. As pointed out earlier, the optimum conditions for any aspect of performance is usually a variable dependent on the intensities of any of several environmental factors.

It should be clear that any prediction as to field performance which is based on tests under controlled conditions must always be tentative. This is not to deny that growth chambers are a boon to the plant physiol- ogist who is unconcerned with ecologic matters, or that studies of pho- toperiodism and temperature tolerance performed under controlled con- ditions have proven useful clues to field behavior, but only to point out that in ecology most growth-chamber studies should be used primarily as an aid in designing critical experiments to be performed in natural environments. Single factors or factor-complexes can indeed be altered in natural environments as well as in controlled environments, by shading, supplying additional water, intercepting precipitation, fertilization, etc. If uncontrolled conditions then affect both experimental and control plants equally, this is a boon rather than a hazard.

Technologically it is possible to multiply the complexity of present-day growth chambers and phytotrons so that nearly all details of field environment are measured with a control system that simultaneously regulated conditions in the chamber to match those in the field (75). But this involves expense beyond the means of most investigators.

The Phytometer Method

The above discussions of the complexity of environment and of plant responses emphasize the inadequacy of instruments that measure single factors like wind and temperature. Even when accurate records of all known transpiration-promoting factors are available, we do not know how to evaluate them in order to predict transpiration rates. Nor can the photosynthetic rate be predicted when all measurable factors are known.

The net influence of a particular environment on the plant is expressed most precisely in the responses of the plant growing in that environment. Consequently the theoretically ideal instrument to evaluate environment in terms of plant growth is a series of plants which are grown in different habitats in order to observe differences in their structure, behavior, or chemical composition. Such experimental plants, or pieces of low vegetation transplanted intact, are called *phytometers* (144). These not only have the advantage of integrating all factors, but also they overcome the problem presented by the fact that instruments are frequently more sensitive to variations in single factors than is the plant on account of the ability of protoplasm to make adjustments. Obviously the plants in the series should be genetically as nearly identical as possible; error due to variability may be reduced still further by using sufficient replications to permit a biometric analysis of results. Phytometers should always be accompanied by adequate batteries of instruments so that quantitative relationships can be established between stimulus and response.

Perhaps the most frequently overlooked limitation to this method lies in the fact that no two species react in exactly the same way to a given set of environmental conditions; therefore the behavior of one species cannot be used to prognosticate the behavior of another on the same series of habitats (213). Also, the fact must be taken into account that, when additional growth is used as a criterion of effectiveness of habitat factors, the amount of new growth is conditioned by the initial size of the organism as well as by environment. Thus it has been observed that the larger the seed the greater the size of the seedling after a given period of development with this effect sometimes persisting into the second year (594).

In North America the widest practical application of the phytometer method has been made by foresters who have used the average rate of height growth of a particular kind of tree as a criterion of the productivity of different habitats for lumber of that species.

In Conclusion

The ecologic approach involves a struggle for perspective which is easy to underestimate both as to importance and complexity. A realization of the multiplicity of interacting factors and the complexity of plant requirements constitutes one of the two most profound biologic principles that has been contributed by autecologic inquiry. The second will be discussed in Chapter 9.

9

Ecologic Adaptation and Evolution

Adaptation

Any feature of an organism or its parts which is of definite value in allowing that organism to exist under the conditions of its habitat may be called an *adaptation*. Such features may insure a degree of success either by allowing the plant to make especially full use of the amounts of nutrients, water, heat, or light available to it or by bestowing a significant amount of protection against some adverse factor, such as extremes of temperature, drouth, or parasites. By accumulating adaptations, organisms utilize the earth's resources ever more efficiently, and after eons of development a great many if not most of the characteristics of each species are adaptive. In fact, it has been said that an organism is "a bundle of adaptations."

Conspicuous adaptations, such as those of carnivorous plants and entomophilous flowers, are perhaps too well known, for they tend to distract attention from inconspicuous but far more numerous adaptations that are to be found everywhere. For example, radial symmetry is particularly appropriate for plants whose moisture, light, and nutrients come from all directions. Likewise, because the absorption of CO_2 by the shoot and water by the root are largely conditioned by the amount of absorptive surface presented, profuse branching of both root and shoot is advantageous. Light penetrates only a short distance into tissues; therefore the prevalence of broad, thin photosynthetic organs throughout the phyla is of significance. The drier or colder the climate, the lower is the osmotic potential of plant saps (314), and the colder the climate, the greater is the calorific value of plant fats (521).

It has been noted that although plants are in general well adapted

355

to the quantity and quality of energy received from the sun, their CO_2 optimum is much higher than that contained in the air. This has been interpreted as indicating that the atmosphere during the evolutionary emergence of modern plants was higher in CO_2.

In addition to such general types of adaptations, special habitats are often accompanied by plant characteristics (succulence, aerating tissue, etc.) that seem of value there. Since the possible means of adaptation to a particular type of environment are not many, taxonomically diverse species may follow converging lines of ecologic adaptation. Thus under desert environment the stem-succulent habit has arisen independently in the Cactaceae, the Euphorbiaceae, and the Asclepiadaceae. These groups retain many of their ancestral characters, especially reproductive structures, even though developing many physiologic and morphologic traits that are similar and bear the same relationship to desert environment. The principle is further illustrated by the root tendrils of *Vanilla* as compared with the ecologically equivalent stem tendrils of the grape, and by aerenchyma as compared with lacunar tissue in various emergent hydrophytes.

It often happens, however, that a single kind of adverse circumstance is met by entirely different types of adaptations which have equivalent value. An excellent example of this is provided by the divergence in physiology and morphology exhibited by desert plants. Here plants with very evident morphologic adaptations grow among others that lack visible manifestations of adaptations but rather are possessed of the necessary physiologic requirements to make their existence possible. Thus adaptation may be purely physiologic, as illustrated by resistance to disease, host preference among parasites, degree of palatability to herbivores, endurance of desiccation, etc. What appear to be purely morphologic adaptions are illustrated by sunken stomata, special pollinating mechanisms, form of shoot, etc. In reality there is much less difference between strictly physiologic and morphologic adaptations than is apparent, for a morphologic feature is but an expression of physiologic processes otherwise unevident.

Morphologic characteristics should always be subjected to physiologic experimentation before they are classified as having survival value, for interpretations of the physiologic significance of morphology based on human logic cannot be relied upon. It was pointed out, for example, that water may be transpired faster through the thick cuticle of one plant than through the thinner cuticle of another (620). Wherever morphology and physiology tell different stories, as in this instance, physiology must provide the correct answer, because morphologic features are assumed to have significance only as far as they influence the physiology of an organ. It is not a particularly comforting fact to note that most statements regarding the merits of assumed adaptations are unsubstantiated by physiologic tests.

All physiologic processes in plants are affected by the supply of water, heat, nutrients, and light. It follows that each plant must be fairly accurately adjusted to its habitat, at least in that it surpasses certain minimum adaptational requirements. To be perfectly adapted a plant would theoretically have to make the fullest use of the available energy and nutrients. It is significant in this connection that the crop plants with the highest carbohydrate production do use nearly all of the favorably warm season. Much of agronomic practice is directed consciously or subconsciously toward fitting the life cycles of crop plants into the prevailing climatic regimen by varietal selection, proper choice of planting dates, irrigation, protection against unseasonal frosts, etc. In these ways natural deficiencies in adaptations are compensated for.

Adaptation at first appears to be a process constantly conferring benefits on organisms. However, in the long run it is frequently disastrous as proved repeatedly by the paleontologic record. Where adaptation steadily pursues a given course without interruption it may eventually lead to such a high degree of specialization as to make survival absolutely dependent upon the maintenance of the environmental complex by which adaptation was guided. The trends of specializations in pollination illustrated by *Yucca* and by the Smyrna fig provide examples of this.

Origin of Adaptations

The correspondence between structure and function, and a particular habitat, obviously is not fortuitous. Man has long been interested in explaining how these adaptations came about, but until quite recently the interpretations have been almost entirely philosophical and not based on a careful analysis of facts.

The Older Viewpoint: Anthropomorphism and Teleology

Early ecologic thought was dominated by *anthropomorphism* and *teleology*. Anthropomorphism is the habit of attributing human characteristics to nonhuman objects, as in the expression that a plant "avoids" acid soils, or has a "strategy." Teleology is the doctrine that adaptation is purposive, as is implied in the statement that the wing of a seed is "for" wider dissemination. It was assumed that such an adaptation arose in response to a particular need, and indeed many phenomena superficially appear to warrant such an hypothesis. For example, root hairs, cuticle, and cork ordinarily do not develop on organs grown under water, and at the same time other tissues become modified in such a manner as to favor aeration. However, close analysis reveals that not all responses of plants to environmental stimuli are beneficial, and experimentation has shown that many responses, whether beneficial or detrimental, can be explained on a physical or chemical basis. Thus there is more evidence in support of mechanism than of vitalism.

The change away from the old viewpoint has been so recent, however, that scientific writing of high calibre still contains such terms as "food storage" and "calciphile," literally "lover of calcium." In general the continued use of such expressions is condoned for the sake of convenience or custom, for the older viewpoint has been thoroughly discredited. Precise expression would of course demand the substitutions of expressions such as "food accumulation" and "calciphyte" for the expressions mentioned above.

The Newer Viewpoint: Genetic Variation and Natural Selection

Most adaptations are now believed to arise by the selective action of environment operating as a sieve on genetic variations, the origins of which are strictly matters of chance. Since not all variations are genetically fixed, it will be necessary at this point to distinguish between two major types of variation. It will be noted that the topic here is variations rather than adaptations, for many if not most variations appear to lack survival value and therefore cannot be considered adaptational.

Environmentally induced variations. (= ontogenetic, or somatic variations; = modifications; = acquired, or noninherited characters). When a series of genetically identical plants is grown in diverse habitats, it becomes apparent that, to a certain extent, the characteristics of an individual develop according to the particular habitat in which the plant grows (527). The degree of plasticity and hence variation ranges from striking differences to only slight changes, depending on the hereditary constitution of the plant. Thus *Polygonum amphibium* (60) and *Callitriche palustre* exhibit very different forms when grown in water and on land, and many less striking variations in structure and function have been enumerated earlier in connection with the supply of water, light, oxygen, salts, etc. Environmentally induced variations of a morphologic nature are acquired only as a result of continued exposure during a long portion of the life cycle, but physiologic adjustments may be brought about in a few days (105).

Frequently variations of this type are beneficial and may be classed as adaptations. For example, exposure to drouth induces drouth resistance, chilling increases cold resistance, and unequal shading stimulates a bending toward brighter light. Le Châtelier's theorem can be restated to fit such biologic phenomena as follows: any intensification of an environmental factor tends to increase the organism's resistance to further intensification of that factor.

However, not all environmentally induced variations appear to be useful. Such phenomena as the change in flower color with change in soil pH in *Hydrangea,* the shade-induced lowering of light-saturation levels for photosynthesis (744), and the swelling of the bases of trunks of *Taxodium* at the pond surface, seem best described as neutral. Still other responses to environment may result in considerable detriment to plants. For example, barrel cacti germinate chiefly in the shade of shrubs, but as they grow the stems lean toward brighter light until they topple over and die (374). Again the tall weak stems produced by cereal crops growing under exceptionally favorable moisture and nutrient conditions render the plants very susceptible to lodging and therefore exemplify an at least potentially detrimental response to environment. Finally, the formation of galls clearly favors gall wasps at the expense of the host (159).

Genetically fixed variations.   (= characters). These are ordinarily irreversible and arise only by changes in the structure of genes, * rearrangements of genes within the chromosome framework, recombinations of genes through hybridization, or irregularities during mitosis or meiosis that change the number of chromosomes per cell. Large changes of this nature occur only at infrequent intervals, but it is quite likely that numerous small changes occur frequently and eventually become important because of cumulative effect.

The origin of genetic variations is subject to the laws of chance, but as the individual develops, the relative merits of the new hereditary materials assert themselves more and more with the result that survival, maturation, and reproduction are definitely not random. If the new character is detrimental the organism is less successful than others of its kind, and hence the change is not likely to be preserved. If the innovation confers an advantage upon the recipient, it has better-than-average chances of survival. Thus natural selection, operating on fortuitous ge-

---

* A *gene* is a unit of inheritance, actually a definite segment of a chromosome which, by interaction with other genes and the environment, governs the manifestation of some character of the organism. Genes occupying identical positions in homologous chromosomes are called *alleles.* Different alleles control the same character but may vary its expression.

netic variations, tends to produce new forms ever more closely adapted to the habitat. Mathematically it has been demonstrated that, if a mutation increased the chances of survival only 1%, it would become established in half the population in about 100 generations. This emphasizes the significance of sex, for without this a species would be unable to exploit useful mutations through recombination, although vegetative reproduction alone will permit the perpetuation of unmodified mutations, as horticultural practices testify abundantly.

Natural selection favoring characters that create the maximum harmony between organism and environment provides the only explanation for the countless instances of ecologic similarity among taxonomically diverse species.

Of the thousands of genes that govern the behavior of an organism, only one need change beyond a certain extent in order to disturb the synchrony of the various functions and thereby prove fatal. The vast majority of genetic variations are probably unsuccessful because the physiologic balance has been upset by the new combination of genes. Thus there is an internal requirement for harmony, in addition to the demand for harmony between the new gene complex and the environment. When viewed in this light the slowness of the evolutionary process becomes understandable.

The principle of natural selection cannot be interpreted to mean that all structures and functions of present-day organisms have survived because they have been beneficial. Many if not most genetic variations must be so inconsequential with respect to natural selection that they must be considered neutral, i.e., lacking either positive or negative survival value. The endless variety of blade shapes and venation patterns in leaves is not known to affect the function of these organs and therefore must have no different survival values. Likewise the night-folding of leaflets which are so common in legumes appear unrelated to the welfare of these plants. On changes of this nature the environmental sieve fails to act with a selective or discarding action, and thus netural traits have a chance of being preserved.

Inconsequential characters, however, may be due to genes linked in the same chromosome with other genes that control small but important physiologic characters, in which event natural selection appears to favor characters for which man can rightly see no value (347). Such small and directly insignificant morphologic characters then become important to the biologist solely because they indicate the presence or absence of physiologic attributes that are of ecologic significance (30). This aspect of linkage deserves much consideration, for small morphologic characters, no matter how trivial they may seem to the taxonomic "lumpers," may provide invaluable criteria of the ecologic status of individuals. On the other hand it is definitely known that the genes governing slight morphologic differences may not be linked to those governing slight physiologic differences, so that the two types of variations can be independent.

From an ecologic standpoint, all trivial characters should be ignored unless they are accompanied by some physiologic character. Since its beginning the subscience of taxonomy has been concerned chiefly with the comparative morphology of sexual reproductive organs, but the ecologist is much more interested in the details of structure and function of vegetative organs, especially of juvenile stages, because of their significance in relation to survival. It is unfortunate that it is so inconvenient to recognize taxonomically such inherent differences among plants as the rate of seedling root penetration, time of flowering, etc., for these small physiologic characters are frequently important in determining failure or success.

There seems to be no doubt but that many attributes have been gained or lost in evolution simply as the result of chance mutations of neutral nature. It has been pointed out that, if adaptation by mutation is a random phenomenon unrelated to the organism's needs, warm-climate vegetation should contain species possessing cold-tolerance, for in warm regions a change in this direction would be neutral from the standpoint of natural selection. This has proved to be true, for many plants have been successfully transplanted into regions 5 to 10°C cooler than their native homes (835). Further evidence bearing out this conclusion is provided by the fact that in the floras of the tropics some species have come to require the short days that prevail there, whereas others are indeterminate in daylength requirements. Mutations from either one of these types to the other would be neutral, but those producing long-day requirements would be fatal.

Any mutation that gives the organism some characteristic of potential value, but which at the time has no survival value, may be called a *preadaptation.* For example, blight-resistant strains of species have been discovered in regions where the blight organism does not occur. Long-day-tolerant and cold-resistant species native to the warm tropics provide additional examples of preadaptation. This phenomenon may explain the fact that different species of the same habitat sometimes move into distinctly different habitats when climates become more diversified (834). Thus, the paleontologic record shows that *Sequoia, Fagus,* and *Glyptostrobus* were members of the same forest in Miocene time, but with subsequent climatic differentiation each came to occupy a different climatic area.

The genetically fixed type of adaptation is the more advantageous where an early appearance of the character is essential in the life cycle, for the younger the organism the greater its sensitivity to adversity and the less time it has had to become conditioned to meet crises. It should be noted that the use of the expresion "genetically fixed" does not mean that such characters are not susceptible of environmentally induced modification, for the effects of environment become superimposed on the effects of heredity as the seedling develops into a mature organism. In addition, somatic and genetic adaptation frequently parallel each other and can never be distinguished without resorting to experimentation.

In the discussion above emphasis has been laid on vascular plants, but the same principles are operative among fungi, bacteria and other cryptogams, where the genetic segregates are often referred to as "physiologic races" (458).

Fate of Adaptations

A perplexing problem which long confronted biologists is that certain structures and functions that appear to be definitely related to one type of environment are occasionally found in an entirely different environment. Examples of such paradoxical situations are provided by the xeromorphic yet hydrophytic pond pine (*Pinus rigida* var. *serotina*), by the succulent *Sedum ternatum* of the mesophytic winter-deciduous forest, by semiaquatic cacti, and by the functional nectaries of the parthenogenetic *Taraxacum officinale*. As long as ecologic thinking was dominated by the assumption that an organism and its environment present a closed system of cause and effect, that each characteristic of form and function bears a direct relationship to habitat, phenomena such as these were indeed mysterious enigmas. However, when the truly scientific methods of reason and experimentation were substituted for the method of speculating upon nonexperimental observations, and the dynamic natures of both organisms and environments were taken into account, it became apparent that the relations between organism and environment often become intelligible only when placed in historical perspective.

It is now believed that genetically fixed adaptational features frequently outlive their usefulness, and, as long as they do not prove definitely detrimental under changed habitat conditions, they may persist indefinitely, although they can no longer be called adaptations. It would be a rare circumstance indeed if one gene change which allowed or demanded the movement of a succulent xerophyte into a mesophytic habitat were always accompanied by other gene changes which simultaneously abolished succulence and all other xeromorphic characters at the same instant. Plants have no ability to discard physiologic or morphologic features that are of no use to them. Constant selection pressure keeps essential structures and functions at approximately their peak of efficiency, but, when under changing conditions a structure or function becomes no longer essential, it merely becomes subject to atrophy through a lack of this selective maintenance. Thus essential functions may be in close adjustment with the present factor complex even though "obsolete adaptations" persist.

The Genecologic Classification

During the nineteenth century the fact became established that a species commonly includes genetically distinct races, but genetics itself had not yet been developed and little progress was made in studying this important phenomenon.

Beginning in 1920, Turesson assembled in an experimental garden at Åkarp, Sweden, groups of 20 or more individual plants belonging to the same species but native to different parts of the species' range. Persistent differences among these groups growing in the same environment verified earlier conclusions that a taxonomic species is not a single ecologic unit but is composed of numerous races which exhibit inherent differences in physiology and often in morphology as well (768). The races were observed to differ in earliness of flowering, height, erectness, thickness of leaves, etc., but none of these differences had been sufficient to warrant taxonomic recognition, and often they showed intergradation.

Turesson further showed that many of these genetic races are correlated with particular habitats, that one may be confined to sunny habitats and another to shade, some to alpine regions and some to lowlands, etc. Thus they are truly ecologic races. In other experiments he observed that, when plants similar in appearance were taken from a severe habitat and planted in his garden, they differentiated into several recognizably distinct forms as they recovered from transplanting.

These studies by Turesson were an important factor in bringing about the major revolution in plant taxonomy which gave the ecologic viewpoint the consideration it deserves in that subscience. On the basis of his experimental results Turesson proposed the following classification of plants; subsequent work by other investigators has in the main verified the correctness of his views (138, 140, 141, 286).

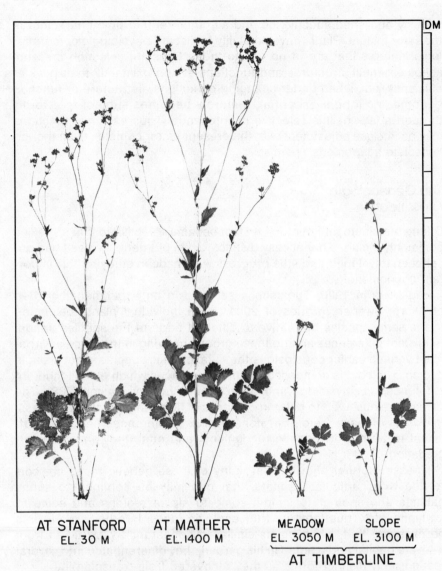

DM

AT STANFORD
EL. 30 M

AT MATHER
EL. 1400 M

MEADOW
EL. 3050 M

SLOPE
EL. 3100 M

AT TIMBERLINE

FIGURE 97. Uniform heredity in varied environment: the appearance of ecophenes developed from clonal divisions of *Potentilla glandulosa* ssp. *hanseni* grown in four markedly different environments: near the seacoast at Stanford, California; in native mid-Sierran habitat at Mather, California; and on two habitats at upper timberline. (Photography by courtesy of J. Clausen and the Carnegie Institution of Washington.)

Ecophenes

(Synonyms: habitat forms, epharmones, plastodemes, ecads, environmentally induced variations, etc.)

These are plants differing in appearance, especially in the size of
vegetative parts, numbers of stems, erectness, and reproductive vigor,
but belonging to essentially homogeneous genetic stock (Fig. 97). Their
distinctness is due entirely to environmental influences, for when different
ecophenes are transplanted into the same habitat these differences
disappear.

As an illustration of this phenomenon, the appearance of a shrubby
species of *Haplopappus* was so strongly affected by soil types that for a
time taxonomists considered the ecophenes as distinct species (*H. ve-
netus* and *H. decumbens*) until transplant studies proved their identity.
Similar examples could be cited in abundance (60).

The ecologic importance of phenotypic plasticity lies in the fact that
it has a bearing on the range of habitats a species can occupy, since
it tends to make the individual adaptable to more than one habitat.
Annuals, and especially weeds, have above-average plasticity (676).

Ecotypes

(Synonyms: ecologic races, physiologic races, ecodemes)

A species is typically composed of a mosaic of populations, each of
which differs in genetically based physiologic (and sometimes morphologic) features having survival value, and is designated as an *ecotype.*
Although much mortality is accidental, individuals possessing gene combinations that are more closely attuned to local peculiarities of environment have higher probabilities of survival, and in this way each variant
of the environmental mosaic tends to modify the characters of all species
that invade it. Opposing this trend are dissemination and pollination
which tend to carry genetic materials back and forth across the borders
of contiguous habitat types. However, hybrids between ecotypes are
incompletely adapted to either habitat, consequently selection pressure
tends to eliminate them and favor homozygosity for those characters
which especially fit a plant for one habitat or the other.

Local differentiation can occur only where natural selection is strong
enough to overcome gene flow. But rain dripping from a wire fence
can contain enough poisonous Zn to result in the selection beneath
of a linear strip of Zn-tolerant ecotypes of grasses in only a few decades
(87)! And in a comparable time span grazing pressure can sort out

low-growing ecotypes that are less damaged by cattle, just in an area that is fenced and subject to grazing (396).

Not only may ecotypes be differentiated by edaphic, microclimatic, or biotic factors (286), but also wherever a species extends across several climatic zones it may evolve a distinct climatic ecotype in each (140). In some species the populations that are encountered along a climatic gradient are restricted to special habitats in each sector, so that the species is composed of a chain of contiguous yet reasonably distinct and homogeneous population segments, which are ecotypes (140). In other species the adaptational pattern appears to be more in the nature of a continuous gradation (an ecocline [286]) which parallels a continuous environmental gradient, so that ecotypes could be recognized only as arbitrarily defined segments. For example, black walnut (*Juglans nigra*) trees grown from nuts obtained in Minnesota are distinctly more winter-hardy than trees grown from nuts obtained in Alabama or Texas, although materials from the two extremities of the range are morphologically indistinguishable (834). Races not morphologically differentiated and hence amenable to nomenclatorial recognition, such as those of *Juglans nigra,* should be indicated by accompanying notes on the region and habitat when living material is collected for cultural purposes.

Since differentiation into ecotypes results from the discriminating selection offered by unlike habitats, it follows that in general the wider the ecologic range of the species the more numerous are its ecotypes. It is possible for environmental selection to produce a complete series of races that are individually sensitive to habitat variations even if they are not macroscopically distinguishable. This may well explain strongly different habitat preferences of the same species in different regions. There are other species in which wide ecologic amplitude * seems due rather to the possession of a large reservoir of alleles without specialization than to differentiation into ecotypes with narrow but complementary requirements.

Studies of ecotypes taken from widely different climatic zones in California have shown that the ecotypes of different species which grow in the same habitat all tend to exhibit certain adaptational characteristics in common. For example, alpine ecotypes of all species tend to be earlier than lowland types when grown on the habitat of the lowland types (140). In some instances the adaptations of corresponding ecotypes of closely related species are so parallel that these ecotypes resemble each other more closely than they resemble other ecotypes within their respective

---

* The range of habitat variation a plant can endure is conveniently expresed as its *ecologic amplitude.*

species. The frequency with which genetic and plastic adaptations parallel each other has long been a phenomenon causing confusion of thought, and one which is still not understood.

A single ecotype may, under suitable stimulus, be represented in several habitats by many ecophenes, and for this reason the critical delineation of ecotypes within a species usually requires experimentation. Each ecotype retains at least some of its distinctive characteristics (such as flower color and shape, habit of branching, venation, and shape of

COASTAL   MID-SIERRAN   ALPINE

FIGURE 98. Varied heredity in uniform environment: the appearance of three distinct ecotypes grown in the same (mid-Sierran) habitat. *Potentilla glandulosa* ssp. *typica* transplanted from coastal habitat; *P. g.* ssp. *reflexa* native of mid-Sierran habitat; *P. g.* ssp. *nevadensis,* transplanted from alpine habitat. (Photograph by courtesy of J. Clausen and the Carnegie Institution of Washington.)

leaves) when transplanted into the same environment with others (Fig. 98), although the ecophenes of one ecotype may closely resemble another ecotype of the same species. Thus the dwarfing of a lowland ecotype transplanted into an alpine habitat may render it superficially similar to another ecotype which is confined to high altitude.

Although an ecotype has a certain degree of homogeneity with respect to ecologically critical alleles, this does not exclude variation due to heterozygosity with respect to other alleles. In fact, genetic variations within ecotypes (each of which is a *biotype*) are theoretically infinite owing to recombinations of a great store of dominant, partially dominant, and recessive genes distributed among the individuals. However, in contrast to the differences among ecotypes, the variations within them appear at random, and because they have almost no survival value these variations are not correlated with environment. Also, because most terrestrial habitats embrace an infinite number of subtle variations, natural selection cannot be expected to eliminate variations below a certain limit set by environmental heterogeneity if an interbreeding population is to remain well represented in an area.

Because ecotypes are interfertile, at least through intermediaries, they are never pure. The differences among ecotypes may be no more than a matter of difference in the frequency of occurrence of certain alleles. Therefore not all members of one ecotype could be accurately classified, even if environmental modifications could be ruled out.

Another consequence of the fact that ecotypes are normally interfertile is that in a region where the ranges of two ecotypes overlap they exchange genes freely, even though outside this region of contact the two types exist unaltered because of constant environmental selection. There is experimental evidence that such hybridization occasionally produces new ecotypes. In Madison, Wisconsin, the natural hybrid between *Lonicera morrowii* and *L. tartarica* has proven to be a highly aggressive and troublesome weed. These are distinct species rather than ecotypes, but the instance is valid as an illustration of the potentialities of crossing. There is much evidence that the survival of new gene combinations resulting from hybridization is rarely favored except when accompanied by the creation of a new kind of environment to which they may be preadapted, such as was provided in geologic history by the development of aridity and by the recession of glaciers to leave large areas of relatively unoccupied soil, or in recent time by the many artificial habitats created by man.

A second way in which new ecotypes may originate is through the slow accumulation of numerous small gene changes which are individually unimportant. This process is not likely in large, freely interbreeding populations, as an innovation is easily lost by the overwhelming numbers of homologous alleles. However, in a small population that is not freely exchanging genes with other populations (owing to habitat discontinuity, or merely to great distance between remote extremities of the species' range) small gene changes have a fair chance of becoming prevalent.

A third manner by which new ecotypes may arise is by chromosome changes. This is indicated by the fact that, in sexual species, polyploids from the same original stock rarely exhibit the same ecologic amplitude (558).

Most changes in hereditary constitution, whether due to hybridization, gene changes, or chromosome changes, probably result in changes in ecologic amplitude. Because new ecotypes originate within the area of their parental types they must of necessity tolerate that environment, but by virtue of their different ecologic amplitudes they are able to extend into new habitats. Such a preadaptation is the more likely to succeed if it originates near the margin of the parental area, especially if such a marginal position is adjacent to a habitat that is favorable to the new entity, for this allows the new ecotype to spread quickly into new territory and thus escape submergence by interbreeding with the parental stock.

The number of ecotypes normally increases when a plant is brought into cultivation. This increase seems explainable by the fact that under cultivation there is relatively little competitive selection, which in nature would promptly eliminate many innovations.

Ecotypes are the basic units of the genecologic classification, but other categories are necessary for grouping these units into a hierarchy showing degrees of kinship and evolutionary relationships.

Ecospecies

An *ecospecies* is a classificatory unit embracing one or more ecotypes which, though freely interfertile, do not cross or at least do not produce offspring strong enough to survive when crossed with ecotypes of another species. The taxonomic species, defined for practical reasons on morphology, frequently cannot be made to correspond with the ecospecies defined on experimental tests of fertility, for sometimes two ecospecies must bear the same taxonomic designation simply because there are no useful morphologic criteria for separation. Also the role of taxonomy seems better served by recognizing as distinct species two population complexes that are geographically isolated and well differentiated, even though they may be completely interfertile.

Coenospecies

Coenospecies, the highest entities of the genecologic system, correspond to the taxonomist's section of a genus, or to all the members of a small genus. Relationships between coenospecies are so remote that genes can seldom be exchanged among them, but on rare occasions they may cross and produce either a sterile hybrid or a fertile hybrid on the level of a new species. Complex coenospecies are represented by a number of ecospecies, but at the other extreme a coenospecies may be represented by a single ecotype.

A concrete example of the application of genecology to a group of plants is provided by studies of *Potentilla* (Table 14).

This summary brings out the point of view adopted by certain well-qualified workers that morphologically indistinguishable ecotypes should not be given Latinized names but should be designated by descriptive

Table 14

Taxonomic and Genecologic Analysis of Part of One Section of *Potentilla*. Five Ecotypes of *P. Glandulosa* Span the Environmental Mosaic (without occupying all components of it) from the Pacific Coast of Central California Across Two Mountain Ranges, Giving the Species a Very Wide Ecologic Amplitude (140). *Achillea* Contains Eleven Ecotypes Across the Same Transect (141).

Coenospecies	Ecospecies	Subspecies	Segment of Environmental Mosaic	Ecotypes
Potentilla, Section Drymocallis	glandulosa	nevadensis	Alpine and upper forest belts (1600–3500 meters)	Alpine
				Subalpine
		hanseni	Mid-altitude meadows of Sierra Nevada	All one
		reflexa	Well-drained slopes of Sierra Nevada (250–2200 meters)	All one
		typica	Coast Range	All one
	fissa	Not analyzed genecologically		
	arguta			

terms and symbols. Also it is generally agreed that the genecologic terminology should never be applied to specific groups of plants until their genetic status has been determined experimentally. Until this is done the homologous taxonomic terms (subspecies, species, and section) should be applied tentatively with the understanding that experimentation will provide the final answer and may necessitate revision of any taxonomic scheme based on morphology alone.

Ecologic Significance of Intraspecific Variations (729)

Ecology, both theoretical and applied, must frequently take genetic variability into consideration in order to explain natural phenomena as well as to avoid error in experimental work.

As indicated above, a taxonomic species may consist of one to many ecotypes. It may have only one ecotype yet occupy wide geographic area if its special habitat (e.g., shallow ponds) is widely distributed. But in general, genetic variability varies directly with extent of range and diversity of habitats occupied.

Species with many ecotypes can fit into a wide range of habitats, whereas genetically impoverished species are commonly adapted to a particular combination of environmental factors and hence have very limited ecologic amplitude. The same variation with respect to adjustment and tolerance extends to ecotypes, for, when plants with identical heredity are grown in different habitats, some ecotypes show little or no ability to succeed in more than one habitat, whereas others are able to tolerate wide variations in environment, even though they exhibit marked differences in appearance (140, 527). Genetic constitution that allows for plasticity of structure and function of the individual probably has more survival value than variability due to multiple alleles, for successive changes in environment can more easily eliminate all of the latter.

The value of a species as an indicator of particular environmental conditions is highly dependent on its degree of ecotypic specialization, for the ecotype, rather than the species, is the fundamental ecologic unit. It follows that an experimenter should file pressed specimens (*vouchers*) as an aid in settling any subsequent questions that may arise about the ecotype which was studied.

The degree of ecologic amplitude has an important bearing on the possibilities of natural or artificial extensions of species ranges. When a plant is moved from one region to another, or its original habitat undergoes change, its physiologic processes must act at a new pace dictated by the new factor intensities. Survival depends on the degree of environmental change in terms of the degree to which the plant's hereditary make-up enables it to accommodate to the change. Weeds appear to be groups with exceptionally broad ecologic amplitude, for, when they are freed of competition and original parasites, and the natural barriers to their distribution are overcome, these plants prosper under a wide variety of environmental conditions. Thus *Bromus tectorum,* a Eurasian weed that has become established throughout the United States, appears to be represented so widely in the new continent by a single ecotype (371). As a rule, however, a vascular plant is less successful in any new habitat. Man has been able to extend the ranges of economic plants so widely only because he has produced new ecotypes, * adopted cultivation practices which compensate for much climatic adversity, and kept the plants free of serious competition from native vegetation.

In forest and range management, however, it is rarely feasible to provide any measure of freedom from competition or other habitat modifications, so that the genetic constitution must be more closely matched with the habitat factors than it is in cultivated plants. The result of a number of experiments warrants the conclusion that among trees, at least, few if any ecotypes are more successful in a region than those native to it (549). Foresters have become so conscious of this fact that seed collections from different regions are labeled and kept separate, and nursery stock is distributed in accordance with this information (36). Attempts have been made, especially with range plants, to gather together in a garden for comparative evaluation a wide selection of ecotypes belonging to economically important species. From the discussions above it should be evident that, theoretically at least, the conditions under which the selections are compared should match those under which they are to be grown, for environment may so condition the expression of hereditary characters as to suppress some and magnify others (683). With cultivated plants this is not so necessary as with forest trees or range grasses which are given no protection after planting. It is significant to note that the comparative plantings of different ecotypes on one habitat gives results applicable to that habitat alone; relative behavior in other environments may differ considerably (140).

Theoretically the introduction of a species into a new locality is most likely to succeed if the greatest possible number of ecotypes is introduced simultaneously (728, 835). Not only does this increase the chances of finding a suitable ecotype among those already in existence, but also the possibility is increased that, by hybridization, a race especially suited

---

* It has been suggested that ecotypes produced by artificial selection be called *agroecotypes.*

to the new habitat may arise. Artificial crossing of ecotypes has shown that it is possible to produce new ecotypes which will grow well in an environment that neither of the parents could tolerate. This process of crossing and selection among ecotypes is the only manner by which plants become genetically acclimated to new environments.

The existence of morphologically indistinguishable ecotypes points clearly to the absolute necessity of using either *clonal lines* or *pure lines* wherever possible in ecologic experimentation. A clone is a group of individuals propagated asexually from a single parent, so that they normally have absolutely uniform heredity. A *ramet* is one member of a clone. A pure line is the offspring of a homozygous, self-fertilized individual.

Isolation
and Evolution

Usually species and ecotypes are populations of individuals in which many if not most homologous genes differ; i.e., they are heterozygous to varying degrees. Crossing within these populations results in countless recombinations of the different genes and allows favorable and neutral innovations to spread and become common characters. However, when a barrier prevents the mingling of new characters as they arise, separated populations sooner or later evolve along divergent lines owing to the independent accumulation and loss of genes within each population. Barriers having such an effect may be extrinsic (ecologic):

1. Geographic: seas, continents, unfavorable climatic areas.
2. Local: differences in habitat preferences.
3. Biologic: areas occupied by other organisms that are inimical to survival, or areas from which essential symbionts are excluded.

or intrinsic:

1. Physiologic: differences in time of pollination, etc.
2. Morphologic: differences in pollinating mechanisms, etc.
3. Cytogenetic: differences in chromosome number, etc.

Intrinsic barriers are permanent, and if different genetic adaptations then develop, the populations so differentiated must be considered ecospecies even if the morphologic differences remain too small to warrant nomenclatorial recognition.

Once extrinsic isolation has allowed separate populations to develop intrinsic barriers, a breakdown of extrinsic isolation may permit the recently divergent ecospecies to mingle on the same habitat without crossing. Morphologically such ecospecies may be so similar that only experimental hybridization or the occurrence of bimodal frequency curves in random measurements of the mixed population may reveal their existence (17).

Frequently species or subspecies represented by a single ecotype confined to a limited area exhibit exceptionally small ecologic amplitude when compared with others which, by means of genetic diversification, are represented in a variety of habitats. The apparent genetic impoverishment of these entities may arise by either of two courses of events.

First, a species may suffer one or more catastrophes which destroy all but a fragment of the total population. The remnants are called *relics.* * Subsequent to the catastrophe the species is represented by only one or at most a few ecotypes which carry but a limited number of contrasting alleles.

Alternatively, an ecotype of narrow ecologic amplitude may result from a very limited crossing of an effective geographic barrier. The population descending from such a single introduction obviously can have no greater genetic heterogeneity than that possessed by the individuals which immigrated. Here specialization would not be the result of environmental selection as is true of most ecotypes, and the terms *geoecotype* (286), and *insular species* have been proposed for these.

It should be clear that both of the above kinds of genetically impoverished entities occupy restricted areas. Thus the generalization can be drawn that the degree of genetic homogeneity is somewhat proportional to the degree of restriction in local distribution and in geographic area.

Species of small range and ecologic amplitude are in precarious positions as far as continued survival is concerned. Because of their genetic homogeneity even small changes in environment can exceed their tolerance and thus bring about extinction. On the other hand, if environment remains favorable they may persist and come to possess variability through the accumulation of mutations. Also it is theoretically possible that favorable environmental changes may come about that would allow isolated colonies to reunite and share variability acquired during the period of isolation or would allow hybridization to produce new combinations.

Free interbreeding is more effective in dispersing genes uniformly through small populations than through large, for in the latter mutations

---

* The term relic(t) is also used in other senses in ecology. Any entity, such as a community, an individual, a species, a genus, that now occupies but a small part of an area or a habitat in which it was once more abundant may be referred to as a relic in that situation. Thus the term may have local as well as geographic application.

are more numerous on account of the laws of chance, yet distance alone retards gene flow. This phenomenon is added to ecotypic differentiation in bringing about a positive correlation between amount of variability and size of a species' range.

Isolated populations have unique opportunities for evolution, for new genes can easily become fixed in the absence of continual dilution. However, this is not necessarily advantageous, for neutral or even detrimental characters can become fixed just as easily. Evolutionary divergence of this nature has been called *drift* in contrast to the divergence that results from the accumulation of adaptive characters. Thus it can be seen that two subspecies may occupy different habitats that are ecologically identical even though discontinuous, and then the subspecies must be considered as still belonging to the same ecotype.

Earlier it was pointed out that adaptational variation could be discontinuous (ecotypes) or continuous (ecoclines). Nonadaptive variation also may be discontinuous, as when an island population develops morphologic distinctiveness through drift, and the term *topotype* has been suggested. If it is continuous, as illustrated by the progressive increase in ratio of scape to spike length in *Plantago maritima* from west to east across North America to Europe, the term *topocline* is appropriate.

In Conclusion

The fact that environment shapes the course of evolution was recognized with the acceptance of Darwin's work, but intensive experimentation on the problem was delayed more than half a century until developments in the field of genetics made it possible. Only within the past half century has it been established that the taxonomic species is not a single ecologic unit, and this has been the second major contribution of autecologic study to science.

Literature Cited

1. Abd-el-Rahman, A. A., and K. H. Natanouny. 1964. Osmotic pressure of desert plants under different environmental conditions, J. Bot. U. A. R. 7:95–107.
2. Abd-el-Rahman, A. A., and M. H. Sharkawi. 1968. Dew condensation under desert conditions. Cairo Univ. Bul. Faculty Sci. 42:87–113.
3. Addoms, R. M., and F. C. Mounce, 1931. Notes on the nutrient requirements and the histology of the cranberry (Vaccinium macrocarpon) with special reference to mycorrhiza. Plant Physiol. 6:653–668.
4. Agerter, S. R., and W. S. Glock. 1965. An annotated bibliography of tree growth and growth rings, 1950–1962. Univ. Ariz. Press, Tucson. 179 p.
5. Ahlgren, I. F., and C. E. Ahlgren. 1961. Ecological effects of forest fires. Bot. Rev. 26:483–533.
6. Ahmadjian, V. 1967. The lichen symbiosis. Blaisdell Publ. Co., Waltham, Mass. 152 p.
7. Aikman, J. M. 1936. The radiometer: A simple instrument for the measurement of radiant energy in field studies. Iowa Acad. Sci. Proc. 43:95–99.
8. Albert, W. B., and O. Armstrong. 1931. Effects of high soil moisture and lack of soil aeration upon fruiting behavior of young cotton plants. Plant Physiol. 6:585–591.
9. Alden, J., and R. K. Hermann. 1971. Aspects of the cold-hardiness mechanism in plants. Bot. Rev. 37:37–142.
10. Allard, H. A. 1942. Lack of available phosphorus preventing normal succession on small areas on Bull Run Mountain in Virginia. Ecol. 23:345–353.
11. Allen, W. E. 1934. What is an optimum? Ecol. 15:218–221.
12. Allison, F. E. 1968. Soil aggregation—some facts and fallacies as seen by a microbiologist. Soil Sci. 106:136–143.
13. Alvey, N. G. 1955. Error in glasshouse experiments. Plant and Soil 6:347–359.
14. Alvim, P. de T. 1960. Moisture stress as a requirement for flowering of coffee. Sci. 132:354.
15. Ames, J. W. and K. Kitsua. 1933. Assimilation of P and K by barley plants grown according to Neubauer procedure and in undiluted soil. Soil Sci. 35:197–207.
16. Anderson, A. J. 1956. Molybdenum as a fertilizer. Adv. in Agron. 8:163–202.
17. Anderson, E., and W. B. Turrill. 1938. Statistical studies on two populations of Fraxinus. New Phytol. 37:160–172.
18. Anderson, L. E. 1943. The distribution of Tortula pagorum (Wilde) de Not in North America. Bryol. 46:47–66.
19. Anderson, Y. O. 1955. Seasonal development in sun and shade leaves. Ecol. 36:430–439.

377

20. Anonymous. 1940. Influence of vegetation and watershed treatments on runoff, silting and streamflow. U.S.D.A. Misc. Publ. 397. 80 p.

21. Anonymous. 1951. Soil survey manual. U.S.D.A. Handbook No. 18. 503 p.

22. Anonymous. 1953. *Acacia mellifera*/grassland cycle. Sudan Report For. Dept. 1951–1952:44.

23. Anonymous. 1958. Alpine snowfields gain moisture from air. Rocky Mtn. Forest and Range Exp. Sta., Ann. Rept. 1957: 3–4.

24. Anonymous. 1970. Salt injury to roadside plantings studied. Shade Tree 43:112.

25. Anonymous. 1971. Increased plant crop damage from air pollutants reported. Western Landscaping News 11(7):1,24.

26. Armiger, W. H., and M. Fried. 1958. Effect of pot size and shape on yield and P uptake by millet. Agron. J. 50:462–465.

27. Arrhenius, O. 1957. Plant food from decayed leaves and conifer needles. Acta Agric. Scandinavica 7:341–340.

28. Arthur, J. M. 1936. Radiation and anthocyanin pigments. *In* B. M. Duggar, Biological effects of radiation. 2:1109–1150.

29. Association of Official Seed Analysts of N. A. 1938. Rules and recommendations for testing seeds. U.S.D.A. Cir. 480. 24 p.

30. Austin, L., et al. 1945. Use of shoot characters in selecting ponderosa pines resistant to resin midge. Ecol. 26:288–296.

31. Bair, R. A. 1942. Climatological measurements for use in the prediction of maize yields. Ecol. 23:79–88.

32. Baker, G. O., and K. H. W. Klages. 1938. Crop rotation studies. Idaho Agric. Exp. Sta. Bul. 227. 34 p.

33. Baker, H. G. 1961. The adaptation of flowering plants to nocturnal and crepuscular pollinators. Quart. Rev. Biol. 36:64–73.

34. Baker, H. G. 1963. Evolutionary mechanisms in pollination biology. Sci. 139:877–883.

35. Balchin, W. G. V., and N. Pye. 1950. Observations on local temperature variations and plant responses. J. Ecol. 38:345–353.

36. Baldwin, H. I. 1942. Forest tree seed. Chronica Botanica Co., Waltham, Mass. 240 p.

37. Baleigh, S. E., and O. Biddulph. 1970. The photosynthetic action spectrum of the bean plant. Plant Physiol. 46:1–5.

38. Barley, K. P. 1954. Effects of root growth and decay on the permeability of a synthetic sandy loam. Soil Sci. 78:205–210.

39. Barley, K. P., and E. L. Greacen. 1967. Mechanical resistance as a soil factor influencing the growth of roots and underground shoots. Adv. Agron. 19:1–43.

40. Barrows, F. L. 1941. Propagation of *Epigaea repens* L. II. The endophytic fungus. Boyce Thomps. Inst. Contrib. 11:431–440.

41. Bartlett, J. L. 1905. The influence of small lakes on local temperature conditions. Mo. Wea. Rev. 33:147–148.

42. Barton, L. V. 1944. Some seeds showing special dormancy. Boyce Thomps. Inst. Contrib. 13:259–271.

43. Bates, C. G. 1911. Windbreaks: their influence and value. U.S. For. Serv. Bul. 86. 100 p.

44. Batten, L. 1918. Observations on the ecology of *Epilobium hirsutum*. J. Ecol. 6:161–177.

45. Baufait, W. R. 1966. An integrating device for evaluating prescribed fires. For. Sci. 12:27–29.

46. Beadle, N. C. W. 1954. Soil phosphate and the delimitation of plant communities in eastern Australia. Ecol. 35:370–375.

47. Beardsley, G. F., and W. A. Cannon. 1930. Note on the effects of a mud-flow at Mt. Shasta on the vegetation. Ecol. 11:326–336.
48. Beatley, Janice C. 1965. Effects of radioactive and non-radioactive dust upon Larrea divaricata Cav., Nevada Test Site. Health Physics 11:1621–1625.
49. Becking, J. H. 1970. Plant-endophyte symbiosis in non-leguminous plants. Plant and Soil 32:611–654.
50. Beeson, C. F. 1946. Forestry, horticulture and the moon. For. Abstr. 8:191–198.
51. Bensin, B. M. 1955. The prospects for various uses for solar reflectors as a means of environmental control of microclimates in agriculture. Proc. conf. on solar energy: The scientific basis. Tucson. 2(Part 1, Sec.A):119–136.
52. Bentley, J. R., and R. L. Fenner. 1958. Soil temperatures during burning related to postfire seedbeds on woodland range. J. For. 56:737–740.
53. Bergen, J. D. 1971. An inexpensive heated thermistor anemometer. Agric. Meteorol. 8:395–405.
54. Bergman, H. F. 1920. The relation of aeration to the growth and activity of roots and its influence on the ecesis of plants in swamps. Ann. Bot. 34:13–33.
55. Bergman, H. F. 1959. Oxygen deficiency as a cause of disease in plants. Bot. Rev. 25:417–485.
56. Bernstein, L., and H. C. Hayward. 1958. Physiology of salt tolerance. Ann. Rev. Plant Physiol. 9:25–46.
57. Berry, S. 1914. Work of the California gray squirrel on conifer seed in the southern Sierras. Soc. Am. For. Proc. 9:95–97.
58. Biebel, J. 1937. Temperature, photoperiod, flowering and morphology in *Cosmos* and China aster. Amer. Soc. Hort. Sci. Proc. 34:635–643.
59. Billings, W. D., and W. B. Drew. 1938. Bark factors affecting the distribution of corticolous bryophytic communities. Amer. Midl. Nat. 20:302–330.
60. Bissell, C. H. 1902. Biological relationships of *Polygonum hartwrightii* to *P. amphibium*. Rhodora 4:104–105.
61. Biswell, H. H., et al. 1953. Frost heaving of grass and brush seedlings on burned chamise brushlands in California. J. Range Man. 6:172–180.
62. Björkman, E. 1949. The ecological significance of the ectotrophic mycorrhizal association in forest trees. Svensk. Bot. Tids. 43:223–262.
63. Björkman, E. 1970. Forest tree mycorrhiza—the conditions for its formation and the significance for tree growth and afforestation. Plant and Soil 32:589–610.
64. Black, M., and P. F. Wareing. 1955. Photoperiodic control of germination in *Betula pubescens* Ehrh. Physiol. Plant. 8:300–316.
65. Blackman, G. E., and J. N. Black. 1959. Physiological and ecological studies in the analysis of plant environment, XI. A further assessment of the influence of shading on the growth of different species in the vegetative phase. Ann. Bot. 23:51–63.
66. Blackman, G. E., and G. L. Wilson. 1951. An analysis of the differential effects of light intensity on net assimilation rate, leaf-area ratio and relative growth rate of different species. Ann. Bot. 15:373–428.
67. Blaisdell, J. P. 1953. Ecological effects of planned burning of sagebrush-grass range on the upper Snake River Plains. U.S.D.A. Tech. Bul. 1075. 39 p.
68. Bliss, L. C. 1962. Caloric and lipid content in alpine tundra plants. Ecol. 43:753–757.
69. Bloomfield, C. A. 1953. A study of podzolization. II. The mobilization of iron and aluminum by the leaves and bark of *Agathis australis* (Kauri). J. Soil Sci. 4:17–23.
70. Bönning, R. H., and C. A. Burnside. 1956. The effect of light intensity on rate of apparent photosynthesis in leaves of sun and shade plants. Am. J. Bot. 43:557–561.

70a. Bolas, B. D. 1934. The influence of light and temperature on the assimilation rate of seedling tomato plants, variety E.S.I. Ann. Rept. Exp. Sta. Nursery & Mark. Gard. Indus. Soc. Cheshunt. 20:86–89.

71. Bond, G., and G. D. Scott. 1955. An examination of some symbiotic systems for fixation of N. Ann. Bot. 19:67–77.

72. Bonner, J. 1962. The upper limit of crop yield. Sci. 137:11–15.

73. Bormann, F. H. 1956. Percentage light readings, their intensity-duration aspects, and their significance in estimating photosynthesis. Ecol. 37:473–476.

74. Borthwick, H. A., et al. 1956. Photoperiodism. In A Hollaender (ed.) Radiation biology. 3:479–517. McGraw-Hill Book Co., N.Y.

75. Bosian, G. 1965. Control of conditions in the plant chamber: fully automatic regulation of wind velocity, temperature and relative humidity to conform to microclimatic field conditions. U.N.E.S.C.O. Arid Zone Research 15:233–238.

76. Boswell, V. R., et al. 1940. A study of rapid deterioration of vegetable seeds and methods for its prevention. U.S.D.A. Techn. Bul. 708. 48 p.

77. Boswell, V. R., et al. 1935. A method for making mechanical analysis of the ultimate natural structure of soils. Soil Sci. 40:481–485.

78. Boswell, V. R., et al. 1950. A practical soil moisture meter as a scientific guide to irrigation practices. Agron. J. 42:104–107.

79. Boswell, V. R., et al. 1951. A recalibration of the hydrometer method for making mechanical analysis of soils. Agron. J. 43:434–438.

80. Bourdeau, P. F., and G. M. Woodwell. 1965. Measurements of plant CO_2 exchange by infrared absorption under controlled conditions and in the field. In F. E. Echhardt (ed.) Methodology of plant ecophysiology. U.N.E.S.C.O. Arid Zone Research 15:283–289.

81. Boyce, S. G. 1954. The salt spray community. Ecol. Mono. 24:29–67.

82. Boyer, J. S. 1969. Measurement of the water status of plants. Ann. Rev. Plant Physiol. 20:351–364.

83. Boyko. H. 1945. On forest types of the semi-arid areas at lower latitudes. Palestine J. Bot., Rehovot Ser. 5:1–21.

84. Boyle, L. W. 1950. Collar rot of peanuts, primarily a heat canker. Phytopath. 41:39.

85. Brackett, F. S. 1936. Measurement and application of visible and near-visible radiation. In B. M. Duggar (ed.), Biological effects of radiation. 1:123–210.

86. Bradford, F. C. 1922. The relation of temperature to blossoming in the apple and the peach. Mo. Agric. Exp. Sta. Res. Bul. 53. 51 p.

87. Bradshaw, A. D., et al. Industrialization, evolution and the development of heavy metal tolerance in plants. In G. T. Goodman, et al. (ed.) Ecology and the industrial state. Blackwell Publ. Co., Oxford. 327–343.

88. Brenchly, W. 1920. The relations between growth and the environmental conditions of temperature and bright sunshine. Ann. Appl. Bot. 6:211–244.

89. Brett, C. H. 1944. An electrically regulated humidity control. J. Econ. Entomol. 37:552–553.

90. Brezeale, J. F. 1926. Alkali tolerance of plants considered as a phenomenon of adaptation. Ariz. Agric. Exp. Sta. Tech. Bul. 11:237–256.

91. Brezeale, J. F. 1928. The effect of one element of plant food on the absorption by plants of another element. Ariz. Agric. Exp. Sta. Tech. Bul. 19:465–480.

92. Briggs, L. J., and H. L. Shantz. 1912. The relative wilting coefficients for different plants. Bot. Gaz. 53:229–235.

93. Briggs, L. J., and H. L. Shantz. 1914. The relative water requirement of plants. J. Agric. Res. 3:1–63.

94. Briggs, L. J., and H. L. Shantz. 1916. Hourly transpiration on clear days as determined by cyclic environmental factors. J. Agric. Res. 5:583–651.
95. Brinley, F. J. 1942. Relation of domestic sewage to stream productivity. Ohio J. Sci. 42:173–176.
96. Brockmann-Jerosch, H. 1919. Tree limits and the climatic character. J. Ecol. 8:63–65 (rev.).
97. Brodie, H. J. 1955. Springboard plant dispersal mechanisms operated by rain. Can. J. Bot. 33:156–167.
98. Brooks, M. G. 1951. Effects of black walnut trees and their products on other vegetation. W. Va. Agric. Exp. Sta. Bul. 347. 31 p.
99. Brown, C. A. 1941. Studies on the isolated praries of Louisiana. Am. J. Bot. 28:16s.
100. Brown, D. S. 1953. The apparent efficiencies of different temperatures on the development of apricot fruit. Am. Soc. Hort. Sci. Proc. 62:173–183.
101. Brown, D. S. 1957. The rest period of apricot flower buds as described by a regression of time of bloom on temperature. Plant Physiol. 32:75–85.
102. Brown, D. S. 1958. A comparison of the temperature of the flower buds of Royal apricot with standard and black bulb thermograph records during the winter. Am. Soc. Hort. Sci. Proc. 72:113–122.
103. Brown, G. W., and J. T. Krygier. 1970. Effects of clear-cutting on stream temperature. Water Resources Res. 6:1133–1139.
104. Brown, I. C. 1943. A rapid method of determining exchangeable hydrogen and total exchangeable bases in soils. Soil Sci. 56:353–357.
105. Brown, W. H. 1912. The relation of evaporation to the water content of the soil at the time of wilting. Plant World 15:121–134.
106. Bryant, A. E. 1934. Comparison of anatomical and histological differences between roots of barley grown in aerated and in nonaerated culture solutions. Plant Physiol. 9:389–391.
107. Buchner, P. 1965. Endosymbiosis of animals with plant microorganisms. John Wiley and Sons, N.Y. 909 p.
108. Burns, G. P. 1923. Measurement of solar radiant energy in plant habitats. Ecol. 4:189–195.
109. Burton, G. W. 1944. Seed production of several southern grasses as influenced by burning and fertilization. Am. Soc. Agron. J. 36:523–529.
110. Busse, W. F. 1930. Effect of low temperatures on germination of impermeable seeds. Bot. Gaz. 89:169–179.
111. Butters, F. K. 1914. Some peculiar cases of plant distribution in the Selkirk Mountains, British Columbia. Univ. Minn. Bot. Studies 4:313–331.
112. Byram, G. M. 1948. Terrestrial radiation and its importance in some forestry problems. J. For. 46:653–658.
113. Cain, S. A., and J. D. O. Miller. 1933. Leaf structure of Rhododendron catawbiense Mich. grown in Picea-Abies forest and in heath communities. Am. Midl. Nat. 14:69–82.
114. Cain, S. A., and J. E. Potzger. 1933. Comparison of leaf tissues of Gaylussacia baccata (Wang.) C. Koch. and Vaccinium vacillans Kalm. grown under different conditions. Am. Midl. Nat. 14:97–112.
115. Calder, K. L., and S. A. Mumford. 1948. Frost control with air circulators. Weatherwise 1:105.
116. Caldwell, M. M. 1968. Solar ultraviolet radiation as an ecological factor for alpine plants. Ecol. Mono. 38:243–268.
117. Caldwell, M. M., and M. L. Caldwell. 1970. A fine wire psychrometer for measurement of humidity in the vegetation layer. Ecol. 51:918–920.

118. Campbell, C. J. 1970. Ecological implications of riparian vegetation management. J. Soil and Water Cons. 25:49–52.

119. Campbell, C. J., and J. E. Strong. 1964. Salt gland anatomy in *Tamarix pentandra* (*Tamaricaceae*). Southw. Nat. 9:232–238.

120. Cannon, Helen L. 1960. Botanical prospecting for ore deposits. Sci. 132:591–598.

121. Cannon, W. A. 1925. Physiological features of roots. Carnegie Inst. Wash. Publ. 368. 168 p.

122. Cannon, W. A. 1940. Oxygen relations in hydrophytes. Sci. 91:43–44.

123. Capalungan, A. V., and H. F. Murphy. 1930. Wilting coefficient studies. Am. Soc. Agron. J. 22:842–847.

124. Caprio, J. M. 1966. Pattern of plant development in the western United States. Mont. Agric. Exp. Sta. Bul. 607. 42 p.

125. Chandler, D. C. 1942. Limnological studies of western Lake Erie. II. Light penetration and its relation to turbidity. Ecol. 23:41–52.

126. Chandler, R. F., Jr. 1939. Cation exchange properties of certain forest soils in the Adirondak section. J. Agric. Res. 59:491–506.

127. Chang, H. T., and W. E. Loomis. 1945. Effect of CO_2 on absorption of water and nutrients by roots. Plant Physiol. 20:221–232.

128. Chapman, A. G. 1935. The effects of black locust on associated species with reference to forest trees. Ecol. Mono. 5:37–60.

129. Chapmen, H. D., et al. 1940. The determination of pH at soil moisture contents approximating field conditions. Soil. Sic. Soc. Am. Proc. 5:191–200.

130. Chapman, V. J. 1942. The new perspective in the halophytes. Quart. Rev. Biol. 17:291–311.

131. Childer, N. F., and D. G. White. 1950. Some physiological effects of excess soil moisture on Stayman Winesap apple trees. Ohio Agric. Exp. Sta. Bul. 694. 36 p.

132. Chopra, R. S. 1940. Experimental afforestation of water-logged areas in the Punjab. Indian For. 66:545–551.

133. Chupp, C. 1946. Soil temperature, moisture, aeration, and pH as factors in disease incidence. Soil Sci. 61:31–36.

134. Clark, J. A., and J. Levitt. 1956. The basis of drouth resistance in the soybean plant. Physiol. Plant. 9:598–606.

135. Clark, J. R. 1969. Thermal pollution and aquatic life. Sci. Amer. 220(3):18–27.

136. Clark, W. M. 1928. The determination of hydrogen ions. Williams and Wilkins, Baltimore. 3rd ed. 717 p.

137. Clarke, S. E. and E. W. Tisdale. 1945. The chemical composition of native forage plants of southern Alberta and Saskatchewan in relation to grazing practices. Can. Dept. Agric. Tech. Bul. 54. 60 p.

138. Clausen, J. 1951. Stages in the evolution of plant species. Cornell Univ. Press, Ithaca. 206 p.

139. Clausen, J., et al. 1939. The concept of species based on experiment. Am. J. Bot. 26:103–106.

140. Clausen, J., et al. 1940. Experimental studies on the nature of species. Carnegie Inst. Wash. Publ. 520. 452 p.

141. Clausen, J., et al. 1948. Experimental studies on the nature of species. III. Environmental responses of climatic races of *Achillea*. Carnegie Inst. Wash. Publ. 581. 129 p.

142. Clements, F. E. 1910. The life history of lodgepole burn forests. U.S.D.A. For. Serv. Bul. 79. 56 p.

143. Clements, F. E. 1938. Climatic cycles and human populations in the Great Plains. Sci. Mo. 47:193–210.

144. Clements, F. E., and G. W. Goldsmith. 1924. The phytometer method in ecology. Carnegie Inst. Wash. Publ. 356. 106 p.

145. Clements, F. E., and F. L. Long. 1923. Experimental pollination. Carnegie Inst. Wash. Publ. 336. 274 p.

146. Clements, J. B. 1941. The introduction of pines into Nyasaland. Nyasaland Agric. Quart. J. 1:5–15.

147. Coile, T. S. 1940. Soil changes associated with loblolly pine succession on abandoned agricultural land of the Piedmont Plateau. Duke Univ. Sch. For. Bul. 5. 85 p.

148. Coile, T. S. 1952. Soil and the growth of forests. Adv. in Agron. 4:329–398.

149. Coit, J. E., and R. W. Hodgson. 1919. An investigation of the abnormal shedding of young fruits of the Washington navel orange. Cal. Univ. Publ. Agric. Sci. 3:283–368.

150. Cole, J. R. 1959. Spanish-moss in pecan trees can be controlled by spraying dormant trees with copper sulfate and calcium arsenate. Plant Dis. Rep. 43:960–961.

151. Colman, E. A. 1947. A laboratory procedure for determining the field capacity of soils. Soil Sci. 63: 277–283.

152. Connell, A. B. 1923. Measuring soil temperature by standard thermometer suspended in iron pipe. Ecol. 4:313–316.

153. Connor, H. E. 1966. Breeding systems in New Zealand grasses, VII. Periodic flowering of snow tussock, *Chionochloa rigida*. N. Z. J. Bot. 4:392–397.

154. Conrad, V., and L. W. Pollak. 1950. Methods in climatology. Harvard Univ. Press, Cambridge, Mass. 459 p.

155. Conway, Verona M. 1940. Aeration and plant growth in wet soils. Bot. Rev. 6:179–189.

156. Cooper, J. P., and N. M. Tainton. 1968. Light and temperature requirements for the growth of tropical and temperate grasses. Herb. Abstr. 38:167–176.

157. Cope, F., and E. S. Trickett. 1965. Measuring soil moisture. Soils and Fert. 28:201–208.

158. Costello, D. F., and G. T. Turner. 1941. Vegetation changes following exclusion of livestock from grazed ranges. J. For. 39:310–315.

159. Coulter, J. M., et al. 1931. Textbook of botany. Vol. 3. Ecology. American Book Co., New York City. 499 p.

160. Cowlishaw, S. J. 1951. The effect of sampling cages on the yield of herbage. J. Brit. Grassl. Soc. 6:179–182.

161. Cox, H. J. 1922. Thermal belts and fruit growing in North America. Mo. Wea. Rev. Suppl. 19:1–98.

162. Craib, I. J. 1929. Some aspects of soil moisture in the forest. Yale Univ. Sch. For. Bul. 25. 62 p.

163. Crawford, R. M. M. 1961. Effect of temperature fluctuations on flowering in *Salvia splendens*. Nature 189:75–76.

164. Crawford, R. M. M., and P. D. Taylor. 1969. Organic acid metabolism in relation to flooding tolerance of roots. J. Ecol. 57:235–244.

165. Crocker, W. 1936. Effect of visible spectrum upon the germination of seeds and fruits. *In* B. M. Duggar. Biological effects of radiation. Vol. I. McGraw-Hill Book Co., N.Y. 791–828.

166. Cumming, B. G., and P. W. Voisey. 1962. Note on a small high intensity light unit for use in plant research. Can. J. Plant Sci. 42:392–395.

167. Currie, P. O., and D. L. Goodwin. 1966. Consumption of forage by blacktailed jackrabbits in salt-desert ranges of Utah. J. Wildl. Man. 30:304–311.

168. Curtis, J. D. 1936. Snow damage to plantations. J. For. 34:613–619.
169. Curtis, J. T. 1952. Outline for ecological life history studies of vascular epiphytic plants. Ecol. 33:550–558.
170. Curtis, J. T., and G. Cottam. 1950. Antibiotic and autotoxic effects in prairie sunflower. Torrey Bot. Club Bul. 77:187–191.
171. Curtis, L. C. 1943. Deleterious effects of guttated fluids on foliage. Am. J. Bot. 30:778–781.
172. Dachnowski-Stokes, A. P. 1935. Peat land as a conserver of rainfall and water supplies. Ecol. 16:173–177.
173. D'Albe, E. M. F. 1958. The modification of microclimates. U.N.E.S.C.O. Arid Zone Research 10:126–146.
174. Darrow, G. M. 1942. Rest period requirements for blueberries. Am. Soc. Hort. Sci. Proc. 41:189–194.
175. Daubenmire, R. 1940. Exclosure technique in ecology. Ecol. 21:514–515.
176. Daubenmire, R. 1943. Soil temperature versus drouth as a factor determining lower altitudinal limits of trees in the Rocky Mountains. Bot. Gaz. 105:1–13.
177. Daubenmire, R. 1943. Some observations on epiphyllous lichens in northern Idaho. Am. Midl. Nat. 30:447–451.
178. Daubenmire, R. 1945. An improved type of precision dendrometer. Ecol. 26:97–98.
179. Daubenmire, R. 1957. Injury to plants from rapidly dropping temperature in Washington and northern Idaho. J. For. 55:581–585.
180. Daubenmire, R. 1960. A seven-year study of cone production as related to xylem layers and temperature in *Pinus ponderosa*. Am. Midl. Nat. 64:187–193.
181. Daubenmire, R. 1968. Ecology of fire in grasslands. Adv. Ecol. Res. 5:209–266.
182. Daubenmire, R., and H. Charter. 1942. Behavior of woody desert legumes at the wilting percentage of the soil. Bot. Gaz. 103:762–770.
183. Davenport, D. C., et al. 1969. Antitranspirants research and its possible application to hydrology. Water Res. 5:735–743.
184. Davies, W. E. 1959. Experiments on the control of broomrape in red clover. Plant Path. 8:19–22.
185. Davis, J. H. The ecology and geologic role of mangroves in Florida. Carnegie Inst. Wash. Publ. 517:303–412.
186. Day, W. R. 1946. Ecology and the study of climate. Nature 157:827–829.
187. Day, W. R. 1946. The pathology of beech on chalk soils. Quart. J. For. 40:72–82.
188. Day, W. R., and T. R. Peace. 1934. The experimental production and the diagnosis of frost injury of forest trees. Oxford For. Mem. 16. 60 p.
189. Deacon, E. L., et al. 1958. Evaporation and the water balance. U.N.E.S.C.O. Arid Zone Res. 10:9–34.
190. Dean, B. E. 1933. Effect of soil type and aeration upon root systems of certain aquatic plants. Plant Physiol. 8:203–222.
191. DeBell, D. S. 1970. Phytotoxins: new problems in forestry? J. For. 68:335–337.
192. Demaree, D. 1932. Submergins experiments with *Taxodium*. Ecol. 13:258–262.
193. Denny, F. E. 1927. Field medhod for determining the saltiness of brackish water. Ecol. 8:106–112.
194. De Silva, B. L. T. 1934. The distribution of "calcicole" and "calcifuge" species in relation to the content of the soil in calcium carbonate and exchangeable calcium and to soil reaction. J. Ecol. 22:532–553.
195. Deters, M. E., and H. Schmitz. 1936. Drouth damage to prairie shelterbelts in Minnesota. Minn. Agric. Exp. Sta. Bul. 329. 28 p.
196. de Villiers, G. D. B. 1943. Research on the influence of climate on deciduous fruit growing. Chron. Bot. 7:388–390.

197. DeWitt, J. B., and J. V. Derby, Jr. 1955. Changes in nutritive value of browse plants following forest fires. J. Wildl. Man. 19:65–70.
198. Dexter, T. S. 1956. The evaluation of crop plants for winter hardiness. Adv. in Agron. 8:203–241.
199. Diebold, C. H. 1938. The effect of vegetation upon snow cover and frost prevention during the March 1936 floods. J. For. 36:1131–1137.
200. Dittmer, H. J. 1937. A quantitative study of the roots and root hairs of a winter rye plant. Am. J. Bot. 24:417–420.
201. Doneen, L. D., and J. H. MacGillivray. 1943. Germination (emergence) of vegetable seeds as affected by different soil moisture conditions. Plant Physiol. 18:524–529.
202. Doorenbos, J. 1953. Review of the literature on dormancy in buds of woody plants. Landb. Hogesch., Wageningen 53:1–24.
203. Downie, D. G. 1940. On the germination and growth of *Goodyera repens*. Bot. Soc. Edinburgh Trans. and Proc. 33:36–51.
204. Downs, R. J. 1958. Photoperiodic control of growth and dormancy in woody plants. p. 529–537. *In* K. V. Thimann (ed.) The physiology of forest trees. Ronald Press, New York. 678 p.
205. Duddington, C. L. 1955. Fungi that attack microscopic animals. Bot. Rev. 21:377–439.
206. Duggar, B. M. 1936. Biological effects of radiation. Vols. 1 and 2. McGraw-Hill Book Co., N.Y. 1342 p.
207. Duley, F. L., and L. L. Kelly. 1939. Effects of soil type, slope, and surface conditions on intake of water. Neb. Agric. Exp. Sta. Res. Bul. 112. 16 p.
208. Duncan, W. H. 1933. Ecological comparison of leaf structures of *Rhododendron punctatum* Andr., and the ontogeny of the pidermal scales. Am. Midl. Nat. 14:83–96.
209. Dunlap, A. A. 1943. Low light intensity and cotton boll-shedding. Sci. 98:568–569.
210. Dutton, H. J., and C. Juday. 1944. Chromatic adaptation in relation to color and depth distribution of freshwater phytoplankton and large aquatic plants. Ecol. 25:273–282.
211. Eaton, F. M. 1924. Assimilation respiration balance as related to length of day reactions in soy beans. Bot. Gaz. 77:311–321.
212. Ebell, L. F. 1967. Cone production induced by drouth in potted Douglas-fir. Can. Dept. For., Bi-m. Res. Notes 23:26–27.
213. Eckert, R. E., and A. T. Bleak. 1960. The nutrient status of four mountain soils in western Nevada and eastern California. J. Range Man. 13:184–188.
214. Edlefsen, N. E., and A. B. C. Anderson. 1943. Thermodynamics of soil moisture. Hilgardia 15:31–298.
215. Edmond, D. B. 1958. The influence of treading on pasture. A preliminary study. N. Zeal. J. Agric. Res. 1:319–328.
216. Edwards. C. A. 1965. Effects of pesticide residues on soil invertebrates and plants. 240–261. *In* G. T. Goodman, et al. (ed.) Ecology and the industrial society. Blackwell Sci. Publs., Oxford. 395 p.
217. Edwards, D. C. 1942. Grass-burning. Empire J. Exp. Gric. 10:219–231.
218. Edwards, T. I. 1932. Temperature relations of seed germination. Quart. Rev. Biol. 7:428–443.
219. Eggert, R. 1944. Cambium temperatures of peach and apple trees in winter. Am. Soc. Hort. Sci. Proc. 45:33–36.
220. Eggert, R. 1946. The construction and installation of thermocouples for biological research. J. Agric. Res. 72:341–355.
221. Ehrler, W. L. 1963. Water absorption of alfalfa as affected by low root temperature and other factors of a controlled environment. Agron. J. 55:363–366.
222. Einarsen, A. S. 1946. Crude protein determination of deer food as an applied management technique. N. A. Wildl. Conf. Trans. 11:309–312.

223. Ekern, P. C. 1964. Direct interception of cloud water at Lanaihale, Hawaii. Soil Sci. Soc. Amer. Proc. 28:419–421.

224. Elliott, Frances H. 1946. *Saintpaulia* leaf spot and temperature differential. Amer. Soc. Hort. Sci. Proc. 47:511–514.

225. Elliott, G. R. B. 1924. Relation between the downward penetration of corn roots and water level in peat soil. Ecol. 5:175–178.

226. Emden, H. F. van 1965. The effect of uncultivated land on the distribution of cabbage aphid (*Brevicoryne brassicae*) on an adjacent crop. J. Appl. Ecol. 2:171–196.

227. Ernest, E. C. M. 1935. Factors rendering the plasmolytic method inapplicable in the estimation of osmotic values of plant cells. Plant Physiol. 10:553–558.

228. Evans, G. C. 1969. The spectral composition of light in the field. 1. Its measurement and ecological importance. J. Ecol. 57:109–126.

229. Evans, L. T. 1963. Extrapolation from controlled environments to the field. 421–437. *In* L. T. Evans (ed.) Environmental control of plant growth. Academic Press, N.Y. 449 p.

230. Evans. L. T. 1953. The ecology of the halophytic vegetation at Lake Ellesmere, New Zealand. J. Ecol. 41:106–122.

231. Evans, M. W. 1931. Relation of latitude to time of blooming of timothy. Ecol. 12:182–187.

232. Evenari, M. 1956. Seed germination. *In* A. Hollaender, Radiation biology. 3:519–549. McGraw-Hill Book Co., N.Y.

233. Faegri, K., and L. van der Pilj. 1971. The principles of pollination ecology. Pergamon Press, N.Y. 291 p.

234. Farnham, R. S., and H. R. Finney. 1965. Classification and properties of organic soils. Adv. Agron. 17:115–162.

235. Featherly, H. I. 1941. The effect of grape vines on trees. Okla. Acad. Sci. Proc. 21:61–62.

236. Federer, C. A., and C. B. Tanner. 1966. Sensors for measuring light available for photosynthesis. Ecol. 47:654–657.

237. Felt, E. P. 1940. Plant galls and gall-makers. Comstock Publ. Co., Ithaca. 364 p.

238. Fernald, M. L. 1919. Lithological factors limiting the ranges of *Pinus banksiana* and *Thuja occidentalis*. Rhodora 21:41–65.

239. Finn, R. G. 1942. Mycorrhizal innoculation of soil of low fertility. Black Rock For. Papers 1:115–117.

240. Finnell, H. H. 1928. Effect of wind on plant growth. Am. Soc. Agron. J. 20:1206–1210.

241. Finnell, H. H. 1929. Heavy plains soil moisture problems. Okla. Agric. Exp. Sta. Bul. 193. 7 p.

242. Fippin, E. O. 1945. Plant nutrient losses in silt and water in the Tennessee River system. Soil Sci. 60:223–239.

243. Fireman, M., and H. E. Hayward. 1952. Indicator significance of some shrubs in the Escalante Desert, Utah. Bot. Gaz. 114:143–155.

244. Fisher, G. M. 1935. Comparative germination of tree species on various kinds of surface-soil material in the western white pine type. Ecol. 16:606–611.

245. Fisher, R. T. 1928. Soil changes and silviculture on the Harvard Forest. Ecol. 9:6–11.

246. Fitton, E. M., and C. F. Brooks. 1931. Soil temperature in the United States. Mo. Wea. Rev. 59:6–16.

247. Fitzgerald, P. D., and D. S. Rickard. 1960. A comparison of Penman's and Thornthwaite's method of determining soil moisture deficits. N. Zeal. J. Agric. Res. 3:106–112.

248. Flowers, S. 1934. Vegetation of the Great Salt Lake region. Bot. Gaz. 95:353–418.

249. Foister, C. E. 1946. The relation of weather to fungus diseases of plants. II. Bot. Rev. 12:548–591.

250. Fowle, F. E. 1927. Smithsonian physical tables. Smithsonian Misc. Coll. Publ. 2539. 458 p.

251. Foy, C. L., et al. 1971. Root exudation of plant-growth regulators. 75–85. *In* Biochemical interactions among plants. U.S. Nat. Acad. Sci. Washington, D.C. 134 p.

252. Frank, E. C., and R. Lee. 1966. Potential solar beam irradiation on slopes. Tables for 30-50° latitude. U.S.F.S. Res. Paper RM-18. 116 p.

253. Freyman, S. 1968. Spectral distribution of light in forests of the Douglas fir zone of southern British Columbia. Can. J. Plant Sci. 48:326–328.

254. Fritsch, F. E. 1936. The role of the terrestrial algae in nature. 195–217. *In* T. H. Goodspeed (ed.) Essays in geobotany in honor of W. A. Setchell. Univ. Calif. Press, Berkeley.

255. Fritts, H. C. 1966. Growth-rings of trees: their correlation with climate. Sci. 154:973–979.

256. Frutiger, H. 1964. Snow avalanches along Colorado mountain highways. U.S. For. Serv. Res. Paper RM-7. 85 p.

257. Fuller H. J. 1948. Carbon dioxide concentration of the atmosphere above Illinois forest and grassland. Amer. Midl. Nat. 39:247–249.

258. Gardner, J. L. 1942. Studies in tillering. Ecol. 23:162–174.

259. Garren, K. H. 1943. Effects of fire on the vegetation of the southeastern United States. Bot. Rev. 9:617–654.

260. Gast, P. R. 1965. Modification and measurement of sun, sky and terrestrial radiation for eco-physiological studies. *In* F. E. Eckardt (ed.) Methodology of plant eco-physiology. U.N.E.S.C.O. Arid Zone Res. 15:29–45.

261. Gates, D. M. 1962. Energy exchange in the biosphere. Harper and Row, N.Y. 151 p.

262. Gates, F. C. 1912. The relation of snow cover to winter killing in *Chamaedaphne calyculata*. Torreya 12:257–262.

263. Geiger, R. 1950. The climate near the ground. Harvard Univ. Press, Cambridge. 459 p.

264. Gemmer, E. W. 1929. A method of recording maximum and minimum temperatures of forest soils. Sci. 70:505–506.

265. Gerdel, R. W. 1948. Penetration of radiation into the snow pack. Am. Geophys. Union Trans. 29:366–374.

266. Giddings, L. A. 1914. Transpiration of *Silphium laciniatum* L. Plant World 17:309–328.

267. Gilchrist, M. 1908. Effect of swaying by the wind on the formation of mechanical tissue. Mich. Acad. Sci. Rept. 10:45.

268. Gilead, M., and N. Rosenan. 1958. Climatological observational requirements in arid zones. U.N.E.S.C.O. Arid Zone Research 10:181–188.

269. Gill, C. J. 1970. The flooding tolerance of woody species—a review. For. Ab. 31:671–688.

270. Gill, L. S., and F. G. Hawksworth. 1961. The mistletoes: a literature review. U.S.D.A. Tech. Bul. 1241. 87 p.

271. Gindel, I. 1966. Attraction of atmospheric moisture by woody xerophytes in arid climates. Commonw. For. Rev. 45:297–321.

272. Gleason, C. H. 1953. Indicators of erosion on watershed land in California. Amer. Geophys. Union Trans. 34:419–426.

273. Glock, W. S. 1951. Cambial frost injuries and multiple growth layers at Lubbock, Texas. Ecol. 32:28–37.

274. Glover, J., and M. D. Gwynne. 1962. Light rainfall and plant survival in east Africa. 1. Maize. J. Ecol. 50:111–118.

275. Gloyne, R. W. 1955. Some effects of shelter-breaks and wind-breaks. Meteorol. Mag., London 84:272–281.

276. Godwin, H. 1935. The effect of handling on the respiration of cherry laurel leaves. New Phytol. 34:403–406.

277. Golden, L. B., et al. 1943. A comparison of methods of determining the exchangeable cations and the exchange capacity of Maryland soil. Soil Sci. Soc. Am. Proc. 7:154–161.

388 Literature Cited

278. Goldsmith, G. and A. L. Hafenrichter. 1932. Anthokinetics. Carnegie Inst. Wash. Publ. 420. 198 p.

279. Goode, J. E. 1956. Soil-moisture relationships in fruit plantations. Ann. Appl. Biol. 44:525–530.

280. Gorham, E. 1953. The development of the humus layer of some woodlands of the English Lake District. J. Ecol. 41:123–152.

281. Gorham, E., and A. G. Gordon. 1960. The influence of smelter fumes upon the chemical composition of lake waters near Sudbury, Ontario, and upon the surrounding vegetation. Can. J. Bot. 38:477–487.

282. Gourley, J. H., and G. T. Nightingale. 1921. The effects of shading some horticultural plants. N.H. Agric. Exp. Sta. Tech. Bul. 18. 22 p.

283. Grable, A. R. 1966. Soil aeration and plant growth. Adv. Agron. 18:58–106.

284. Grainger, J. 1939. Studies upon the time and flowering of plants. Anatomical, floristic and phenologic aspects of the problem. Ann. Appl. Biol. 26:684–704.

285. Greenwood, D. J. 1969. Effect of oxygen distribution in the soil on plant growth. 202–223. In W. J. Whittington (ed.) Root growth. Butterworths, London. 450 p.

286. Gregor, J. W. 1942. The units of experimental taxonomy. Chron Bot. 7:193–196.

287. Gregor, J. W., 1946. Ecotypic differentiation. New Phytol. 45:254–270.

288. Gressitt, J. L., et al. 1965. Flora and fauna on back of large Papuan moss-forest weevils. Sci. 150:1833–1835.

289. Gries, G. A. 1943. The effect of plant-decomposition products on root diseases. Phytopath. 33:1111–1112.

290. Griggs, R. F. 1933. The colonization of Katmai ash, and new inorganic "soil." Am. J. Bot. 20:92–113.

291. Grime, J. P. 1966. Shade avoidance and shade tolerance in flowering plants. 187–207. In R. Bainbridge, et al. (ed.) Light as an ecological factor. J. Wiley and Sons, N.Y. 452 p.

292. Grime, J. P., and D. W. Jeffrey. 1965. Seedling establishment in vertical gradients of sunlight. J. Ecol. 53:621–642.

293. Grinnell, J. 1923. The burrowing rodents of California as agents in soil formation. J. Mammal. 4:137–149.

294. Grinnell, J. 1936. Up-hill planters. The Condor 38:80–82.

295. Groom, P., and E. Wilson. 1925. On the pneumatophores of paludal species of *Amoora, Carapa,* and *Heritiera.* Ann. Bot. 39:9–24.

296. Grubb, P. J., et al. 1969. The ecology of chalk heath: its relevance to the calcicole-calcifuge and soil acidification problems. J. Ecol. 57:175–212.

297. Guppy, H. B. 1917. Plants, seeds, and currents in the West Indies and Azores. Williams and Norgate, London. 531 p.

298. Haase, E. F. 1970. Environmental fluctuations on south-facing slopes in the Santa Catalina Mountains of Arizona. Ecol. 51:959–974.

299. Hagan, R. M., et al. 1957. Relationships of soil moisture stress to different aspects of growth in ladino clover. Soil Sci. Soc. Am. Proc. 21:360–365.

300. Haig, I. T., et al. 1941. Natural regeneration in the western white pine type. U.S.D.A. Tech. Bul. 767. 98 p.

301. Haines, E. H. 1922. Influence of varying soil conditions on night-air temperatures. Mo. Wea. Rev. 50:363–366.

302. Hale, M. E., Jr. 1970. The biology of lichens. Edward Arnold, London. 184 p.

303. Halket, A. C. 1931. The flowers of *Silene saxifraga* L.; an inquiry into the cause of

their day closure and the mechanism concerned in effecting their periodic movements. Ann. Bot. 45:15–37.

304. Halkais, N. A., et al. 1955. Determining water needs for crops from climatic data. Hilgardia 24:207–233.

305. Hambidge, C. (ed.). 1941. Hunger signs in crops. Washington, D.C. 327 p.

306. Hamilton, E. L. 1954. Rainfall sampling on rugged terrain. U.S.D.A. Tech. Bul. 1096. 41 p.

307. Handley, W. R. C. 1939. The effect of prolonged chilling on water movement and radial growth in trees. Ann. Bot. 3:803–813.

308. Handley, W. R. C. 1954. Mull and mor formation in relation to forest soils. Brit. For. Comm. Bul. 23. 115 p.

309. Hansen, T. S., et al. 1923. A study of the damping-off disease of coniferous seedlings. Minn. Agric. Exp. Sta. Tech. Bul. 15. 35 p.

310. Hanson, H. C. 1917. Leaf structure as related to environment. Am. J. Bot. 4:533–560.

311. Hanson, H. C. 1939. Fire in land use and management. Am. Midl. Nat. 21:415–434.

312. Harley, J. L. 1939. The early growth of beech seedlings under natural and experimental conditions. J. Ecol. 27:384–401.

313. Harrington, G. T. 1923. Use of alternating temperatures in the germination of seeds. J. Agric. Res. 23:295–332.

314. Harris, J. A. 1917. Physical chemistry in the service of phytogeography. Sci. 46:25–30.

315. Harris, R. F., et al. 1966. Dynamics of soil aggregation. Adv. Agron. 18:107–169.

316. Harris, T. M. 1946. Zinc poisoning of wild plants from wire netting. New Phytol. 45:50–55.

317. Hart, F. C. 1937. Precipitation and runoff in relation to altitude in the Rocky Mountain region. J. For. 35:1005–1010.

318. Hart, T. S. 1946. Notes on the identification and growth of certain dodder-laurels. Victorian Nat. 63:12–16.

319. Hartley, C., et al. 1919. Moulding of snow-smothered nursery stock. Phytopath. 9:521–531.

320. Harvey, R. B. 1925. Red as a protective color in vegetation. J. For. 23:179–180.

321. Hassler, A. D. 1938. Fish biology and limnology of Crater Lake, Oregon. J. Wildl. Man. 2:94–103.

322. Hassler, J. F., et al. 1948. Protection of vegetation from frost damage by use of radiant energy. Mich. Agric. Exp. Sta. Quart. Bul. 30:339–360.

323. Hatch, A. B. 1937. The physical basis of mycotrophy in *Pinus*. Black Rock For. Bul. 6. 168 p.

324. Hawksworth, D. L., and F. Rose. 1970. Qualitative scale for estimating sulphur dioxide air pollution in England and Wales using epiphytic lichens. Nature 227:145–148.

325. Hayes, G. L. 1941. Influence of altitude and aspect on daily variations in factors of forest fire danger. U.S.D.A. Cir. 591. 39 p.

326. Hayward, H. E., and C. H. Wadleigh. 1949. Plant growth on saline and alkali soils. Adv. Agron. 1:1–38.

327. Heald, F. D., and R. A. Studhalter. 1914. Birds are carriers of chestnut-blight fungus. J. Agric. Res. 2:405–422.

328. Heath, G. W., and M. K. Arnold. 1966. Studies in leaf litter breakdown. 1. Breakdown rates of leaves of different species. Pedobiol. Jena 6:1–12.

329. Hedgecock, G. G. 1912. Winter-killing and smelter-injury in the forests of Montana. Torreya 12:25–30.

330. Heiberg, S. O., and D. P. White. 1951. Potassium deficiency of reforested pine and spruce stands in northern New York. Soil Sci. Soc. Amer. Proc. 15:369–376.

331. Hein, M. A., and P. R. Henson. 1942. Comparison of the effect of clipping and grazing treatments on the botanical composition of permanent pasture mixtures. Am. Soc. Agron. J. 34:566–573.

332. Heinicke, D. R. 1963. The micro-climate of fruit trees. 1. Light measurements with uranyl oxalate actinometers. Can. J. Plant Sci. 43:561–568.

333. Heller, V. G. 1938. The chemical content of Oklahoma rainfall. Okla. Agric. Exp. Sta. Tech. Bul. 1. 23 p.

334. Helmers, A. E. 1943. The ecological anatomy of ponderosa pine needles. Am. Midl. Nat. 29:55–71.

335. Helmers, A. E. 1954. Precipitation measurements on wind-swept slopes. Am. Geophys. Union Trans. 35:471–474.

336. Henry, S. M. (ed.). 1966. Symbiosis, its physiological and biochemical significance. Vol. 1. Associations of microorganisms and plants, and marine organisms. 478 p. Vol. 2. Associations of invertebrates, birds, ruminants, and other biota. Academic Press, N.Y. 443 p.

337. Hepting, G. H. 1941. Prediction of cull following fire in Appalachian oaks. J. Agric. Res. 62:109–120.

338. Herbert, D. A. 1928. Nutritional exchange between lianas and trees. Roy. Soc. Queensl. Proc. 39:115–118.

339. Heslop-Harrison, J. 1964. Forty years of genecology. Adv. Ecol. Res. 2:159–264.

340. Hewitt, E. J. 1966. Sand and water culture methods used in the study of plant nutrition. Bur. Hort. Plantat. Crops, E. Malling. Tech. Commun. 22 (2nd. ed) 547 p.

341. Hewitt, E. J., et al. 1954. The production of copper, zinc, and molybdenum deficiencies in crop plants grown in sand culture with special reference to some effects of water supply and seed reserves. Plant and Soil 5:205–222.

342. Hey, G. L., and J. E. Carter. 1931. The effect of ultra-violet light radiation on the vegetative growth of wheat seedlings and their infection by *Erysiphe graminis*. Phytopath. 21:695–699.

343. Heyligers, P. C. 1963. Vegetation and soil of a white-sand savanna in Surinam. Verhandl. Konik. Nederl. Akad. Wetensch., Afd. Natuurk. Tweede Reeks 54(3):1–148.

344. Heyward, F. 1938. Soil temperatures during forest fires in the longleaf pine region. J. For. 36:478–491.

345. Heyward, F., and R. M. Barnette. 1934. Effect of frequent fires on chemical composition of forest soils in the longleaf pine region. Fla. Agric. Exp. Sta. Bul. 265. 39 p.

346. Hiesey, W. M. 1953. Growth and development of species and hybrids of Poa under controlled temperatures. Am. J. Bot. 40:205–221.

347. Hiesey, W. M., and H. W. Milner. 1965. Physiology of ecological races and species. Ann. Rev. Plant Physiol. 16:203–216.

348. Higgs, D. E. B., and D. B. James. Comparative studies on the biology of upland grasses. 1. Rate of dry matter production and its control in four grass species. J. Ecol. 57:553–563.

349. Highkin, H. R. 1958. Temperature-induced variability in peas. Am. J. Bot. 45:626–631.

350. Hoffman, G. R. 1966. Ecological studies of *Funaria hygrometrica* Hedw. in eastern Washington and northern Idaho. Ecol. Mono. 36:157–180.

351. Hoffman, M. B., and G. R. Schlubatis. 1928. The significance of soil variation in raspberry culture. Mich. Agric. Exp. Sta. Spec. Bul. 177. 20 p.

352. Holttum, R. E. 1968. The response of plants to climatic change in Singapore. J. Ecol. 56:5p.

353. Hoover, J. W. 1940. Agricultural meteorology: a statistical study of conservation of precipitation by summer fallowed soil tanks at Swift Current, Saskatchewan. Can. J. Res. Sec. C Bot. Sci. 18:388–400.

354. Hoover, M. D., et al. 1954. Soil sampling for pore space and percolation. U.S. Forest Serv. Southeast For. Exp. Sta. Paper 42. 29 p.

355. Hopkins, A. D. 1918. Periodical events and the natural law as guides to agricultural research and practice. Mo. Wea. Rev. Suppl. 9. 42 p.

356. Hopkins, B. 1962. The measurement of available light by the use of *Chlorella*. New Phytol. 61:221–223.

357. Hopkins, B. 1965. Observations on savanna burning in the Olokemiji Forest Reserve. J. Appl. Ecol. 2:367–381.

358. Hopkins, D. M., and R. S. Sigafoos. 1950. Frost action and vegetation patterns on Seward Peninsula, Alaska. U.S.G.S. Bul. 974-C. 101 p.

359. Hori, T. (ed.) 1953. Studies on fogs in relation to fog-preventing forest. Tanne Trading Co., Ltd. Sapporo, Hokkaido. 399 p.

360. Horton, R. E., and J. S. Cole. 1934. Compilation and summary of the evaporation ercords of the Bureau of Plant Industry, U.S.D.A. 1921–1932. Mo. Wea. Rev. 62:77–89.

361. Hosegood, P. H. 1963. The root distribution of kikuyu grass and wattle trees. E. Afr. Agric. For. J. 29:60–61.

362. Hosokawa, T., and N. Odani. 1957. The daily compensation period and vertical ranges of epiphytes in a beech forest. J. Ecol. 45:901–915.

363. Hough, A. F. 1945. Frost pockets and other microclimates in forests of the northern Allegheny Plateau. Ecol. 26:235–250.

364. Hovin. A. W. 1957. Bulk emasculation by high temperatures in annual bluegrass, Poa annua L. Agron. J. 49:463.

365. How, S. Y., and F. G. Merkle. 1950. Chemical composition of certain calcifugous plants. Soil Sci. 69:471–486.

366. Howard, R. A. 1970. The "alpine" plants of the Antilles. Biotropica 2:24–28.

367. Huberman, M. A. 1941. Why phenology? J. For. 39:1007–1013.

368. Huberman, M. A. 1943. Sunscald of eastern white pine, *Pinus strobus* L. Ecol. 24:456–471.

369. Hudson, J. P. 1965. Gauges for the study of evapotranspiration rates. *In* F. E. Eckardt (ed.) Methodology of plant eco-physiology. U.N.E.S.C.O. Arid Zone Research 15:443–451.

370. Hughes, F. E. 1965. Tension wood, a review of literature. For. Ab. 26:2–9,179–186.

371. Hulbert, L. C. 1955. Ecological studies of *Bromus tectorum* and other annual brome-grasses. Ecol. Mono. 25:181–213.

372. Hulbert, L. C. 1969. Fire and litter effects in undisturbed bluestem prairie in Kansas. Ecol. 50:874–877.

373. Humm, H. J. 1944. Bacterial leaf nodules. N.Y. Bot. Gard. J. 45:193–198.

374. Humphrey, R. R. 1936. Growth habits of barrel cacti. Madroño 3:348–352.

375. Hungerford, K. E. 1957. A portable instrument shelter for ecological studies. Ecol. 38:150–151.

376. Hunter, A. S., and O. J. Kelley. 1946. The extension of plant roots into dry soil. Plant Physiol. 21:445–451.

377. Hursh, C. R., and H. C. Pereira. 1953. Field moisture balance in the Shimba Hills, Kenya. E. Afr. Agric. J. 18:139–145.

378. Hutchins, L. M. 1926. Studies on the oxygen-supplying power of the soil together with quantitative observations on the oxygen-supplying power requisite for seed germination. Plant Physiol. 1:95–150.

379. Hutt, W. N. 1922. Thermal belts from the horticultural viewpoint. Mo. Wea. Rev. Suppl. 19:99–106.
380. Hynes, H. B. N. 1965. A survey of water pollution problems. p. 49–63. *In* G. T. Goodman, et al. (ed.) Ecology and the industrial society. Blackwell Sci. Publs., Oxford. 395 p.
381. Iizumi, S., and Y. Iwanami. 1965. Some measurements of burning temperatures on the mountain grassland at Kawatabi in northeastern Japan. Tohoku Univ. Sci. Rept. Resp. Inst. D16:33–46.
382. Iljin, W. S. 1957. Drouth resistance in plants and physiological processes. Ann. Rev. Plant Physiol. 8:257–274.
383. Ingham, G. 1950. The mineral content of air and rain and its importance to agriculture. J. Agric. Sci. 40:55–61.
384. Isaac, L. A. 1930. Seed flight in the Douglas fir region. J. For. 28:492–499.
385. Isaac, L. A., and H. G. Hopkins. 1937. The forest soil of the Douglas fir region, and changes wrought upon it by logging and slash burning. Ecol. 18:264–279.
386. Isanogle, I. T. 1944. Effects of controlled shading upon the development of leaf structure in two deciduous tree species. Ecol. 25:404–413.
387. Jackson, W. T. 1955. The role of adventitious roots in recovery following flooding of the original root systems. Am. J. Bot. 42:816–819.
388. Jacobs, M. R. 1954. The effect of wind sway on the form and development of Pinus radiata D. Don. Austral. J. Bot. 2:35–51.
389. James, W. O. 1958. Succulent plants. Endeavour 17:90–95.
390. Jameson, D. A. 1963. Responses of individual plants to harvesting. Bot. Rev. 29:532–594.
391. Janzen, D. H. 1967. Fire, vegetation structure, and the ant × *Acacia* interaction in Central America. Ecol. 48:26–35.
392. Janzen, D. H., 1971. Seed predation by animals. Ann. Rev. Ecol. System. 2:462–492.
393. Jarvis, P. G. 1964. The adaptability to light intensity of seedlings of *Quercus petraea* (Matt.) Liebl. J. Ecol. 52:545–571.
394. Johnson, D. S., and H. H. York. 1915. The relation of plants to tide levels. Carnegie Inst. Wash. Publ. 206. 162 p.
395. Johnson, I. M. 1941. Gypsophily among Mexican desert plants. Arnold Arbor. J. 22:145–170.
396. Jones, E. T. 1956. Some observations on grassland research and grassland husbandry. Brit. Grassl. Soc. J. 11:73–81.
397. Jones, L. R. 1938. Relation of soil temperature to chlorosis of *Gardenia*. J. Agric. Res. 57:611–621.
398. Jones, L. R., and W. B. Tisdale. 1922. The influence of soil temperature upon the development of flax wilt. Phytopath. 12:409–413.
399. Junge, C. E. 1963. Air chemistry and radioactivity. Academic Press, Inc., N.Y. 382
400. Katznelson, H., et al. 1948. Soil microorganisms and the rhizosphere. Bot. Rev. 14:543–587.
401. Kaufert, F. H. 1933. Fire and decay in the southern bottomland hardwoods. J. For. 31:64–67.
402. Keen, B. A., and E. J. Russell. 1921. The factors determining soil temperature. J. Agric. Sci. 11:211–239.
403. Kelley, A. P. 1950. Mycotrophy in plants: lectures on the biology of mycorrhizae and related structures. Chronica Botanica Co., Waltham, Mass. 223 p.
404. Kerfoot, O. 1968. Mist precipitation on vegetation. For. Ab. 29:8–20.
405. Kern, F. D. 1921. Observations on the dissemination of the barberry. Ecol. 2:211–214.
406. Kezer, A., and D. W. Robertson. 1927. The critical period of applying irrigation water to wheat. Am. Soc. Agron. J. 19:80–116.

407. Kilmer, V. J., and L. T. Alexander. 1949. Methods of making mechanical analyses of soils. Soil Sci. 68:15–24.

408. Kincaid, R. H. 1935. Effect of certain environmental factors on germination of Florida cigar-wrapper tobacco seeds. Fla. Agric. Exp. Sta. Bul. 277. 47 p.

409. Kitchen, H. B. (ed.). 1948. Diagnostic technique for soils and crops. Amer. Potash Inst., Wash., D.C. 308 p.

410. Kittredge, J., Jr. 1938. The magnitude and regional distribution of water losses influenced by vegetation. J. For. 36:775–778.

411. Kittredge, J., Jr. 1955. Some characteristics of forest floors from a variety of forest types in California. J. For. 53:645–647.

412. Klages, K. H. W. 1942. Ecological crop geography. Macmillan Co., New York. 615 p.

413. Knight, R. O. 1966. The plant in relation to water. Dover Publications, N.Y. 147 p.

414. Knudson, L. 1929. Physiological investigations on orchid seed germination. Internat. Congr. Plant Sci. Proc. 1926. 2:1783–1789.

415. Knuth, P. E. O. W. 1906–1909. Handbook of flower pollination. Clarendon Press, Oxford. 3 vols.

416. Koch, L. W. 1955. The peach replant problem in Ontario. Can. J. Bot. 33:450–460.

417. Kohnke, H., et al. 1940. A survey and discussion of lysimeters. U.S.D.A. Misc. Publ. 372. 67 p.

418. Korstian, C. F. 1923. Control of snow molding of coniferous nursery stock. J. Agric. Res. 24:741–748.

419. Korstian, C. F. 1925. Some ecological effects of shading coniferous nursery stock. Ecol. 6:48–51.

420. Korstian, C. F., and N. J. Fetherolf. 1921. Control of stem girdle of spruce transplants caused by excessive heat. Phytopath. 11:485–490.

421. Kramer, P. J. 1934. Effects of soil temperature on the absorption of water by plants. Sci. 79:371–372.

422. Kramer, P. J. 1938. Root resistance as a cause of the absorption lag. Am. J. Bot. 25:110–113.

423. Kramer, P. J. 1950. Effects of wilting on the subsequent intake of water by plants. Am. J. Bot. 37:280–284.

424. Kramer, P. J. 1951. Causes of injury to plants resulting from flooding of the soil. Plant Physiol. 26:722–736.

425. Kramer, P. J., and T. S. Coile. 1940. An estimation of the volume of water made available by root extension. Plant Physiol. 15:743–747.

425. Kramer, P. J., and J. P. Decker. 1944. Relation between light intensity and rate of photosynthesis of loblolly pine and certain hardwoods. Plant Physiol. 19:350–358.

426. Kramer, P. J., and H. Brix. 1965. Measurement of water stress in plants. U.N.E.S.C.O. Arid Zone Research 25:343–351.

427. Kramer, P. J., et al. 1952. Gas exchange of cypress knees. Ecol. 33:117–120.

428. Kramer, P. J., et al. 1952. Survival of pine and hardwood seedlings in forest and open. Ecol. 33:428–430.

429. Kuijt, J. 1969. The biology of parasitic flowering plants. Univ. Cal. Press, Berkeley. 246 p.

430. Kullenberg, B. 1950. Investigations on the pollination of *Ophrys* species. Oikos 2:1–19.

431. Kurz, H., and D. Demaree. 1934. Cypress buttresses and knees in relation to water and air. Ecol. 15:36–41.

432. Laessle, A. M. 1961. A micro-limnological study of Jamaican bromeliads. Ecol. 42:499–517.

433. La Garde, R. V. 1929. Non-symbiotic germination of orchids. Mo. Bot. Gard. Ann. 16:499–514.

434. Laing, H. E. 1940. The composition of the internal atmosphere of *Nuphar advenum* and other water plants. Am. J. Bot. 27:861–868.

435. Lamb, D. 1965. Horizontal interception of precipitation. Austral. For. Res. 1:37–59.

436. Lamb, J., Jr., and J. E. Chapman. 1943. Effect of surface stones on erosion, evaporation, soil temperature, and soil moisture. Am. Soc. Agron. J. 35:567–578.

437. Landsberg, H. 1932. Is the "growing season" a significant climatological element? Am. Meteorol. Soc. Bul. 16:169–170.

438. Landsberg, H. 1944. Physical climatology. Pennsylvania State College, State College, Penn. 283 p.

439. Landsberg, H. 1953. The origin of the atmosphere. Sci. Am. 189(2):82–86.

440. Landsberg, H. 1970. Man-made climatic changes. Sci. 170:1265–1274.

441. Lang, A. 1963. Achievements, challenges, and limitations of phytotrons. p. 405–419. *In* L. T. Evans (ed.) Environmental control of plant growth. Academic Press, N.Y. 449 p.

442. Lange, O. L. 1965. Leaf temperatures and methods of measurement. p. 203–209. *In* F. E. Eckardt (ed.) Methodology of plant eco-physiology. U.N.E.S.C.O. Arid Zone Research 15:203–209.

443. Langheim, D. C. 1941. Effect of light on growth habit of plants. Sci. 93:576–577.

444. Larsen, J. A. 1922. Effect of removal of the virgin white pine stand upon the physical factors of site. Ecol. 3:302–305.

445. Larson, W. E., et al. 1959. Equipment and methods for measurement of soil temperature. U.S.D.A., Agric. Res. Serv. Publ. 41–27. 32 p.

446. Launchbaugh, J. L. 1954. A simple puller for soil tubes. J. Range Man. 7:182.

447. Laurie, A. 1931. Photoperiodism—practical application to greenhouse culture. Am. Soc. Hort. Sci. Proc. 27:319–322.

448. Lauritzen, J. I., et al. 1946. Influence of light and temperature on sugar cane and *Erianthus*. J. Agric. Res. 72:1–18.

449. Lawrence, D. B. 1939. Some features of the vegetation of the Columbia River Gorge with special reference to asymmetry in forest trees. Ecol. Mono. 19:217–257.

450. Lawrence, D. B., et al. 1947. Data essential to completeness of reports on seed germination of native plants. Ecol. 28:76.

451. Leamer, R. W., and B. Shaw. 1941. A simple apparatus for measuring non-capillary porosity on an extensive scale. Am. Soc. Agron J. 33:1003–1008.

452. LeBarron, R. K. 1939. The role of forest fires in the reproduction of black spruce. Minn. Acad. Sci. Proc. 7:10–14.

453. LeBarron, R. K., and J. R. Neetzel. 1942. Drainage of forested swamps. Ecol. 23:457–465.

454. Lehenbauer, P. A. 1914. Growth of maize seedlings in relation to temperature. Physiol. Res. 1:247–288.

455. Lemon, P. C. 1961. Forest ecology of ice storms. Torrey Bot. Club. Bul. 88:21–29.

456. Leonard, O. A., and J. A. Pinckard. 1946. Effect of various oxygen and carbon dioxide concentrations on cotton root development. Plant Physiol. 21:18–36.

457. Leopold, A., and S. E. Jones. 1947. A phenologic record of Sauk and Dane Counties, Wisconsin, 1935–1945. Ecol. Mono. 17:81–122.

458. Levisohn, I. 1956. Strain differentiation in a root-infecting fungus. Nature 183:1065–1066.

459. Levitt, J. 1972. Responses of plants to environmental stresses. Academic Press, N.Y. 732 p.

460. Levy, E. B. 1937. The conversion of rainforest to grassland in New Zealand. Internat. Grassl. Congr. Rept. 4th:71–77.
461. Lewis, C. E. 1965. Firebreak is verastile necessity. Fla. Cattleman & Livestock J. 29(8):38–39.
462. Lieth, H. 1970. Phenology in productivity studies. p. 29–46. *In* J. Jacobs, et al. (ed.) Ecological studies. Springer Verlag, N.Y. 304 p.
463. Lieth, H., and D. H. Ashton. 1961. The light compensation points of some herbaceous plants inside and outside deciduous woods in Germany. Can. J. Bot. 39:1255–1259.
464. Lindsey, A. A., and J. E. Newman. 1956. Use of official weather data in spring time-temperature analysis of an Indiana phenological record. Ecol. 37:812–823.
465. List, R. J. 1958. Smithsonian meteorological tables. Ed. 6., rev. Smiths. Inst., Wash., D.C. 527 p.
466. Livingston, B. E. 1916. Physiological temperature indices for the study of plant growth in relation to climatic conditions. Physiol. Res. 1:399–420.
467. Livingston, B. E. 1934. Environments. Sci. 80:569–576.
468. Livingston, B. E. 1935. Atmometers of porous porcelain and paper, their use in physiological ecology. Ecol. 16:438–472.
469. Livingston, B. E., and F. W. Haasis. 1929. The measurement of evaporation in freezing weather. J. Ecol. 17:315–320.
470. Lloyd, F. E. 1905. The artificial induction of leaf-formation in the ocotillo. Torreya 5:175–179.
471. Lloyd, F. E. 1942. The carnivorous plants. Chronica Botanica Co., Waltham, Mass. 352 p.
472. Lloyd, P. S. 1971. Effects of fire on the chemical status of herbaceous communities of the Derbyshire Dales. J. Ecol. 59:261–273.
473. Lodewiick, J. E. 1928. Seasonal activity of the cambium in some northeastern trees. N.Y. State Coll. For. Tech. Publ. 23. 52 p.
474. Loehwing, W. F. 1930. The effect of light intensity on tissue fluids in wheat. Iowa Acad. Sci. Proc. 37:107–110.
475. Loehwing, W. F. 1934. Physiological aspects of the effect of continuous soil aeration on plant growth. Plant Physiol. 9:567–584.
476. Long, D. H. 1940. Spruce regeneration in Canada. The maritimes. For. Chron. 16:6–9.
477. Loomis, W. E. 1965. Absorption of radiant energy of leaves. Ecol. 46:14–17.
478. Loomis, W. E., and C. A. Shull. 1937. Methods in plant physiology. McGraw-Hill Book Co., N.Y. 472 p.
479. Long, Frances L. 1934. Application of calorimetric methods to ecological research. Plant Physiol. 9:323–337.
480. Lorch, J. 1958. Analysis of windbreak effects. Res. Council Israel Bul. Sec. D 6:211–220.
481. Lorenz, R. W. 1939. High temperature tolerance of forest trees. Minn. Agric. Exp. Sta. Tech. Bul. 141. 25 p.
482. Lloyd, M. G., et al. 1959. A study of spore dispersion by use of silver-iodide particles. Am. Meteorol. Soc. Bul. 40:305–309.
483. Loveless, A. R. 1962. Further evidence to support a nutritional interpretation of sclerophylly. Ann. Bot. 26:551–561.
484. Low, A. J. 1964. Compression wood in conifers: a review of literature. For. Abstr. 25(3):xxxv–xliii.
485. Lowdermilk, W. C. 1925. Factors affecting reproduction of Engelmann spruce. J. Agric. Res. 30:995–1009.

486. Lucas, C. E. 1947. The ecological effects of external metabolites. Biol. Rev. 22:270–295.

487. Ludwig, J. W., and J. L. Harper. 1958. The influence of the environment on seed and seedling mortality. VIII. The influence of soil colour. J. Ecol. 46:381–391.

488. Ludwig, J. W., et al. 1957. The influence of environment on seed and seedling mortality. III. The influence of aspect on maize germination. J. Ecol. 45:205–224.

489. Lugo-Lopez, M. A. 1952. Comparative value of various methods of approximating the permanent wilting percentage. Puerto Rico Univ. J. Agric. 36:122–133.

490. Lull, H. W., and K. G. Reinhart. 1955. Soil-moisture measurement. U.S. Forest Service, Southern For. Exp. Sta. Occas. Paper 140. 56 p.

491. Lundegardh, H. 1935. The influence of the soil upon the growth of the plant. Soil Sci. 40:89–101.

492. Lunt, H. A. 1951. Liming and twenty years of litter raking and burning under red (and white) pine. Soil Sci. Soc. Amer. Proc. 15:381–390.

493. Lutz, H. J. 1943. Injury to trees by *Celastrus* and *Vitis.* Torrey Bot. Club Bul. 70:436–439.

494. Lutz, H. J. 1944. Determination of certain physical properties of forest soils. I. Methods of utilizing samples collected in metal cylinders. Soil. Sci. 57:475–487.

495. Lutz, H. J. 1944. Determination of certain physical properties of forest soils. II. Methods of utilizing loose samples collected from pits. Soil Sci. 58:325–333.

496. Lyles, L., and N. P. Woodruff. 1960. Moisture control techniques for experimental field plots. Agron J. 52:298–299.

497. Lynch, J. J. 1941. The place of burning in the management of gulf coast wildlife refuges. J. Wildl. Man. 5:454–457.

498. Lyon, G. L., et al. 1971. Calcium, magnesium and trace elements in the New Zealand serpentine flora. J. Ecol. 59:421–429.

499. McAtee, W. L. 1939. Wildfowl food plants. Their value, propagation, and management Collegiate Press, Ames, Iowa. 141 p.

500. McAtee, W. L. 1947. Distribution of seeds by birds. Am. Midl. Nat. 38:214–223.

501. McCalla, T. M. 1971. Studies on phytotoxic substances from soil microorganisms and crop residues at Lincoln, Nebraska. p. 39–43. *In* Biochemical interactions among plants. U.S. Nat. Acad. Sci., Washington, D.C.

502. McCool, M. M., and G. J. Bouyoucos. 1929. Causes and effects of soil heaving. Mich. Agric. Exp. Sta. Spec. Bul. 192. 11 p.

503. McCool, M. M., and A. N. Johnson. 1938. Nitrogen and sulfur content of leaves of plants at different distances from industrial centers. Boyce Thompson. Inst. Contrib. 9:371–380.

504. MacDougal, D. T. 1907. Factors affecting the seasonal activities of plants. Plant World 10:217–237.

505. MacDougal, D. T. 1924. Dendrographic measurements. Carnegie Inst. Wash. Publ. 350. 88 p.

506. MacDougal, D. T., and E. Spaulding. 1910. The water-balance of succulent plants. Carnegie Inst. Wash. Publ. 141. 77 p.

507. MacDougal, D. T., and E. B. Working. 1921. Another high-temperature record for growth and endurance. Sci. 54:152–153.

508. MacDougal, W. B. 1918. The classification of symbiotic phenomena. Plant world 21:250–256.

509. McEwan, L. C., and D. R. Dietz. 1965. Shade effects on chemical composition of herbage in the Black Hills. J. Range Man. 18:184–190.

510. MacFadyen, A. 1956. The use of a temperature integrator in the study of soil temperatures. Oikos 7:56–81.

511. MacHattie, L. B. 1963. Winter injury of lodgepole pine foliage. Weather (London) 18:301–307.

512. McIlrath, W. J. 1956. Cotton stem intumescences as a result of flooding. Plant Dis. Rept. 40:65–67.

513. McIlrath, W. J. 1965. Radiation screens for air temperature measurement. Ecol. 46:533–538.

514. McIntyre, D. S. 1970. The platinum microelectrode method for soil aeration measurement. Adv. Agron. 22:235–283.

515. Maciolek, J. A. 1954. Artificial fertilization of lakes and ponds; a review of the literature. U.S. Fish and Wildl. Serv., Spec. Sci. Rept., Fisheries No. 113. 41 p.

516. Mack, A. 1960. A bath for soil temperature control in pot culture. Agron. J. 52:299.

517. McLintock, T. F. 1959. A method of obtaining soil-sample volumes in stony soils. J. For. 57:832–834.

518. McMillan, C. 1959. The role of ecotypic variation in the distribution of the central grassland of North America. Ecolo. Mono. 29:285–308.

519. McMillan, C. 1971. Environmental factors affecting the establishment of the black mangrove on the central Texas coast. Ecol. 52:927–930.

520. Macmillan, H. G. 1923. Cause of sunscald of beans. Phytopath. 13:376–380.

521. McNair, J. B. 1945. Plant fats in relation to environment and evolution. Bot. Rev. 11:1–59.

522. Magistad, O. C. 1945. Plant growth on saline and alkali soils. Bot. Rev. 11:181–230.

523. Mail, G. A. 1935. Soil temperature apparatus for field work. Soil Sci. 40:285–286.

524. Malyuga, D. P. 1964. Biochemical methods of prospecting. Plenum Publ. Corp., N.Y. 210 p.

525. Mani, M. S. 1963. The ecology of plant galls. W. Junk, The Hague. 434 p.

526. Marlatt, W. E., et al. 1961. A comparison of computed and measured soil moisture under snap beans. J. Geophys. Res. 66:535–541.

527. Marsden-Jones, E. M., and W. B. Turrill. 1945. Sixth report of the transplant experiments of the British Ecological Society at Potterne, Wiltshire. J. Ecol. 33:57–81.

528. Marshall, R. 1927. Influence of precipitation cycles on forestry. J. For. 25:415–429.

529. Martinelli, M., Jr. 1964. Watershed management in the Rocky Mountains alpine and subalpine zones. U.S.F.S. Res. Note RM-36. 7 p.

530. Martin, E. V., and F. E. Clements. 1935. Studies of the effects of artificial wind on growth and transpiration in *Helianthus annuus*. Plant Physiol. 10:613–636.

531. Martin, J. H. 1930. The comparative drouth resistance of sorghums and corn. Am. Soc. Agron. J. 22:993–1003.

532. Martin, M. H. 1968. Measurement of soil aeration. p. 181–190. *In* R. M. Wadsworth (ed.) The measurement of environmental factors in terrestrial ecology. Blackwell Sci. Publ., Oxford, Engl. 314 p.

533. Marvin, C. F. 1941. Psychrometric tables for obtaining the vapor pressure, relative humidity, and temperature of the dew point. U.S. Weather Bur. Bul. 235. 87 p.

534. Mather, J. R. 1954. The determination of soil moisture from climatic data. Am. Meteorol. Soc. Bul. 35:63–68.

535. Meineke, E. P. 1925. An effect of drouth in the forests of the Sierra Nevada. Phytopath. 15:549–553.

536. Meinzer, O. E. 1927. Plants as indicators of ground water. U.S.G.S. Water Supply Paper 577. 95 p.

537. Meinzer, O. E. 1942. Hydrology. McGraw-Hill Book Co., N.Y. 700 p.

538. Melin, E. 1953. Physiology of mycorrhizal relations in plants. Ann. Rev. Plant Physiol. 4:325–346.

539. Melin, E. 1954. Growth factor requirements of mycorrhizal fungi of forest trees. Svensk. Bot. Tidskr. 48:86–94.

540. Mes, Margaretha G. 1954. Excretion (secretion) of P and other mineral elements by leaves under the influence of rain. S. Afr. J. Sci. 50:167–172.

541. Metz, L. J., et al. 1966. Sampling soil and foliage in a pine plantation. Soil Sci. Soc. Am. Proc. 30:387–399.

542. Meyers, B. S., et al. 1943. Effect of depth of immersion on apparent photosynthesis in submerged vascular aquatics. Ecol. 24:393–399.

543. Middleton, H. E. 1920. The moisture equivalent in relation to the mechanical analysis of soils. Soil Sci. 9:159–167.

544. Middleton, W. E. K., and A. F. Spilhaus. 1953. Meteorological instruments. 3rd ed. Univ. Toronto Press, Ontario. 286 p.

545. Miller, F. J. 1938. The influence of mycorrhizae on the growth of shortleaf pine seedlings. J. For. 36:526–527.

546. Miller, S. A., and A. P. Mazurak. 1958. Relationships of soil particle and pore sizes to the growth of sunflowers (*Helianthus annuus* L.) Soil Sci. Soc. Am. Proc. 22:275–278.

547. Montieth, J. L., et al. 1964. Crop photosynthesis and the flux of carbon dioxide below the canopy. J. Appl. Ecol. 1:321–337.

548. Morinaga, T. 1926. The favorable effect of reduced oxygen supply upon the germination of certain seeds. Am. J. Bot. 13:159–166.

549. Morris, W. C. 1934. Heredity tests of Douglas fir seed and their application to forest management. J. For. 32:351.

550. Morrissey, S. 1955. Chloride ions in the secretion of the pitcher plant. Nature 176:1220–1221.

551. Mortimer, C. H. 1953. A review of temperature measurement in limnology. Internat. Assoc. Limnol. Commun. 1:1–25.

552. Moss, A. E. 1940. Effect of wind-driven salt water. J. For. 38:421–425.

553. Moss, D., et al. 1969. Carbon dioxide compensation points in related plant species. Sci. 164:187–188.

554. Moss, E. H. 1936. The ecology of *Epilobium angustifolium* with particular reference to rings of periderm in the wood. Am. J. Bot. 23:114–120.

555. Mount, A. B. 1969. Eucalypt ecology as related to fire. Tall Timbers Fire Ecol. Conf. Proc. 9:75–108.

556. Muelder, D. W., et al. 1963. Measurement of potential evaporation rates in ecology and silviculture with particular reference to the Piche atmometer. J. For. 61:840–845.

557. Mueller, I. M., and J. E. Weaver. 1942. Relative drouth resistance of seedlings of dominant prairie grasses. Ecol. 23:387–398.

558. Müntzing, A. 1936. The evolutionary significance of autopolyploidy. Hereditas 21:263–278.

559. Muller, C. H. 1965. Inhibitory terpenes volatilized from *Salvia* shrubs. Torrey Bot. Club Bul. 92:38–45.

560. Munn, R. E. 1966. Descriptive micrometeorology. Academic Press, N.Y. 245 p.

561. Murphy, J. L. 1970. Research looks at air quality and forest burning. J. For. 68:530–535.

562. Muscatine, L., and C. Hand. 1958. Direct evidence for the transfer of materials from symbiotic algae to the tissues of a coelenterate. U.S. Nat. Acad. Sci. Proc. 44:1259–1263.

563. Neumann, J. 1953. Some microclimatological measurements in a potato field. Israel Meteolor. Serv. Misc. Papers, Ser. C. No. 6.

564. Newcombe, F. C. 1922. Significance of the behavior of sensitive stigmas. Am. J. Bot. 9:99–120.

565. Nichol, H. 1955. The pH/pC concept. Ecol. 36:506.

566. Nichols, G. E. 1913. A simple revolving table for standardizing porous cup atmometers. Bot. Gaz. 55:249–251.

567. Nielsen, K. F., and E. C. Humphries. 1966. Effect of root temperature on plant growth. Soils and Fert. 29:1–7.

568. Njoku, E. 1958. The photoperiodic response of some Nigerian plants. W. Afr. Sci. Assoc. J. 4:99–111.

569. Nobuhara, H., and M. Toyohara. 1964. Observations on the damages of the coastal vegetation. II. Jap. J. Ecol. 14:195–200.

570. Nobuhara, H., et al. 1962. Observations on the damages of the coastal vegetation. I. Jap. J. Ecol. 12:101–107.

571. Nordhagen, R. 1959. Remarks on some new or little known myrmechorous plants from North America and east Asia. Bul. Res. Council Israel, Sec. E. Bot. 7D: 184–201.

572. Norman, A. G. 1951. Role of soil organisms in nutrient availability. p. 167–183. *In* E. Truog (ed.) Mineral nutrition of plants. Univ. Wisc. Press, Madison. 469 p.

573. Norris, D. O. 1956. Legumes and the rhizobium symbiosis. Empire J. Exptl. Agric. 24:247–270.

574. Nuttonson, M. Y. 1955. Wheat-climate relationships and the use of phenology in ascertaining the thermal and photo-thermal requirements of wheat. Amer. Inst. Crop Ecol., Wash., D.C. 388 p.

575. Oberlander, G. T. 1956. Summer fog precipitation on the San Francisco Peninsula. Ecol. 37:851–852.

576. O'Connor, T. C. 1955. On the measurement of global radiation using black and white atmometers. Geofiz. pura e Appl. 30:130–136.

577. Ogura, Y. 1940. On the types of abnormal roots in mangrove and swamp plants. Bot. Mag. Tokyo 54:389–404.

578. Oliver, W. R. B. 1930. New Zealand epiphytes. J. Ecol. 18:1–50.

579. Olmsted, C. E. 1944. Photoperiodic responses in twelve geographic strains of side-oats grama. Bot. Gaz. 106:46–74.

580. Ooosting, H. J. 1933. Physical-chemical variables in a Minnesota lake. Ecol. Mono. 3:493–533.

581. Oosting, H. J. 1954. Ecological processes and vegetation of the maritime strand in the southeastern United States. Bot. Rev. 20:226–262.

582. Oppenheimer, H. R. 1960. Adaptation to drouth; xerophytism. U.N.E.S.C.O. Arid Zone Res. 15:105–138.

583. Orshan, G. 1963. Seasonal dimorphism of desert and Mediterranean chamaephytes and its significance and a factor in their water economy. p 206–222. *In* A. J. Rutter and F. H. Whitehead (ed.). The water relations of plants. Blackwell Sci. Publ., London. 394 p.

584. Osterhhout, W. J. V. 1918. Endurance of extreme conditions and its relation to the theory of adaptation. Am. J. Bot. 5:507–510.

585. Ovington, J. D. 1955. Ecological conditions of different woodland types. Linn. Soc. London Proc. 165:103–105.

586. Ozanne, P. G. 1955. The effect of light on zinc deficiency in subterranean clover (*Trifolium subterraneum* L.). Austral. J. Bot. 8:344–353.

587. Pady, S. M., and L. Kapica. 1955. Fungi in air over the Atlantic Ocean. Mycologia 47:34–50.

588. Page, J. B., and G. B. Bodman. 1951. The effect of soil physical properties on nutrient availability. p 133–166. *In* E. Truog (ed.) Mineral nutrition of plants. Univ. Wisc. Press, Madison. 469 p.

589. Palmer, W. C., and A. V. Havens. 1958. A graphical technique for determining evapotranspiration by the Thornthwaite technique. Mo. Wea. Rev. 86:123–128.

590. Parr, R. G. 1947. A hot-wire anemometer for low wind speeds. J. Sci. Instr. 24:317–319.

591. Parsons, R. F. 1968. The significance of growth-rate comparisons for plant ecology. Am. Nat. 102:595–597.

592. Parsons, R. F., and A. M. Gill. 1968. The effects of salt spray on vegetation at Wilson's Promontory, Victoria, Australia. Roy. Soc. Vict. Proc. 81:1–10.

593. Patton, R. T. 1930. The factors controlling the distribution of trees in Victoria. Roy. Soc. Vict. Proc. 42:154–210.

594. Pauley, S. S., et al. 1955. Seed source trials of eastern white pine. For. Sci. 1:244–256.

595. Pavari, A. 1949. Control of mountain torrents and avalanches through establishment and maintenance of forest cover. U.N. Sci. Conf. Cons. Util. Nat. Res., Lake Success 5:168–170.

596. Pavlychenko, T. K. 1936. The soil block washing method in quantitative root study. Can. J. Res. 15:33–57.

597. Pearsall, W. H. 1920. The aquatic vegetation of the English lakes. J. Ecol. 8:163–199.

598. Pearsall, W. H. 1954. Growth and production. Adv. Sci. 42:1–10.

599. Pearse, C. K., and S. B. Wooley. 1936. The influence of range plant cover on the rate of absorption of surface water by soils. J. For. 34:844–847.

600. Pearson, G. A. 1924. Studies in transpiration of coniferous tree seedlings. Ecol. 5:340–347.

601. Pearson, G. A. 1931. Forest types in the southwest as determined by climate and soil. U.S.D.A. Tech. Bul. 247. 143 p.

602. Pearson, G. A. 1931. Recovery of western yellow pine seedlings from injury by grazing animals. J. For. 29:876–894.

603. Pearson, G. A. 1936. Some observations on the reactions of pine seedlings to shade. Ecol. 17:270–276.

604. Pearson, G. A., and A. S. McIntyre. 1935. Slash disposal in ponderosa pine forests of the southwest. U.S.D.A. Cir. 357. 28 p.

605. Peattie, R. 1936. Mountain geography: a critique and field study. Harvard Univ. Press, Cambridge. 257 p.

606. Peele, T. C., and O. W. Beale. 1950. Relation of moisture equivalent to field capacity and moisture retained at 15 atmospheres pressure to the wilting percentage. Agron. J. 42:604–607.

607. Pendleton, R. L. and D. Nickerson. 1951. Soil colors and special Munsell soil color charts. Soil Sci. 71:35–43.

608. Penfound, W. T. 1932. The anatomy of the castor bean as conditioned by light intensity and soil moisture. Am. J. Bot. 19:538–546.

609. Penfound, W. T. 1934. Comparative structure of wood in the "knees," swollen bases, and normal trunks of the tupelo gum (*Nyssa aquatica* L.). Am. J. Bot. 21:623–631.

610. Penfound, W. T., and F. G. Deiler. 1947. On the ecology of Spanish moss. Ecol. 28:455–458.

611. Penfound, W. T., and E. S. Hathaway. 1938. Plant communities in the marshlands of southeastern Louisiana. Ecol. Mono. 8:1–56.

612. Penman, H. L. 1956. Estimating evaporation. Am. Geophys. Union Trans. 37:43–46.

613. Perry, T. O., et al. 1950. Estimation of photosynthetically active radiation under a forest canopy with chlorophyll extracts and from basal area measurements. Ecol. 50:39–44.

614. Pessin, L. J. 1922. Epiphyllous plants of certain regions in Jamaica. Torrey Bot. Club Bul. 49:1–14.

615. Pewe, T. L. 1957. Permafrost and its effect on life in the north. Oregon State Coll., Biol. Coll. 1957:12–25.

616. Pfeiffer, N. E. 1928. Anatomical study of plants grown under glass transmitting light of various ranges of wavelengths. Bot. Gaz. 85:427–436.

617. Philip, C. B. 1952. Notes on tabanid flies and other victims caught by the carnivorous plant *Sarracenia flava*. Fla. Entomol. 35:151–155.

618. Phillips, J. F. V. 1929. The influence of *Usnea* sp. (near *barbata* Fr.) upon the supporting tree. Roy. Soc. S. Afr. Trans. 17:101–107.

619. Phillips, W. S. 1963. Depth of roots in soil. Ecol. 44:424.

620. Pieniazek, S. A. 1944. Physical characters of the skin in relation to apple fruit transpiration. Plant Physiol. 19:529–536.

621. Pierce, L. T. 1934. Temperature variations along a forested slope in the Bent Creek Experimental Forest, North Carolina. Mo. Wea. Rev. 62:8–12.

622. Pisek, A. 1960. The nature of the temperature optimum of photosynthesis. Bul. Res. Council Israel, Sec. D Bot. 8D: 285–289.

623. Platt, R. B., and J. N. Wolf. 1950. General uses and methods of thermistors in temperature investigations with special reference to a technique for high sensitivity contact temperature measurements. Plant Physiol. 25:507–512.

624. Plummer, A. P. 1943. The germination and early seedling development of twelve range grasses. Am. Soc. Agron. J. 35:19–34.

625. Post, K. 1937. Further responses of miscellaneous plants to temperature. Am. Soc. Hort. Sci. Proc. 34:627–629.

626. Preece, T. F., and C. H. Dickinson (ed.) 1971. Ecology of leaf surface microorganisms. Academic Press, N.Y. 640 p.

627. Price, K. R. 1965. A field method for studying root systems. Health Physics 11:1521–1525.

628. Proctor, J., and S. R. J. Woodell. 1971. The plant ecology of serpentine. 1. Serpentine vegetation of England and Scotland. J. Ecol. 59:375–395.

629. Pruitt, W. O., Jr. 1952. A method of mounting thermistors for field use. Ecol. 33:550.

630. Purer, Edith A. 1942. Plant ecology of the coastal salt marshlands of San Diego County, California. Ecol. Mono. 12:81–111.

631. Purvis, O. N. 1934. An analysis of the influence of temperature during germination on the subsequent development of certain winter cereals and its relation to the effect of length of day. Ann. Bot. 48:919–955.

632. Quick, C. R. 1959. *Ceanothus* seeds and seedlings on burns. Madroño 15:79–81.

633. Raber, O. 1937. Water utilization by trees, with special reference to the economic forest species of the north temperate zone. U.S.D.A. Misc. Publ. 257. 97 p.

634. Rackham, O. 1966. Radiation, transpiration, and growth in a woodland. p 167–185. *In* R. Bainbridge, et al. (ed.) Light as an ecological factor. John Wiley & Sons, N.Y. 452 p.

635. Rajan, A. K., et al. 1971. Interrelationships between the nature of the light source, ambient air temperature, and the vegetative growth of diffierent species within growth cabinets. Ann. Bot. 35:323–343.

636. Ranwell, D. S. (ed.). 1967. Sub-committee report on landscape improvement advice and research. J. Appl. Ecol. 4:1p–8p.

637. Rao, P. V. 1938. Effect of artificial wind on growth and transpiration in the Italian millet, *Setaria italica*. Torrey Bot. Club Bul. 65:229–232.

638. Rather, J. B. 1917. An accurate loss-on-ignition method for the determination of organic matter in soils. Ark. Agric. Exp. Sta. Tech. Bul. 140. 16 p.

639. Raunkiaer, C. 1934. The life forms of plants and statistical plant geography. Clarendon Press, Oxford. 632 p.

640. Rawlins, S. L., and F. N. Dalton. 1967. Psychrometric measurement of soil water potential without precise temperature control. Soil Sci. Soc. Amer. Proc. 31:297–301.

641. Read, R. A. 1964. Tree windbreaks for the central Great Plains. U.S.D.A. Handb. 250. 68 p.

642. Redmond, D. R. 1954. Variations in development of yellow birch roots in two soil types. For. Chron. 30:401–406.

643. Rediske, J. H., et al. 1963. Anthracene technique for evaluating canopy density following application of herbicides. For. Sci. 9:339–343.

644. Redway, J. W. 1931. Thermometer shelters. Ecol. 12:618–620.

645. Reed, H. S., and E. T. Bartholomew. 1930. The effects of desiccating winds on citrus trees. Calif. Agric. Exp. Sta. Bul. 484. 59 p.

646. Reed, J. F., and R. W. Cummings. 1945. Soil reaction-glass electrode and colorimetric methods for determining pH values of soils. Soil Sci. 59:97–104.

647. Rees, A. R. 1964. The flowering behaviour of *Clerodendrum incisum* in southern Nigeria. J. Ecol. 52:9–17.

648. Rensburg, H. J. van. 1952. Grass burning experiments on the Msima River stock farm, southern highlands, Tanganyika. E. Afr. Agric. J. 17:119–129.

649. Rhind, D. 1935. A note on photoperiodism in *Sesamum*. Indian J. Agric. Sci. 5:729–736.

650. Richards, L. A. 1942. Soil moisture tensiometer materials and construction. Soil Sci. 53:241–248.

651. Richards, L. A. (ed.). 1954. Diagnosis and improvement of saline and alkali soils. U.S.D.A. Handb. 60. 160 p.

652. Richards, L. A., et al. 1949. Some freezing point depression measurements on cores of soil on which cotton and sunflower plants were wilted. Soil Sci. Soc. Amer. Proc. 14:47–50.

653. Rieley, J. O., et al. 1969. The measurement of microclimatic factors under a vegetation canopy—a reappraisal of Wilm's method. J. Ecol. 57:101–108.

654. Ritchie, J. C., and S. Lichti-Federovitch. 1967. Pollen dispersal phenomena in Arctic-subarctic Canada. Rev. Palaeobot. Palynol. 3:255–266.

655. Robbins, W. W. 1917. Native vegetation and climate of colorado in their relation to agriculture. Colo. Agric. Exp. Sta. Bul. 224. 56 p.

656. Roberts, E. G. 1936. Germination and survival of longleaf pine. J. For. 34:884–885.

657. Roberts, R. H. 1943. The role of night temperature in plant performance. Sci. 98:265.

658. Robins, J. S., and H. R. Haise. 1961. Determination of consumptive use of water by irrigated crops in the western United States. Soil. Sci. Soc. Am. Proc. 25:150–154.

659. Rodhe, W. 1948. Environmental requirements of freshwater plankton algae. Symbol. Bot. Upsal. 10:1–149.

660. Ronco, F. 1969. Assimilation chamber for outdoor measurements of photosynthesis of tree seedlings. U.S. For. Serv. Res. Note RM-142. 4 p.

661. Rosenfels, R. S. 1940. Spread of white-top seed in the droppings of grazing cattle. Nev. Agric. Exp. Sta. Bul. 152. 5 p.

662. Rowe, J. S. 1964. Environmental preconditioning, with special reference to forestry. Ecol. 45:399–403.

663. Ruebel, E. 1935. The replaceability of ecological factors and the law of the minimum. Ecol. 16:336–341.

664. Runyon, E. H. 1936. Relation of water content to dry weight in leaves of the creosote bush. Bot. Gaz. 97:518–553.

665. Russell, M. B. 1949. Methods of measuring soil structure and aeration. Soil Sci. 68:25–35.

666. Russell, M. B. 1952. Soil aeration and plant growth. p 253–301. *In* B. T. Shaw (ed.) Soil physical conditions and plant growth. Academic Press, N.Y. 491 p.

667. Rutter, N. 1968. Tattering of flags at different sites in relation to wind and weather. Agric. Met. Amst. 5:163–181.

668. Salisbury, F. B. 1954. Some chemical and biological investigations of materials derived from hydrothermally altered rock in Utah. Soil Sci. 78:277–294.

669. Salter, P. J., and J. B. Williams, 1965. The influence of texture on moisture characteristics of soils. II. Available water capacity and moisture release characteristics. J. Soil Sci. 16:310–317.

670. Samish, R. M. 1954. Dormancy in woody plants. Ann. Rev. Plant Physiol. 5:183–204.

671. Sasaki, N. 1950. A new method for surface-temperature measurement. Rev. Sci. Instr. 21:1–3.

672. Satchell, J. E. 1958. Earthworm biology and soil fertility. Soils and Fert. 21:209–219.

673. Satoo, T. 1949. Influence of wind on evaporation from combined evaporating surfaces. A note on the studies of the effect of wind on transpiration of trees. Tokyo Univ. Forests Bul. 37:31–40.

674. Sayre, J. D. 1920. Relation of hairy leaf coverings to transpiration. Ohio J. Sci. 20:55–86.

675. Schaffner, J. H. 1922. Control of the sexual state in *Arisaema triphyllum* and *A. draconitum.* Am. J. Bot. 9:72–78.

676. Schery, R. W. 1965. This remarkable Kentucky bluegrass. Mo. Bot. Gard. Ann. 52:444–451.

677. Schlich, W. 1904. Manual of forestry. Vol. 2. Silviculture. Bradbury, Agnew & Co., London, 393 p.

678. Scholander, P. F., et al. 1955. Gas exchange in the roots of mangroves. Am. J. Bot. 42:92–98.

679. Schomer, H. A. 1934. Photosynthesis of waterplants at various depths in the lakes of northern Wisconsin. Ecol. 15:217–218.

680. Schramm, J. R. 1966. Plant colonization studies on black wastes from anthracite mining in Pennsylvania. Am. Philos. Soc. Trans. 56:1–194.

681. Schreiner, O. 1923. Toxic organic constituents and the influence of oxidation. Am. Soc. Agron. J. 15:270–276.

682. Schulman, E. 1954. Longevity under adversity of conifers. Sci. 119:296–399.

683. Schultz, H. K. 1941. A study of methods of breeding orchard grass, *Dactylis glomerata* L. Am. Soc. Agron. J. 33:546–558.

684. Schuster, C. E., and R. E. Stephenson. 1940. Soil moisture, root distribution and aeration as factors in nut production in western Oregon. Ore. Agric. Exp. Sta. Bul. 372. 32 p.

685. Schuurman, J. J., and M. A. J. Goedwaagen. 1955. A new method for the simultaneous preservation of profiles and root systems. Plant and Soil 6:373–381.

686. Scott, D. H., and H. Wagner. 1888. On the floating roots of *Sesbania aculeata.* Ann. Bot. 1:307–314.

687. Sellschop, J. P. F., and S. C. Salmon. 1928. The influence of chilling, above the freezing point, on certain crop plants. J. Agric. Res. 37:315–338.

688. Semmens, E. S. 1947. Chemical effects of moonlight. Nature 159:613.

689. Schachori, A. Y., and A. Michaeli. 1965. Water yields of forest, maquis and grass covers in semi-arid regions; a literature review. U.N.E.S.C.O. Arid Zone Research 15:467–477.

690. Sharpe, C. F. S. 1938. Landslides and related phenomena. Columbia Univ. Press, N.Y. 137 p.

691. Shields, Lora M. 1951. Leaf xeromorphy in dicotyledon species from a gypsum sand deposit. Am. J. Bot. 38:175–190.

692. Shields, Lora M., and L. W. Durrell. 1964. Algae in relation to soil fertility. Bot. Rev. 30:92–128.

693. Shirley, H. L. 1929. The influence of light intensity and light quality upon the growth and survival of plants. Am. J. Bot. 16:354–390.

694. Shirley, H. L. 1930. A thermoelectric radiometer for ecological use on land and in water. Ecol. 11:61–71.

695. Shirley, H. L. 1932. Does light burning stimulate aspen suckers? J. For. 30:419–420.

696. Shirley, H. L. 1934. Observations on drouth injury in Minnesota forests. Ecol. 15:42–48.

697. Shirley, H. L., and L. J. Meuli. 1939. Influence of moisture supply on drouth resistance of conifebs. J. Agric. Res. 59:1–21.

698. Show, S. B., and E. I. Kotok. 1924. The role of fire in the California pine forests, U.S.D.A. Bul. 1294. 80 p.

699. Shreve, Edith B. 1916. An analysis of the causes of variations in the transpiring power of cacti. Physiol. Res. 2:73–127.

700. Shreve, F. 1911. The influence of low temperature on the distribution of the giant cactus. Plant World 14:136–146.

701. Shreve, F. 1914. The direct effects of rainfall on hygrophilous vegetation. J. Ecol. 2:82–98.

702. Shreve, F. 1931. Physical conditions in sun and shade. *Ecol.* 12:96–104.

703. Shreve, F., and T. D. Mallory. 1933. The relation of caliche to desert plants. Soil Sci. 35:99–112.

704. Sideris, C. P. 1955. Effects of sea water sprays on pineapple plants. Phytopath. 45:590–594.

705. Sifton, H. B., 1945. Air-space tissue in plants. Bot. Rev. 11:108–143.

706. Sigafoos, R. C., and D. M. Hopkins. 1951. Frost-heaved tussocks in Massachusetts. Amer. Jour. Sci. 249:312–317.

707. Siggins, H. W. 1933. Distribution and rate of fall of conifer seeds. Jour. Agr. Res. 47:119–128.

708. Silen, R. R. 1956. Use of temperature pellets in regeneration research. Jour. For. 54:311–312.

709. Sinclair, J. G. 1922. Temperatures of the soil and air in a desert. Mo. Wea. Rev. 49:142–144.

710. Sinclair, W. A. 1969. Polluted air: Potent new selective force in forests. J. For. 67:305–309.

711. Slavik, B., and J. Catsky. 1965. Colorimetric determination of CO_2 exchange in field and laboratory. U.N.E.S.C.O. Arid Zone Res. 25:291–298.

712. Sloover, J. de, and F. Le Blanc. 1968. Mapping of atmospheric pollution on the basis of lichen sensitivity. Recent Adv. Trop. Ecol. 1:42–56.

713. Small, J. 1918. The origin and development of the Compositae. IX. Fruit dispersal. New Phytol. 17:200–230.

714. Small, J. 1954. Modern aspects of pH, with special reference to plants and soils. Balliere, Tindall & Cox, London. 247 p.

715. Smith, A. 1926. A contribution to the study of interrelations between the temperature of the soil and of the atmosphere and a new type of thermometer for such study. Soil Sci. 22:447–456.

716. Smith, C. F., and S. E. Aldous. 1947. The influence of mammals and birds in retarding artificial and natural reseeding of coniferous forests in the United States. Jour. For. 45:361–369.

717. Smith, E. C. 1936. The effects of radiation on fungi. *In* B. M. Duggar, Biological effects of radiation 2:889–918.

718. Smith, E. P. 1930. Flower colors as natural indicators. Bot. Soc. Edinburgh Trans. and Proc. 30:230–238.

719. Smith, H. W., and C. D. Moodie. 1955. Syllabus for soil classification. (Mimeographed.) Dept. Agron., State Coll. Wash., Pullman.

720. Smith, N. F. 1948. Controlled burning in Michigan's forest and game management programs. Soc. Amer. For. Proc. 1947:200–205.

721. Smith, W. H. 1947. Control of low-temperature injury in the Victoria plum. Nature (London) 159:541–542.

722. Soil Sur. Staff., U.S.D.A. 1951. Soil survey manual. U.S. Govt. Print Off., Washington, D.C. 503 p.

723. Spaith, J. N., and C. H. Diebold. 1939. Some interrelationships among water tables, soil temperature and snow cover in the forest and adjacent open areas in south-central New York. N.Y. (Cornell) Agric. Exp. Sta. Mem. 213. 76 p.

724. Sparling, J. H., and M. Alt. 1966. The establishment of carbon dioxide concentration gradients in Ontario woodlands. Can. J. Bot. 44:321–329.

725. Sprague, H. B. (ed.). 1964. Hunger signs in crops, ed. 3. D. McKay Co., N.Y. 461 p.

726. Sproull, W. T. 1970. Air pollution and its control. Exposition Press, N.Y. 106 p.

727. Spurway, C. H. 1941. Soil reaction (pH) preference of plants. Mich. Agric. Exp. Sta. Spec. Bul. 306. 36 p.

727. Stanhill, G. 1957. The effect of differences in soil moisture status on plant growth: A review and analysis of soil-moisture regime experiments. Soil Sci. 84:205–214.

728. Stapledon, R. G. 1928. Cocksfoot grass (*Dactylis glomerata* L.) ecotypes in relation to the biotic factor. J. Ecol. 16:71–104.

729. Stebbins, G. L., Jr. 1950. Variation and evolution in plants. Columbia Univ. Press, N.Y. 658 p.

730. Stelfox, H. B. 1957. Two types of cages found satisfactory for pasture studies. J. Range Man. 10:230–231.

731. Stephens, E. P. 1956. The uprooting of trees: A forest process. Soil Sci. Soc. Amer. 20:113–116.

732. Stewart, H. W. 1927. The effect of texture of sandy soils on the moisture supply for corn during seasons of favorable and unfavorable distribution of rainfall. Soil Sci. 24:231–240.

733. Stewart, W. P. D. 1966. Nitrogen fixation in plants. Oxford Univ. Press, N.Y. 168 p.

734. Stewart, W. P. D. 1967. Nitrogen-fixing plants. Sci. 158:1426–1432.

735. Stickel, P. W., and H. F. Marco. 1936. Forest fire damage studies in the northeast. J. For. 34:420–423.

736. Stoeckeler, J. H., and C. G. Bates. 1939. Shelterbelts: The advantages of porous soils for trees. J. For. 37:205–331.

737. Stoeckeler, J. H., and E. J. Dortignac. 1941. Snowdrifts as a factor in growth and longevity of shelterbelts in the Great Plains. Ecol. 22:117–124.

738. Strain, B. R. 1966. The effect of a late spring frost on the radial growth of variant quaking aspen biotypes. For. Sci. 12:334–337.

739. Strausbaugh, P. D. 1921. Dormancy and hardiness in the plum. Bot. Gaz. 71:337–357.

740. Studhalter, R. A., and W. S. Glock. 1955. Tree growth. Bot. Rev. 21:1–188.

741. Sunderland, N. 1960. Germination of the seeds of angiospermous root parasites. p. 83–93. *In* J. L. Harper (ed.). The biology of weeds. Blackwell Sci. Publ., Oxford. 256 p.

742. Swan, F. R. 1968. Loss-on-ignition properties of some burned New York State soils. W. Va. Acad. Sci. Proc. 40:33–38.

743. Sweet, A. T. 1929. Subsoil as an important factor in the growth of apple trees in the Ozarks. U.S.D.A. Cir. 95: 12 p.

744. Talling, J. F. 1961. Photosynthesis under natural conditions. Ann. Rev. Plant Physiol. 12:133–154.

745. Tamm, C. O. 1951. Removal of plant nutrients from tree crowns by rain. Physiol. Plant. 4:184–188.

746. Taylor, C. F. 1956. A device for recording the duration of dew deposits. Plant Dis. Rep. 40:1025–1028.

747. Taylor, D. F. 1963. Mortarboard psychrometer. U.S.F.S. Res. Paper SE-5. 12 p.

748. Tevis, L., Jr. 1953. Effect of vertebrate animals on seed crop of sugar pine. J. Wildl. Man. 17:128–131.

749. Theis, T., and L. Calpouzos. 1957. A seven-day instrument for recording periods of rainfall and dew. Phytopath. 47:746–747.

750. Thoday, D. 1931. The significance of reduction in the size of leaves. J. Ecol. 19:297–303.

751. Thomas, M. D. 1955. Effect of ecological factors in photosynthesis. Ann. Rev. Plant Physiol. 6:135–156.

752. Thomas, M. D. 1965. The effects of air pollution on plants and animals. p. 11–33. In G. T. Goodman, et al. (ed.) Ecology and the industrial society. Blackwell Sci. Publ., Oxford. 404 p.

753. Thompson, H. C. 1939. Temperature in relation to vegetative and reproductive development in plants. Am. Soc. Hort. Sci. Proc. 37:672–679.

754. Thompson, R. C. 1938. Dormancy in lettuce seed and some factors affecting its germination. U.S.D.A. Tech. Bul. 655. 20 p.

755. Thornton, H. G., and J. Meiklejohn. 1957. Soil microbiology. Ann. Rev. Microbiol. 11:123–148.

756. Thornthwaite, C. W. 1948. An approach toward a rational classification of climate. Geogr. Rev. 38:55–94.

757. Timmons, F. L. 1942. The dissemination of prickly pear seed by jackrabbits. Am. Soc. Agron. J. 34:513–520.

758. Timonin, M. I. 1941. Effect of by-products of plant growth on activity of fungi and actinomyces. Soil. Sci. 52:395–410.

759. Ting, I. P., and W. M. Duggar, Jr. 1968. Non-autotrophic carbon dioxide metabolism in cacti. Bot. Gaz. 129:9–15.

760. Tint, H. 1945. An apparatus for the growth of plants under controlled temperature levels. Phytopath. 35:511–516.

761. Toole, E. H. et al. 1956. Physiology of seed germination. Ann. Rev. Plant Physiol. 7:299–324.

762. Toumey, J. W. 1929. Initial root habit in American trees and its bearing on regeneration. Internat. Congr. Plant. Sci. Proc. 1926:713–728.

763. Transeau, E. N. 1904. On the development of palisade tissue and resinous deposits in leaves. Sci. 19:866–867.

764. Transeau, E. N. 1905. Forest centers of eastern America. Am. Nat. 39:875–889.

765. Truog, E. 1947. Soil reaction influence on availability of plant nutrients. Soil Sci. Soc. Am. Proc. 11:305–308.

766. Tschudy, R. H. 1934. Depth studies on photosynthesis of the red algae. Am. J. Bot. 21:546–556.

767. Turberville, H. W., and A. F. Hough. 1939. Errors in age counts of suppressed trees. J. For. 37:417–418.

768. Turesson, G. 1922. The genotypical response of the plant species to habitat. Hereditas 3:211–350.

769. Turesson, G. 1927. Contributions to the genecology of glacial relics. Hereditas 9:81–101.

770. Turnage, W. V. 1937. Note on accuracy of soil thermographs. Soil Sci. 43:475–476.

771. Turnage, W. V., and A. L. Hinckley. 1938. Freezing weather in relation to plant distribution in the Sonoran desert. Ecol. Mono. 8:529–550.

772. Uggla, E. 1957. Temperatures during controlled burning. The effect of fire on the vegetation and the humus cover. Norrl. Skogsvards. Tidskr. 1957:443–499.

773. Ulrich, A., and C. M. Johnston. 1959. Analytical methods for use in plant analysis. Cal. Agric. Exp. Sta. Bul. 766. 78 p.

774. Ungar, I. A. 1966. Salt tolerance of plants growing in saline areas of Kansas and Oklahoma. Ecol. 47:154–155.

775. U.S. Dept. Agric. 1938. Soils and men. Govt. Print. Off., Washington. 1232 p.

776. U.S. For. Serv. 1971. Prescribed burning symposium. S.E. For. Exp. Sta., Asheville, N.C. 160 p.

777. Uphof, J. C. T. 1942. Ecologic relations of plants with ants and termites. Bot. Rev. 8:536–598.

778. Van Bavel, C. H. M., et al. 1956. Soil moisture measurement by neutron moderation. Soil Sci. 82:29–41.

779. Van Cleef, E. 1908. Is there a type of storm path? Mo. Wea. Rev. 36:56–58.

780. van der Pijl, L. 1956. Remarks on pollination by bats in the genera Freycinetia, Duabanga and Haplophragma and on chiropterophily in general. Acta Bot. Neerl. 5:135–144.

781. Veihmeyer, F. J. 1938. Evaporation from soils and transpiration. Am. Geophys. Union Trans. 1938:612–619.

782. Veihmeyer, F. J., and A. H. Hendrickson. 1938. Soil moisture as an indication of root distribution in deciduous orchards. Plant Physiol. 13:169–178.

783. Veihmeyer, F. J., and A. H. Hendrickson. 1948. The permanent wilting percentage as a reference for the measurement of soil moisture. Am. Geophys. Union Trans. 29:887–891.

784. Veihmeyer, F. J., and A. H. Hendrickson. 1949. Methods of measuring field capacity and permanent wilting percentage of soils. Soil Sci. 68:75–94.

785. Veihmeyer, F. J., and A. H. Hendrickson. 1955. Does transpiration decrease as the soil moisture decreases? Am. Geophys. Union Trans. 36:425–448.

786. Veihmeyer, F. J., et al. 1928. Some factors affecting the moisture equivalent of soils. Internat. Congr. Soil Sci., Proc. and Papers 1:512–534.

787. Venning, F. D. 1948. Stimulation by wind motion of collenchyma formation in celery petioles. Bot. Gaz. 110:511–514.

788. Verall, A. F. 1943. Fungi associated with certain ambrosia beetles. J. Agr. Res. 66:135–144.

789. Verner, L. 1934. A simplified method of determining freezing-point depressions of apple tissue with the Beckmann apparatus. Am. Soc. Hort. Sci. Proc. 31:33–34.

790. Vezina, P. E. 1963. The field performance of ten Bellani radiation integrators. For. Chron. 39:401–402.

791. Visnon, C. G. 1923. Growth and chemical composition of some shaded plants. Am. Soc. Hort. Sci. Proc. 20:293–294.

792. Viro, P. J. 1959. Stoniness of forest soils in Finland, Commun. Inst. Forest. Fenn. 49(4):1–45.

793. Visher, S. 1943. Some climatic influences of the Great Lakes, latitude and mountains. Am. Soc. Hort. Sci. Proc. 20:293–294.

794. Vogel, S. 1968. "Sun leaves" and "shade leaves": differences in convective heat dissipation. Ecol. 49:1203–1204.

795. Voigt, G. K. 1960. Distribution of rainfall under forest stands. For. Sci. 6:2–10.

796. Voth, P. D., and K. C. Hamner. 1940. Response of *Marchantia polymorpha* to nutrient supply and photoperiod. Bot. Gaz. 102:169–205.

797. Wadleigh, C. H., and L. A. Richards. 1951. Soil moisture and the mineral nutrition of plants. p 410–450. *In* E. Truog (ed.) Minderal nutrition of plants. Univ. Wisc. Press, Madison. 469 p.

798. Wagener, W. W. 1957. The limitation of two leafy mistletoes of the genus *Phoradendron* by low temperatures. Ecol. 38:142–145.

799. Wagg, J. W. B. 1963. Notes on food habits of small mammals of the white spruce forest (in Canada). For. Chron. 39:436–445.

800. Wagg, J. W. B. 1964. Design of small mammal exclosures for forest seedling studies. Ecol. 45:199–200.

801. Waggoner, P. E. 1971. Plants and polluted air. Bioscience 21:455–459.

802. Waisel, Y. 1959. Endurance of drouth period beyond the wilting point. Israel Res. Council Bul. 70:44–47.

803. Waisel, Y., and G. Polak. 1969. Estimation of water stresses in the root zone by the double-root technique. Israel J. Bot. 18:123–128.

804. Wakabayashi, S. 1925. The injurious effect of submergence on the cranberry plant. N.J. Agric. Exp. Sta. Bul. 420. 26 p.

805. Wallis, G. W., and S. A. Wilde. 1957. Rapid methods for the determination of carbon dioxide evolved from forest soils. Ecol. 38:359–361.

806. Wallwork, J. A. 1970. Ecology of soil animals. McGraw-Hill Book Co., N.Y. 283 p.

807. Walton, G. S. 1969. Phytotoxicity of NaCl and $CaCl_2$ to Norway maples. Phytopathol. 59:1412–1415.

808. Warcup, J. H. 1951. The ecology of soil fungi. Brot. Mycol. Soc. Trans. 34:376–399.

809. Warington, K. 1936. The effect of constant and fluctuating temperature on the germination of the weed seeds in arable soils. J. Ecol. 24:185–204.

810. Warne, L. G. G. 1942. The supply of water to transpiring leaves. Am. J. Bot. 29:875–884.

811. Warnick, C. C. 1953. Experiments with windshields for precipitation gages. Am. Geophys. Union Trans. 34:379–388.

812. Wassink, E. C., and J. A. J. Stolwijk. 1956. Effects of light quality on plant growth. Ann. Rev. Plant Physiol. 7:373–400.

813. Waters, S. J. P., and C. D. Pigott. 1971. Mineral nutrition and calcifuge behaviour in *Hypericum*. J. Ecol. 59:179–188.

814. Watson, A., et al. 1966. Winter browning of heather (*Calluna vulgaris*) and other moorland plants. Bot. Soc. Edinburgh Trans. 40:195–203.

815. Weaver, H. 1947. Fire—nature's thinning agent in ponderosa pine stands. J. For. 45:437–444.

816. Weaver, H. A., and V. C. Jamison. 1951. Limitations in the use of electrical resistance soil moisture units. Agron. Jour. 43:602–605.

817. Weaver. J. E. 1919. The ecological relations of roots. Carnegie Inst. Wash. Publ. 292. 151 pp.

818. Weaver, J. E., and F. W. Albertson. 1943. Resurvey of grasses, forbs, and underground plant parts at the end of the great drouth. Ecol. Mono. 13:64–117.

819. Weaver, J. E., and W. J. Himmel. 1930. Relation of increased water content and decreased aeration to root development in hydrophytes. Plant Physiol. 5:69–92.

820. Weaver, J. E., and A. Mogensen. 1918. Relative transpiration of coniferous and broad-leaved trees in autumn and winter. Bot. Gaz. 68:393–424.

821. Weaver, J. E., and N. W. Rowland. 1952. Effects of excessive natural mulch on development, yield, and structure of native grassland. Bot. Gaz. 114:1–19.

822. Weidman, R. H. 1920. The windfall problem in the Klamath region. J. For. 18:837–843.

823. Weimer, J. L. 1930. Alfalfa root injuries resulting from freezing. J. Agric. Res. 40:121–143.

824. Weiser, C. J. 1970. Cold resistance and injury in woody plants. Sci. 169:1269–1278.

825. Welch, P. S. 1952. Limnology. McGraw-Hill Book Co., N.Y. 538 p.

826. Went, F. W. 1943. Effect of the root system on tomato stem growth. Plant Physiol. 18:51–65.

827. Went, F. W. 1943. The air-conditioned greenhouses at the California Institute of Technology. Am. Jour. Bot. 30:157–163.

828. Went, F. W. 1944. Thermoperiodicity in growth and fruiting of the tomato. Am. Jour. Bot. 31:135–150.

829. Went, F. W. 1948. Thermoperiodicity. Lotsya 1:145–157.

830. Went, F. W. 1953. The effects of rain and temperature on plant distribution in the desert. In Israel Res. Council, Desert Research. p. 230–240. Intersci. Publ. Co., New York. 641 p.

831. Went, F. W. 1955. Physiological variability in connection with experimental procedures and reproducibility. In W. Ruhland (ed.). Encyclopedia of plant physiology. 1:58–68. Springer-Verlag, Berlin. 850 p.

832. Went, F. W. 1955. Fog, mist, dew and other sources of water. In U.S.D.A. Yearbook, 1955. p. 103–109. Washington, D.C. 751 p.

833. Went, F. W., et al. 1952. Fire and biotic factors affecting germination. Ecol. 33:351–364.

834. White, O. E. 1926. Geographical distribution and the cold-resisting character of certain herbaceous perennial and woody plant groups. Brooklyn Bot. Gard. Rec. 15:1.

835. White, O. E., 1942. Temperature reaction, mutation, and geographical distribution in plant groups. Am. Sci. Congr., Proc. 8th 3:287–294.

836. White, W. N. 1932. A method of estimating ground-water supplies based on discharge by plants and evaporation from soil. U.S.D.I. Geol. Surv. Water. Supply Paper 659:1–105.

837. Whitehead, F. H. 1959. Vegetational changes in response to alterations of surface roughness on Mt. Maiella, Italy. J. Ecol. 47:603–606.

838. Whitford, L. A. 1960. The current effect and growth of freshwater algae. Am. Micro. Soc. Trans. 79:302–309.

839. Whitman, W. C. 1941. Seasonal changes in bound water content of some prairie grasses. Bot. Gaz. 103:38–63.

840. Wiant, H. V., Jr. 1964. The concentration of carbon dioxide at some forest microsites. J. For. 62:817–819.

841. Wiebe, H. H., et al. 1971. Measurement of plant and soil water status. Utah Agric. Exp. Sta. Bul. 484. 71 p.

842. Wilcox, J. C., et al. 1957. Relation of elevation of a mountain stream to reaction and salt content of water and soil. Can. J. Soil Sci. 37:11–20.

843. Wilde, S. A., et al. 1953. Influence of forest cover on the state of the ground water table. Soil Sci. Soc. Am. Proc. 17:65–67.

844. Williams, Stella S. 1951. Microenvironment in relation to experimental techniques. Brit. Grassl. Soc. J. 6:207–217.

845. Williams, T. E., and H. K. Baker. 1957. Studies on the root development of herbage plants. I. Techniques of herbage root investigations. Brit. Grassl. Soc. J. 12:49–55.

846. Williams, W. T., and D. A. Barber. 1961. The functional significance of aerenchyma in plants. p. 132–144. *In* F. L. Milthorpe (ed.) Mechanisms in biological competition. Sympos. Soc. Exptl. Biol. 15. 365 p. Cambridge Univ. Press.

847. Wilner, J. 1952. The effect of seasonal and cultural variations on maturity of woody plants commonly grown on the Canadian prairies. Sci. Agric. 32:568–573.

848. Wilson, H. A. 1957. Effect of vegetation upon aggregation in strip mine spoils. Soil Sci. Soc. Am. Proc. 21:637–640.

849. Wilson, J. D. 1930. A modified form of non-absorbing valve for porous-cup atmometers. Sci. 71:101–103.

850. Wilson, J. W. 1959. Notes on wind and its effects in arctic-alpine vegetation. J. Ecol. 47:415–427.

851. Winston, P. W., and D. H. Bates. 1960. Saturated solutions for the control of humidity in biological research. Ecol. 41:232–237.

852. Wodehouse, R. P. 1935. Pollen grains: their structure, identification and significance in science and medicine. McGraw-Hill Book Co., N.Y. 574 p.

853. Woledge, Jane. 1971. The effect of light intensity during growth on the subsequent rate of photosynthesis of leaves of tall fescue. Ann. Bot. 35:311–322.

854. Wolfenbarger, D. O. 1946. Dispersion of small organisms; distance dispersion rates of bacteria, spores, seeds, pollen and insects; incidence rates of disease and injuries. Am. Midl. Nat. 35:1–152.

855. Wolff, T. 1950. Pollination and fertilization of the fly ophrys, *Ophrys insectifera* L., in Allindelille Fredskov, Denmark. Oikos 2:20–59.

856. Wood, F. A., and D. D. Davis. 1969. Sensitivity to ozone determined for trees. Aci. Agric. 17:4–5.

857. Wood, R. 1951. The significance of managed water levels in developing the fisheries of large impoundments. Tenn. Acad. Sci. J. 26:214–235.

858. Woodwell, G. M., and F. T. Martin. 1964. Persistence of DDT in soils of heavily sprayed forest stands. Sci. 145:481–483.

859. Worall, G. A. 1960. Patchiness in vegetation in the northern Sudan. J. Ecol. 48:107–115.

860. Work, R. A., and M. R. Lewis. 1936. The relation of soil moisture to pear tree wilting in a heavy clay soil. Am. Soc. Agron. J. 28:124–134.

861. Wright, J. G. 1943. Measurement of the degree of shading or crown canopy density in forest sites. For. Chron. 19:183–185.

862. Wright, J. W. 1953. Pollen dispersion studies: some practical applications. J. For. 51:114.

863. Wylie, R. B. 1949. Variations in leaf structure among *Adiantum pedatum* plants growing in a rock cavern. Am. J. Bot. 36:282–287.

864. Yapp, R. H. 1909. On stratification in the vegetation of a marsh, and its relations to evaporation and temperature. Ann. Bot. 23:275–319.

865. Yapp, R. H. 1922. The concept of habitat. J. Ecol. 10:1–17.

866. Young, F. D. 1941. Frost and the prevention of frost damage. U.S.D.A. Farmer's Bul. 1588. 65 p.

867. Zeevaart, J. A. D. 1962. Physiology of flowering. Sci. 137:723–731.

868. Zimmerman, P. W., and A. E. Hitchcock. 1956. Susceptibility of plants to hydrofluoric acid and sulfur dioxide gases. Boyce Thomps. Inst. Contr. 18:263–279.

Index

411